Advances in Light Water Reactor Technologies

Takehiko Saito • Junichi Yamashita
Yuki Ishiwatari • Yoshiaki Oka
Editors

Advances in Light Water Reactor Technologies

Springer

Editors
Takehiko Saito
University of Tokyo
Hongo 7-3-1
113-8656 Tokyo
Bankyo-ku
Japan
saitotake@gmail.com

Junichi Yamashita
University of Tokyo
Hongo 7-3-1
113-8656 Tokyo
Bankyo-ku
Japan
junichi@nuclear.jp

Yuki Ishiwatari
University of Tokyo
Dept. Nuclear Engineering and
Management
Hongo 7-3-1
113-8656 Tokyo
Bunkyo-ku
Japan
ishi@nuclear.jp

Yoshiaki Oka
Waseda University
Joint Department of
Nuclear Energy
Building 51, 11F-09B
3-4-1 Ohkubo,
Shinjuku-ku,
Tokyo, 169-8555
Japan
okay@waseda.jp
Emeritus professor
University of Tokyo

ISBN 978-1-4419-7100-5 e-ISBN 978-1-4419-7101-2
DOI 10.1007/978-1-4419-7101-2
Springer New York Dordrecht Heidelberg London

Library of Congress Control Number: 2010938361

Printed on acid-free paper

Springer is part of Springer Science+Business Media (www.springer.com)

Preface

In December 1951, electric power was generated for the first time by a nuclear reactor called EBR-1 (Experimental Breeder Reactor-1) located at Idaho, USA. Subsequently in *1954*, a small-scale (5 MWe) graphite-moderated, water-cooled reactor Nuclear Power Plant (NPP) began operation at Obninsk in the former USSR (present-day Russia), followed by the first commercial Gas-Cooled Reactor NPP at *Calder Hall*, UK in 1956 and the first commercial Pressurized Water Reactor NPP at Shippingport, PA, USA in 1957.

Since then, many NPPs have been constructed worldwide. According to the IAEA Power Reactor Information System data (updated on December 16, 2009), 436 NPPs are currently in operation with a total net installed capacity of 370,304 MWe. Light water reactors (LWRs) have been most widely used and 88.3% (326,860 MWe) of the world's total nuclear power generation are by 356 LWR NPPs.

The number of NPPs rapidly increased until the Three Mile Island accident in 1979 and the Chernobyl accident in 1986; these events led to a slow down or stoppage in the construction of subsequent plants. However, even during the years of setback that followed, considerable R&D efforts for improving the design of LWRs continued. Thanks to these tireless efforts, evolutionary LWR NPPs have been developed in recent years, and some are already in operation and many are under construction or being planned worldwide.

To build a bridge between fundamental research and practical applications in LWR plants, the University of Tokyo organized the first International Summer School of Nuclear Power Plants at Tokai-mura, Ibaraki Prefecture, Japan, from July 28 to August 5, 2009. The School was hosted by the Executive Committee and was cosponsored by the GoNERI Program of the University of Tokyo and the Japan Atomic Energy Agency, in cooperation with the Atomic Energy Society of Japan. The School presented state-of-the-art technologies, methods, and research studies on NPPs to young researchers and engineers from universities, R&D institutes, and industries working in nuclear science and technology. A total of 57 participants (14 from Japan, 28 from China, 9 from the USA, and 6 from the Republic of Korea), 22 lecturers (invited from internationally renowned manufacturers, research institutes, and universities), and 14 executive committee members and staff joined the School at the Tokai-mura venue. The participants benefited greatly from lectures delivered by the world's top experts who stayed a few days following their lectures to allow intensive exchange of knowledge between lecturers and participants.

In 2004, the IAEA published TECDOC-1391, "Status of Advanced Light Water Reactor Designs," which is an overview of evolutionary LWR design. However, there is no textbook which explains basic research linked to practical LWR applications. To fill this gap, this publication includes 10 selected lectures of the International Summer School and the authors further refined them and elaborated them into a textbook style. Most of the authors are technical experts from manufacturers and their experiences are the key elements of the book. The editors hope the contents will be useful to engineers and researchers at manufacturers, utilities, regulatory bodies, and research institutes as well as to graduate students and professors in the nuclear engineering field.

As for specific evolutionary LWRs, the ABWR, APWR, EPR, and APR1400 have been selected. Relevant studies and research on the safety of these reactors – such as the use of probabilistic safety analysis (PSA) in design and maintenance of the ABWR (Chap.1), development of an advanced accumulator (a new passive ECCS component) of the APWR (Chap.2), studies on severe accident mitigation for the APR1400 (Chap.3), and development of a core catcher for the EPR (Chap.4) – are presented. Current LWR development and severe accident research in China are summarized in Chap.5. Other important advances in LWR technologies – such as full MOX core design, application of CFD in design of LWRs (BWRs), next-generation digital I&C technologies, use of advanced CAD and computer models in design and construction of LWR (ABWR), and advances in seismic design and evaluation of LWR (the new Japanese safety guide on seismic design and seismic PSA) – are given in Chaps.6, 7, 8, 9, and 10, respectively.

Many individuals and organizations have contributed to the realization of this book. The publication of the book and the International Summer School were supported by the Ministry of Education, Culture, Sports, Science, and Technology of Japan through the University of Tokyo Global COE (Center of Excellence) Program "Nuclear Education and Research Initiative," known as GoNERI. In addition to the invited lecturers, sincere appreciation goes to the advisory and international organizing committee members who helped organize the International Summer School. The book was assembled by Ms. Misako Watanabe. The editors are also grateful for the editing assistance of Dr. Carol Kikuchi.

Executive Committee Members of "The First Summer School of Nuclear Power Plant"

Yoshiaki Oka, Chair, University of Tokyo
Yuki Ishiwatari, University of Tokyo
Takaharu Fukuzaki, University of Tokyo
Satoshi Ikejiri, University of Tokyo
Shinichi Morooka, Toshiba/(University of Tokyo)
Takehiko Saito, Nuclear Safety Commission/(University of Tokyo)
Jun Sugimoto, Japan Atomic Energy Agency (JAEA)
Junichi Yamashita, Hitachi-GE/(University of Tokyo)
Zenko Yoshida, Japan Atomic Energy Agency (JAEA)
Advances in Light Water Reactor Technologies
By Yoshiaki Oka, Takehiko Saito, Junichi Yamashita & Yuki Ishiwatari (Editors)

Abbreviations

ABWR	Advanced boiling water reactor
AFWS	Auxiliary feed water system
APRM	Average power range monitor
APWR	Advanced pressurized water reactor
ATWS	Anticipated transient without scram
BWR	Boiling water reactor
CAE	Computer aided engineering
CCS	Containment spray system
CDF	Core damage frequency
CFD	Computational fluid dynamics
CFS	Cavity flooding system
CHF	Critical heat flux
CHRS	Containment heat removal system
DBA	Design-basis accident
DBEGM	Design basis earthquake ground motion
DCH	Direct containment heating
ECCS	Emergency core cooling system
FCI	Fuel coolant interaction
FMCRD	Fine motion control rod drive
HMI	Human machine interface
HMS	Hydrogen mitigation system
HPCS	High pressure core spray system
I&C	Instrumentation and control
IRWST	In-containment refueling water storage tank
IVR	In-vessel retention
LOCA	Loss of coolant accident
LOFW	Loss of feedwater
LPRM	Local power range monitor
LWR	Light water reactor
MCCI	Molten core concrete interaction

MCPR	Minimal critical power ratio
MCR	Main control room
MLHGR	Maximum linear heat generation rate
NPP	Nuclear power plant
NSSS	Nuclear steam supply system
PAR	Passive autocatalytic recombiner
PCCS	Passive containment safety system
PCV	Primary containment vessel
PRNM	Power range neutron monitor
PSA	Probabilistic safety analysis
RCCV	Reinforced concrete containment vessel
RCS	Reactor coolant system
RIP	Reactor internal pump
RPV	Reactor pressure vessel
SA	Sever accident
SCC	Stress corrosion cracking
SG	Steam generator
SIS	Safety injection system
SLC	Standby liquid control
SPSA	Seismic probabilistic safety assessment
SRNM	Startup range neutron monitor
SRV	Safety relief valve

Contents

Contributors

Dietmar Bittermann
AREVA Nuclear Power GmbH, Erlangen, Germany

Xu Cheng
Shanghai Jiao Tong University, Shanghai, China

Manfred Fischer
AREVA Nuclear Power GmbH, Erlangen, Germany

Masahiko Fujii
Toshiba Corporation, Tokyo, Japan

Toshifumi Hayashi
Toshiba Corporation, Tokyo, Japan

Hideaki Heki
Toshiba Corporation, Tokyo, Japan

Junichi Kawahata
Hitachi-GE Nuclear Energy, Ltd, Tokyo, Japan

Sang-Baik Kim
Korea Atomic Energy Research Institute, Daejeon, Korea

Tatsuyuki Maekawa
Toshiba Corporation, Tokyo, Japan

Shinichi Morooka
Toshiba Corporation, Tokyo, Japan

Shohei Motohashi
Japan Nuclear Energy Safety Organization, Tokyo, Japan

Akira Nishimura
Global Nuclear Fuel-Japan Co., Ltd, Tokyo, Kanayawa, Japan

Chapter 1
Application of Probabilistic Safety Analysis in Design and Maintenance of the ABWR

Masahiko Fujii, Shinichi Morooka, and Hideaki Heki

1.1 ABWR Design

1.1.1 ABWR Development

A brief history of the development of nuclear reactor in Japan is summarized in Fig. 1.1. In the 1960s, nuclear reactor technology was introduced mainly from the United States. But in this era, the capacity factor of Japanese boiling water reactors (BWRs) is low because of initial problems such as stress corrosion cracking (SCC). A program to improve the nuclear reactor performance was started. In the 1970s, phases-I and -II of this program was carried out for the purpose of improvement, standardization, and localization of conventional light water reactors (LWRs). The final stage of this program was carried out in the 1980s to develop advanced reactors (both ABWR and APWR), which had to meet the following objectives.

- Provide solutions to technical problems
- Incorporate the latest R&D results, the fruits of experience in plant construction and operation, the world's most advanced BWR technologies and the latest instrumentation & control (I&C) technologies
- Achieve higher plant availability and capacity factor
- Establish a world standard for an LWR

The ABWR was established through this program and it has got worldwide deployment as shown in Fig. 1.2. There are four units operating in Japan and four units are under construction in Taiwan and Japan as of April 2009. An additional ten units are now being planned in Japan and the United States.

M. Fujii (✉), S. Morooka, and H. Heki
Toshiba Corporation, Tokyo, Japan
e-mail: masahiko.fujii@toshiba.co.jp

T. Saito et al. (eds.), *Advances in Light Water Reactor Technologies*,
DOI 10.1007/978-1-4419-7101-2_1, © Springer Science+Business Media, LLC 2011

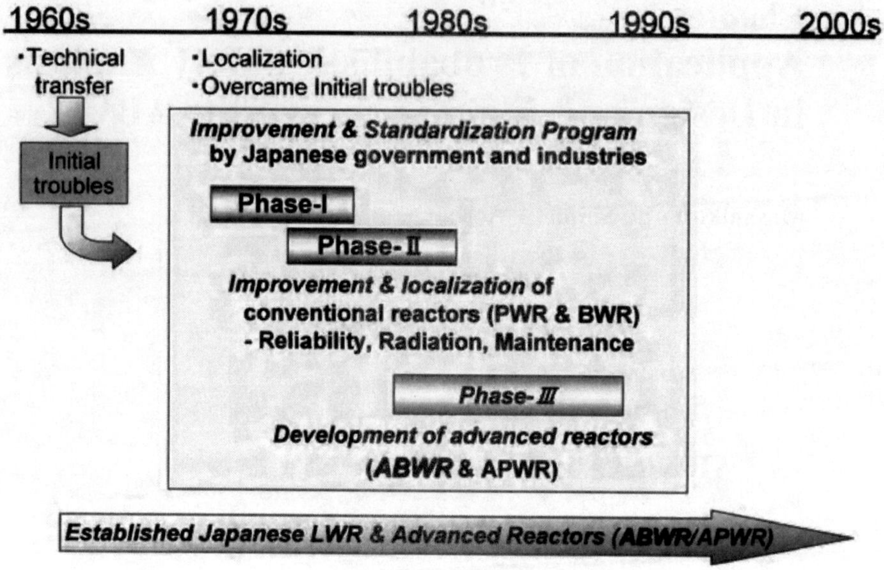

Fig. 1.1 Nuclear reactor development, history in Japan

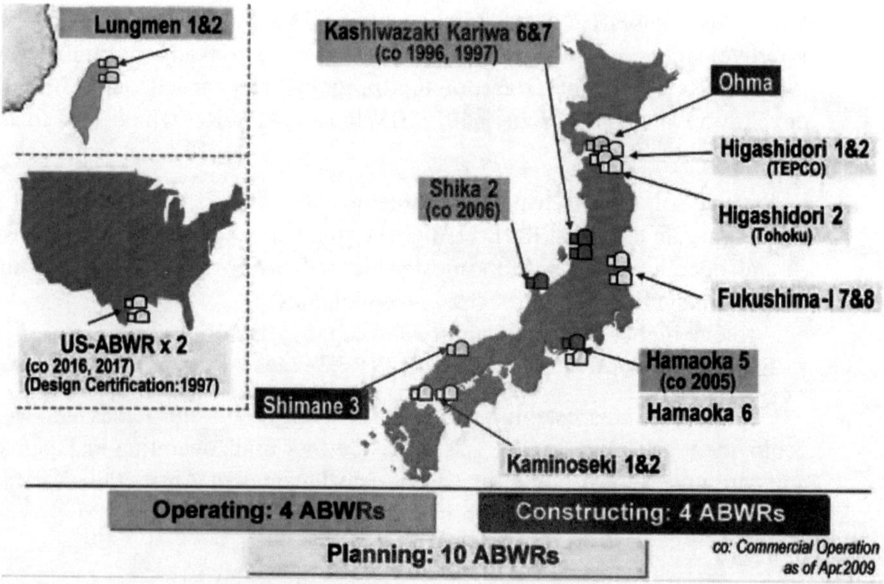

Fig. 1.2 ABWR construction experiences

1.1.2 ABWR Technical Features

The basic technical features of the ABWR are described in Ref. [1] and summarized here as follows:

- Good self-regulation and natural circulation core-cooling capabilities for the reactor.
- Simplified and highly reliable reactor system because the direct cycle is used.
- Good operability of the reactor system and simple adjustment of recirculation flow assures easy control of power output.
- Compact primary containment vessel (PCV) because pressure restriction is done using a suppression chamber.

In addition to these basic features, the ABWR design adopts a safe, reliable nuclear steam supply system, which offers the following features:

- Improved core
- Recirculation system using reactor internal pumps (RIPs)
- Fine motion control rod drive (FMCRD)
- Three-division emergency core cooling system (ECCS)
- Reinforced concrete containment vessel (RCCV)

Table 1.1 lists the main specifications of the ABWR in comparison to the BWR-5 and Table 1.2 compares prominent features of the two types. Figure 1.3 shows the key design features of the ABWR.

The ABWR also adopts the latest I&C technologies, which offer enhanced plant control performance; a highly efficient large capacity turbine/generator system with reheater and an enhanced radioactive waste treatment system that minimizes radwaste. The ABWR design is aimed at optimizing the total plant both by incorporating the new technologies introduced above, and by considering the existing system and equipment designs, and by achieving a compact layout and building design.

The following sub-sections consider the main features of the new technologies in some more detail.

1.1.2.1 Reactor Pressure Vessel and Internals

The shape of the bottom head of the reactor pressure vessel (RPV) was changed from an orb to a disc, and the design of the internals was optimized to allow such changes as adoption of a shorter stand-pipe for the steam separator. The result of these efforts is a 21-m high RPV; about 1 m shorter than that of the 1,100 MWe class BWR.

The number of vent pipes in the PVC was cut by reducing the amount of coolant loss in the event of a loss of coolant accident (LOCA). This was achieved by relocating the main steam restrictor from the main steam piping to the main steam nozzle.

Table 1.1 Main specifications of the ABWR and BWR-5

Item		ABWR	BWR-5
Electrical output		1,356 MWe	1,100 MWe
Thermal output		3,926 MWt	3,293 MWt
Reactor pressure		7.17 MPa	7.03 MPa
Main steam flow		7,640 t/h	6,410 t/h
Feed-water temperature		215°C	215°C
Rated core flow		52×10^6 kg/h	48×10^6 kg/h
Number of fuel bundles		872	764
Number of control rods		205	185
Core average power ratio		50.5 kW/l	50.0 kW/l
Reactor pressure vessel	Inner diameter	7.1 m	6.4 m
	Height	21.0 m	22.2 m
Reactor récirculation system (number of pumps)		Reactor internal pump (10)	External recirculation pump (2) jet pump (20)
Control rod drive	Normal operation	Electrical	Hydraulic
	Scram	Hydraulic	Hydraulic
Emergency core cooling system		Div I: RCIC+LPFL(RHR) Div II: HPCF+LPFL(RHR) Div III: HPCF+LPFL(RHR) ADS	Div I: LPCI+LPCS, ADS Div II: LPCI+LPCI, ADS Div III: HPCS
Residual heat removal system		3 Divisions	2 Divisions
Primary containment vessel		Reinforced concrete containment vessel with steel liner	Free-standing steel containment vessel
Turbine		TC6F-52" (2 stage reheat)	TC6F-41"/43" (non-reheat)

RCIC reactor core isolation cooling system; *LPFL* low-pressure flooder; *RHR* residual heat-removal system; *LPCI* low-pressure core injection system; *LPCS* low-pressure core spray; *HPCF* high-pressure core flooder system; *ADS* automatic depressurization system; *HPCS* high-pressure core spray

Table 1.2 Comparison of prominent features of the ABWR and BWR-5

ITEM	ABWR	BWR-5
Reduction of building volume	0.7	1.0
Enhanced thermal power efficiency (%)	35	33
Excellent operability	A-PODIATM[a]	PODIATM
Enhanced control performance (h)[b]	5	12
Shorter construction period (months)	48	53
Lower construction cost	0.8	1.0
Reduced radwaste (drums/reactor·year)	100	800
Less occupational exposure (Man·Sv/yr)	0.36	1.0
Shorter periodic inspection	45	55
Lower fuel cycle cost	0.8	1.0
Enhanced reliability (times/reactor·yr)[c]	0.1	0.4
Enhanced capacity factor (%)	87	75

[a] Advanced-plant operation by displayed information and automation
[b] Reactor automatic rapid start-up
[c] Scram occurrence

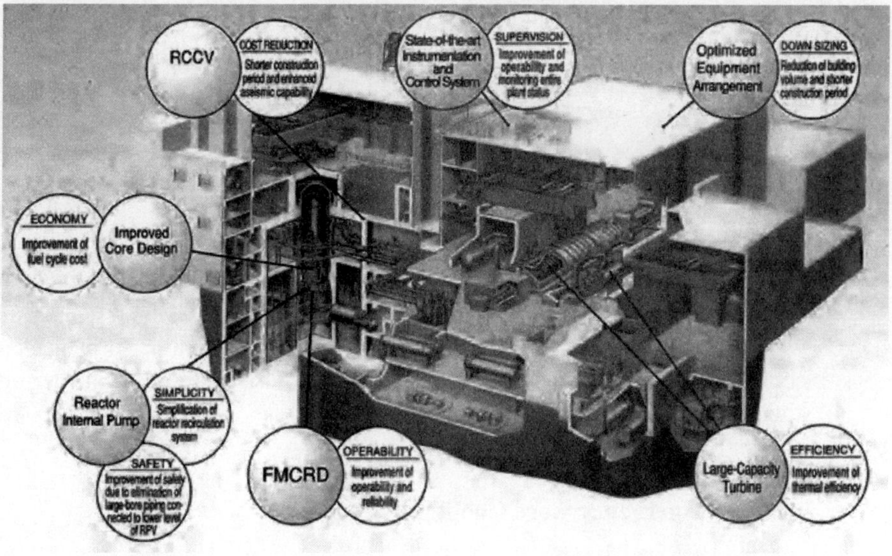

Fig. 1.3 System configuration of ABWR

1.1.2.2 Reactor Internal Pumps

The RIPs are installed directly at the bottom of the RPV, a system design enabling elimination of an external recirculation pump and piping. A small capacity ECCS is able to provide sufficient coolant as there is no need to consider the risk of a large piping rupture. The number of welds that require periodical inspection is reduced, resulting in lower occupational exposure. The smaller PVC allows the overall size of the reactor building to be smaller.

The maximum core flow at the rated thermal power needs 10 pumps in operation. However, the rated core flow can be obtained with only 9 pumps in operation.

1.1.2.3 Fine Motion Control Rod Drive

The FMCRD has two drive systems: a step motor for normal drive and a hydraulic drive for scram. Adoption of the FMCRD brought numerous advantages: higher reliability, more support for automated operation of the plant, a shorter plant start-up time with gang operation of multiple control rods, improved operability, and improved flatness of core power distribution.

To make the drive system maintenance free, a labyrinth seal is applied so that there is no seal against the moving surface inside of the drive system, and the spool piece can be removed at the intermediate flange and inspected, without removing the CRD. To simplify the hydraulic scram accumulator system, two CRDs are driven by a single accumulator.

The FMCRD was designed for an ABWR based on the German KWU design. After prototype testing, 1.5 years of in-plant testing was performed in the United States, at the LaSalle Unit No. 1 Nuclear Power Plant.

1.1.2.4 Improved Core

The ABWR core design focused on flexibility toward application of the latest improved fuel types, such as high burn-up fuel. Enlargement of the distance between the bundles from 12 to 12.2 in. increased the water vs. uranium ratio, which improved the cold shutdown margin and other core performances assuring economic long-term operation. The core was also designed considering future application using plutonium as a fuel.

1.1.2.5 Emergency Core Cooling System

The ECCS design was optimized as a three-division high-pressure system considering the characteristics of the RIPs. The ECCS also includes low-pressure systems. The ECCS maintains core cooling performance during both the short- and long-term cooling periods in the event of a LOCA. Cooling performance was confirmed by testing using a full-size model. Analysis using several computer codes produced the same results as the test.

1.1.2.6 Reinforced Concrete Containment Vessel

Adoption of a cylindrical RCCV, built as part of the ABWR reactor building, instead of the conventional steel PCV reduced the volume of steel required. Effective utilization of the RCCV structure reduced overall costs, and construction of the reactor building and RCCV at the same time cut the construction period.

Thanks to the RIP and enhanced RPV, the RCCV is compact, with a lower center of mass that enhances seismic performance. The adequacy of the design method and the integrity of structure against a combination of loads (internal load, including temperature effect and seismic load) were confirmed by tests using a 1/6 scale model of the RCCV and fuel pool.

1.1.2.7 State-of-the-Art I&C Technologies

The ABWR control room has a main operating console and a large display panel. The compact operating console, incorporating CRTs and flat panel displays, supports operators with automatic CR control and automatic operation after scram. The concentrated and categorized annunciators and the large display panel provide important information to all operators at the same time.

The digital control system and optical fiber network employed is more reliable and has greater performance than an analog system. Conventional plants utilize only a few digital systems, among them the recirculation flow control system, the RW system, and turbine control system. Digital systems are used throughout the ABWR for all plant systems, including safety-related systems.

In the safety protection system, two-out-of-four logic is applied for 4-divisional trip channels. The reliability of the safety-system software logic is assured by design review and verification & validation (V&V) work based on industry standards.

In the instrumentation systems, reliability, operability, and economy are all enhanced. For example, the startup range neutron monitor (SRNM) can monitor neutron flux at both the source range and the intermediate range with a single monitor.

1.1.2.8 Turbine System

The ABWR turbine system uses expertise gained over many years of operation of conventional BWRs to achieve a larger capacity and increased efficiency. Major improvements include: the low-pressure turbine with a 52-in. last-stage blade (the turbine itself can support the larger capacity, as its last stage annulus area is 40% greater than that of a standard 41-in. blade), adoption of the moisture separator reheater, higher turbine inlet pressure, adoption of the heater-drain pump-up system, which returns heater drain to the feedwater lines, and replacement of the combined angle valve for the low-pressure turbine inlet intercept and intermediate valves with butterfly valves, which enhance maintainability, reduce pressure loss, and add to thermal efficiency. Together these modifications give the ABWR a thermal efficiency exceeding that of the 1,100 MWe BWR by 2%.

1.1.2.9 Radioactive Waste Treatment System

The ABWR utilizes the heater-drain pump-up system. This reduces the flow rate of condensate and results in a smaller capacity cleanup system, the main source of low-level radioactive waste. Other measures include adoption of a hollow fiber filter, which does not use a filter aid, and nonregenerative use of the ion-exchange resin in the condensate demineralizer.

Concentrated waste is solidified and spent resin with low-level radioactivity and combustible miscellaneous solid waste are incinerated, reducing the volume of the radioactive waste.

1.1.2.10 Features of ABWR General Arrangement

The basic planning of the Kashiwazaki-Kariwa (K-K) Units 6 and 7 nuclear power plants, i.e., the world's first two ABWR units, sought to improve cost-efficiency and achieve a rational design. It made full use of advances in ABWR technologies and

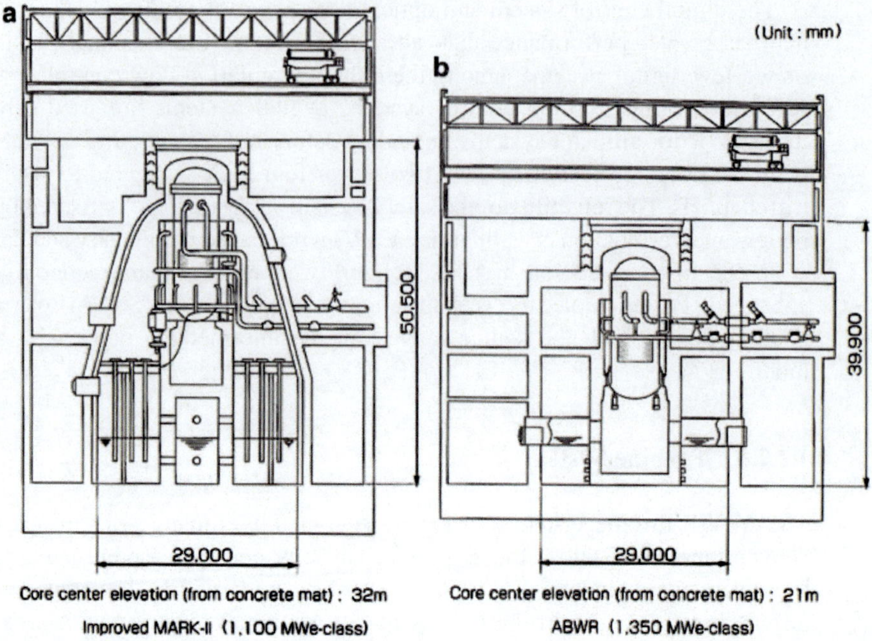

Fig. 1.4 Cross section of the reactor buildings for 1,100 MWe-class BWR and Kashiwazaki-Kariwa Unit 6

design, the economy of scale offered by increased capacity, and the merits of twin-plant construction.

The design integrates the RCCV with the reactor building, achieving a compact structure. The use of RIPs produces an RPV with a lower elevation, giving the resulting building a lower center of mass. The overall result is a more compact design and increased seismic capability, because the height of a building is about 10 m lower than that required to house the 1,100 MWe BWR.

Figure 1.4 shows cross sections of the reactor buildings for the 1,100 MWe BWR (Improved Mark-II) and K-K Unit 6. The turbine building also achieves a smaller volume through design rationalization. It reduces the main piping space for use in the side entry method by arranging the main steam piping on the side of the high pressure turbine.

The main control room, the radwaste building and the service building are shared by K-K Units 6 and 7, and are located between the two. A wind tunnel is used to determine how best to integrate the stack with the reactor building and reduce material volume. Considering maintainability, the floor of the radwaste building provides a shared turbine-maintenance space, and the building provides a route for the turbine crane to run between the two units. This wide-scale rationalization brought the total volume of the Unit 6 buildings (m^3/kWe) to 70% that of the 1,100 MWe BWR.

1.2 Application of PSA in Design and Maintenance of ABWR

The safety design of the ABWR was created using probabilistic safety analysis (PSA). Thorough discussions and details related to the safety design of the ABWR including basic policy, conceptual design process, and actual approach are described in Refs. [2] and [3]. This section gives an outline of them. (Although Ref. [2] discusses the TOSBWR, the discussions are applicable to the ABWR as well. Indeed, the ABWR safety design was conducted according to the concept described in Ref. [3] with the exception that the high-pressure core spray systems (HPCSs) were replaced by high-pressure core flooder systems (HPCFs). Therefore in this section, the TOSBWR described in Ref. [3] is referred to as the ABWR.)

1.2.1 Safety Features of Conventional BWRs

1.2.1.1 Conventional ECCS Design

There is a large piping system in the external recirculation line of conventional BWRs. Figure 1.5 shows the reactor design and the ECCS configuration of conventional BWRs, i.e. BWR-4 and BWR-5. The design-basis accident (DBA) LOCA is a large guillotine break of the recirculation pipe. If a DBA LOCA occurs, a large amount of coolant blows down, and the reactor pressure falls rapidly. Therefore, large-capacity ECCSs are provided. The ECCS pump head is, however, generally

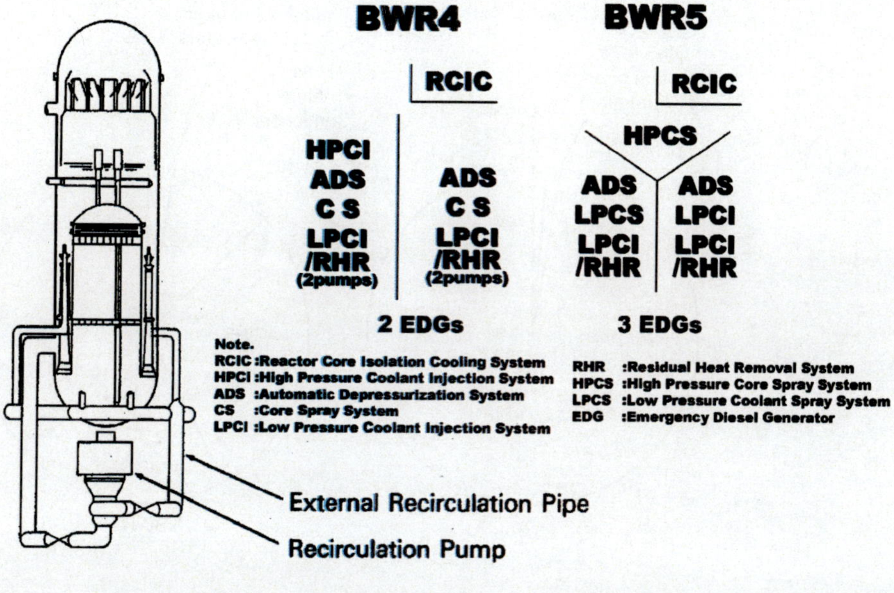

Fig. 1.5 ECCS configuration of conventional BWRs

very low because a large amount of high-head ECCS capacity would be unacceptably costly as high-pressure ECCS pumps are more expensive than low-pressure ECCS pumps of the same capacity. Therefore, conventional BWRs have one high-pressure ECCS (HPCI/HPCS) and four low-pressure ECCSs (CS/LPCS/LPCI). This low-pump-head ECCS design is based on the expectation that the reactor pressure must go down if a large pipe break occurs. Safety regulations require that a plant must cope with the DBA LOCA under the condition of loss of off-site power and a single failure with sufficient margin based on the classical deterministic safety philosophy.

1.2.1.2 Characteristics of the Conventional BWR Risk Profile

Figure 1.6 shows the results of level 1 PSA for internal events at full power for conventional Japanese BWR-4 and BWR-5 plants. These dominant sequences are all related to multiple failures of safe shutdown capabilities after a transient as shown in Fig. 1.7.

Dominant sequences of BWRs are transients followed by multiple failures as follows.

- Loss of feedwater transient followed by multiple failures of the RCIC and HPCS/HPCI systems. This pattern is called the TQUX sequence in PSAs.

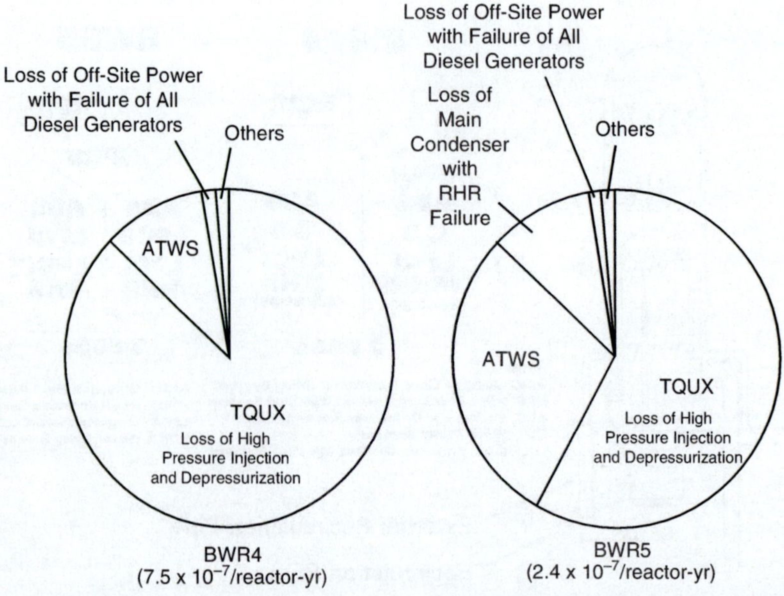

Fig. 1.6 Level 1 PSA results for internal events at full power for Japanese conventional BWRs

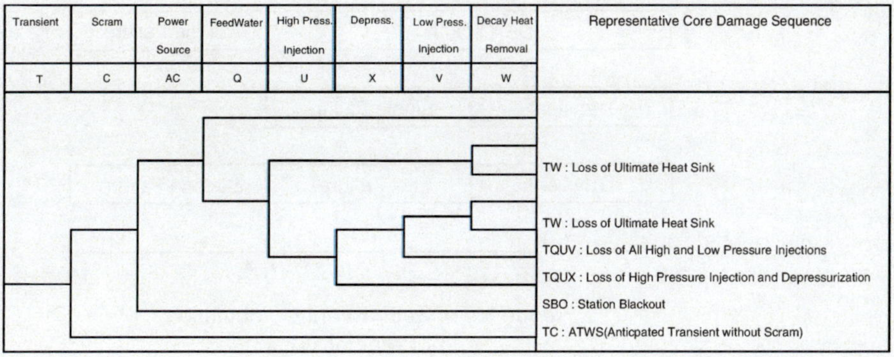

Fig. 1.7 Typical BWR transient-initiated sequences. (Taken from [2] and used with permission from ANS)

- Loss of main condenser followed by multiple failures of both residual heat removal (RHR) trains. Improvement of the RHR system was one of the unresolved safety issues of the US Nuclear Regulatory Commission (NRC).
- Transient followed by common-mode failures of the scram system. This sequence is called an anticipated transient without scram (ATWS). The ATWS was another unresolved safety issue of the NRC.
- Loss of off-site power transient followed by common-mode failures of emergency diesel generators. This sequence is called a station blackout. Station blackout was a third unresolved safety issue.

A transient has two characteristics: multiple failures and a high-pressure sequence. These two characteristics are not seen in a DBA LOCA, where only a single failure is assumed, and the reactor pressure is rapidly depressurized due to the large break itself.

In addition, there are important precursors that could lead to a severe core damage accident as well as unresolved safety issues. These safety issues are all based on experience and relate to actual plant safety performance. However, they cannot be recognized or assessed by the classical deterministic philosophy of safety assessment. This is because it is assumed deterministically that they do not happen. In reality, however, they do occur, and their implications can be assessed by a PSA.

The LOCA is not a dominant sequence in BWRs because the frequency of a DBA LOCA is limited to $\sim 10^{-4}$/reactor yr. Therefore, a combination with only a single failure, which has a probability of $\sim 10^{-2}$/demand, can result in a total frequency of $\sim 10^{-6}$/reactor yr, as illustrated in Fig. 1.8. This value is considered as a limit below which the event need not be considered in a plant design. This is the main reason why it is unnecessary to consider multiple failures in a DBA LOCA assessment. Thus, the use of the single-failure criterion is justified because of the effort not only to maintain the high reliability of safety systems but also to keep the frequency of a DBA LOCA very low.

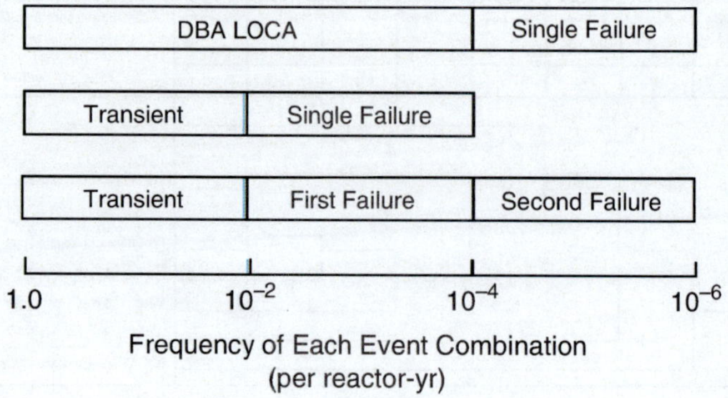

Fig. 1.8 Comparison of the frequencies of event combinations. (Taken from [2] and used with permission from ANS)

For example, if a transient is assumed to be an initiating event, the situation becomes quite different from that of a DBA LOCA. This is because transients occur much more frequently than a DBA LOCA. Figure 1.8 compares the frequencies of different event combinations. Because transients have a higher frequency, ~10^{-2}/reactor yr, the sequence frequency can only be reduced to ~10^{-4}/reactor yr by assuming a single failure. It is necessary to assume additional failures of the safety systems to make the sequence frequency as low as that for the DBA LOCA case. The additional failures include not only independent failures but also common-mode failures. This is because even common-mode failures usually have some probability, for example, from 0.1 to 0.001 in the form of a beta factor. Therefore, they can still reduce the total frequency of an event combination. If the beta factor is close to 1.0, the common-mode failure is a fully dependent failure. The design itself must be improved to avoid this fully dependent failure mode.

The important difference between the two sequences is that DBA LOCAs are rare, but transients occur frequently. A DBA LOCA can be a representative event from the standpoint of the initial effect to a plant. DBAs, however, must also subsume all the other events from the standpoint of demand frequency of safety systems. From this standpoint, a DBA LOCA does not represent all other events. This is because the ECCS also has a very important role in the safe shutdown after a transient.

1.2.2 Philosophies of ABWR Safety Design

The ABWR safety design is based on two important philosophies, i.e., the constant risk philosophy and the positive cost reduction philosophy. The former seeks a uniform distribution of plant risk and the latter aims to improve the cost-effectiveness of safety design.

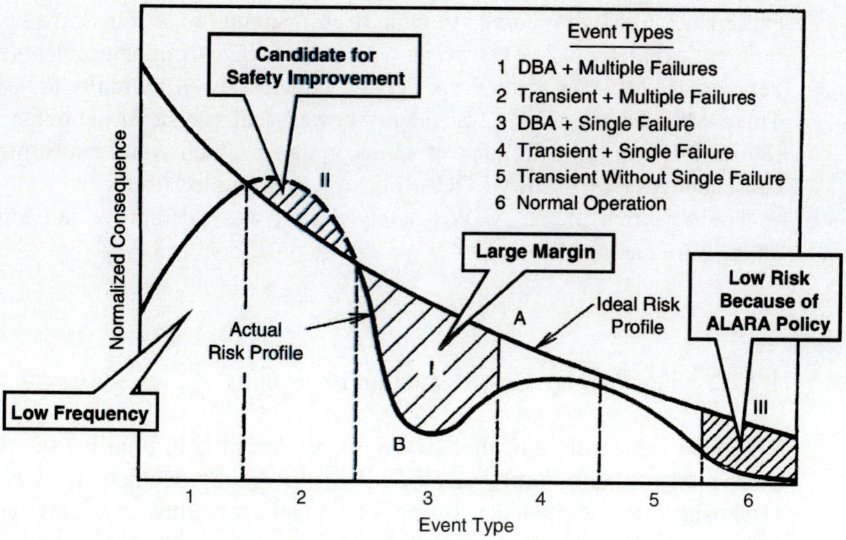

Fig. 1.9 Example of the ideal and the actual risk profile curves in a conventional BWR. (Taken from [2] and used with permission from ANS)

1.2.2.1 The Constant Risk Philosophy

The constant risk philosophy is explained in ANSI-52.1. The concept itself is very basic and classical: for safety of nuclear power plants, plant risk must not be excessively dominated by a few limited prevailing events. In other words, if the probability of a certain event is high and difficult to reduce, the consequences of the event must be limited. On the other hand, if the consequence of a certain event is significant and difficult to reduce, the probability of the event must be limited. By doing so, a plant can be designed that has a constant risk distribution over many events.

Figure 1.9 gives an example of the risk profiles of a conventional BWR. The abscissa shows event types in ascending order of frequency; the abscissa also corresponds to the frequencies of events on a logarithmic scale. The ordinate shows the normalized consequences on a linear scale. The ordinate can represent radiological dose rate or the corresponding death rates. Curve A shows an ideal risk profile. Along this curve, the consequence level decreases as event frequency increases. Therefore, plant risk, defined as the product of the ordinate and the abscissa along curve A, can be maintained as nearly constant.

Curve B shows an actual risk profile of a conventional BWR. Region I corresponds to a DBA plus a single failure, event type 3. In this region, a conventional plant design has a large safety margin, and the actual plant risk level is much lower than the ideal risk level. Region III corresponds to normal operation, event type 6. In this region, the actual plant risk is also much lower than the ideal risk curve because of the ALARA policy. In region II, however, the actual plant risk could exceed the ideal risk curve. Region II corresponds to a transient plus multiple failures, event type 2. This event type also includes minor accidents such as a very small LOCA or a stuck-open relief valve followed by multiple failures. The Three Mile Island (TMI)-2 accident showed that transients or minor accidents followed by multiple failures of safety systems which could cause more severe consequences than a DBA LOCA followed by a single failure.

The objective of the ABWR safety design was to improve the actual safety taking core damage frequency as a measure of safety.

1.2.2.2 The Positive Cost Reduction Philosophy

The other basic policy of the ABWR safety design is to improve safety and also plant total economy simultaneously. Any safety improvement set forth must be established in accordance with the total plant cost reduction. This type of cost reduction is termed "positive cost reduction" because of the positive net increase in cost-effectiveness that it brings about. On the other hand, normal cost reduction is termed "negative cost reduction" because it brings about a small cost reduction at the sacrifice of a large amount of safety and results in a negative net increase in cost-effectiveness. If this philosophy is incorporated into plant design, plant value and attractiveness deteriorate.

If the cost invested in region I in Fig. 1.9 can be reduced and some of the savings can be reinvested in region II, a positive cost reduction in safety design will be attained. In the ABWR safety design, two important values have enabled positive cost reduction: elimination of the risk of a large-break LOCA and incorporation of a constant risk philosophy and probabilistic risk assessment PRA insights.

The most important characteristic of the ABWR safety design is the adoption of internal pumps. These internal pumps can directly impel the water in the reactor vessel and make it possible to eliminate the external recirculation piping system in the ABWR. The ABWR installed the RIPs and eventually eliminated the external recirculation loops resulting in the most simplified primary system that has no large pipes connected below the core.

A detailed comparison of pipe locations connected to the reactor vessel in the ABWR and conventional BWR-5 is depicted in Fig. 1.10. The conventional BWR-5 has a large piping system in the external recirculation system below the core. However, any of the major pipes can be located above the core level, and the pipe size itself is much smaller in the ABWR. There are no large pipes below the top of the active fuel level in the ABWR. This improves the inherent safety of the design against a DBA LOCA. The effect of a DBA LOCA is much reduced, and the ECCS pump capacity can be smaller than in conventional BWRs.

The ABWR has a shorter RPV than the BWR-5, which means that the former has a shallower water depth above the core. Despite this shallower depth, the ABWR was able to achieve no core uncovering at a DBA LOCA. Elimination of large pipes below the core and the three-division high-pressure ECCS contributed to no core

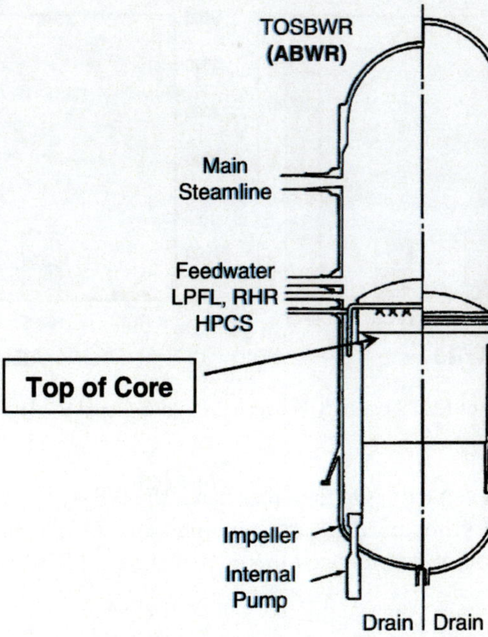

Fig. 1.10 Comparison of pipe locations connected to the reactor vessel in the ABWR (*left side*) and conventional BWR-5 (*right side*). (Taken from [2] and used with permission from ANS)

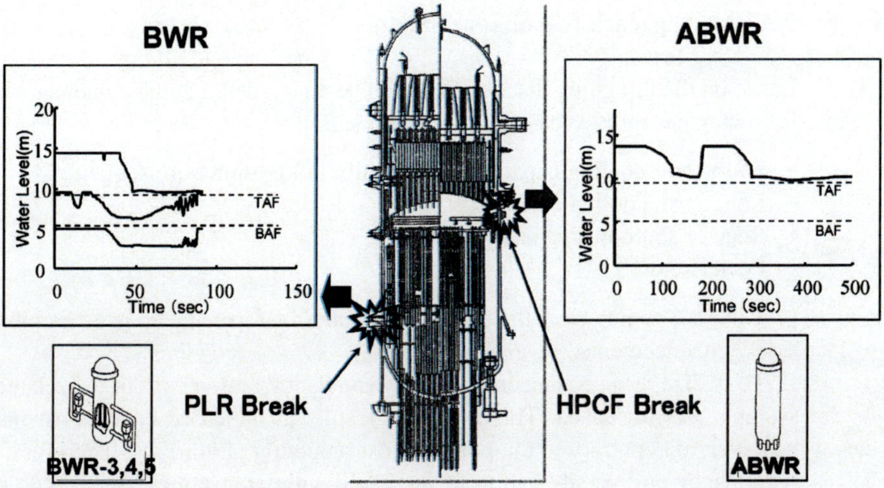

Fig. 1.11 Comparison of ECCS performance between the conventional BWR-5 and the ABWR. (Taken from [3] and used with permission from AESJ)

uncovering. Figure 1.11 compares ECCS performance between the conventional BWR and the ABWR.

Figure 1.12 compares ECCS capacity between the ABWR and BWR-5. The BWR-5 could still experience a large pipe break DBA LOCA and it needs a large

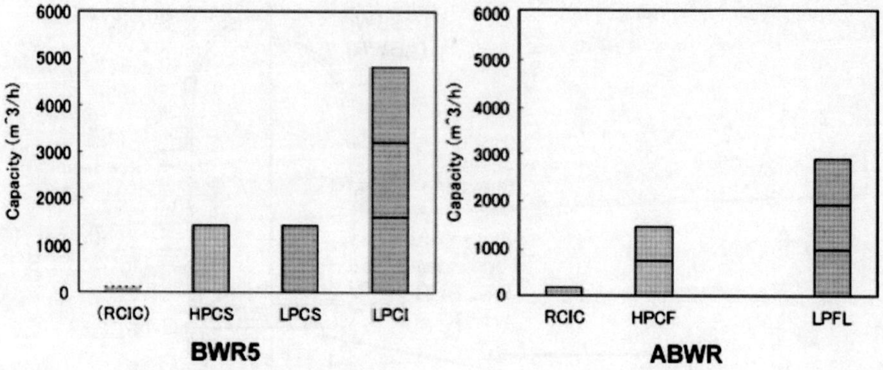

Fig. 1.12 Comparison of ECCS capacity between the conventional BWR-5 and the ABWR

low pressure ECCS capacity. On the other hand, the ABWR does not have a large pipe break and does not need a large low-pressure ECCS capacity. The low-pressure ECCS capacity was reduced to ~60% that of the BWR-5.

1.2.3 Concrete Measures to Enhance Safety

1.2.3.1 Approach to Enhance Safety

Based on the foregoing discussion, it can be concluded that the enhancement of the following capabilities can improve plant safety.

- Short-term cooling capability, especially in high-pressure sequences
- Long-term cooling capability
- Reactor shutdown capability
- Power sources

Table 1.3 summarizes the actual approaches that were taken to accomplish these safety enhancements.

To realize these enhancements, the redundancy or diversity of the related safety systems was increased. This normally results in increased costs. Cost increases, however, are contrary to the positive cost reduction philosophy. Therefore, system redundancy or diversity had to be increased without cost increases. To do this, the safety systems were subdivided. The merit of subdividing the safety systems is an increase in redundancy without a total increase in system capacity. It should be noted, however, that this is true only when 50% capacity is sufficient to fulfill the safety requirement. Otherwise, subdivided safety systems will result in lower total system reliability. This is because not one but two subsystems are required to fulfill the same safety function. Therefore, to take full advantage of the system

Table 1.3 Actual approaches to enhance safety (taken from [2] and used with permission from ANS)

Dominant sequences	Sequence patterns/ precursors	Actual approaches of safety enhancement
TQUX	Loss of feedwater	Enhancement of short-term cooling capability: Subdivide HPCS into $2 \times 50\%$ small HPCS and increase reliability of high-pressure makeup systems
	+ RCIC failure	
	+ HPCI failure	
	Hatch unit 2	
Loss of main condenser with RHR failure	Transient	Enhancement of long-term cooling capability: Subdivide RHR into $3 \times 50\%$ or $4 \times 50\%$ small RHR and increase reliability of long-term cooling
	+ Power conversion system failure + RHR failure	
	Browns Ferry Unit 1	
ATWS	ATWS	Enhancement of reactor shutdown capability: Utilize FMCRD motors to insert control rods and increase reliability of reactor shutdown capability
	Browns Ferry Unit 3	
Loss of off-site power with failure of all diesel generators	Station blackout	Enhancement of power source: incorporate three- or four-division diesel generators
	Quad City Units 1 and 2	

Later: HPCS changed to HPCF and adopted $3 \times 50\%$ RHRs and three-division diesel generators

subdividing technique, the minimum capacity requirement must be reduced to <50%. For a DBA LOCA, the minimum ECCS capacity requirement could be reduced to <50% because the external recirculation pipes are eliminated. For transients, the minimum capacity requirement was originally <50% of that for a DBA LOCA because there is no break in the case of transients.

For the RHR heat exchanger capacity, however, it was impossible to reduce the minimum requirement because it is basically a linear function depending on the amount of decay heat. Therefore, RHR system reliability decreases in the order of $4 \times 50\% > 2 \times 100\% > 3 \times 50\%$. This is because these configurations correspond to two-out-of-four, one-out-of-two, and two-out-of-three success criteria, respectively. Among these configurations, two-out-of-four has the highest reliability, and two-out-of-three has the lowest. Table 1.4 compares reliability assuming a 10^{-2}/demand failure probability for each RHR train.

Table 1.4 Reliability comparison among RHR configurations (taken from [2] and used with permission from ANS)

		RHR configurations	
Case		ABWR 3 × 50%	BWR 2 × 100%
Conservative as licensing	U^a	${}_3C_2 \times P^2 (1 - P) = 3 \times 10^{-4}$	${}_2C_2 \times P^2 = 10^{-4}$
	SC	2 out of 3	1 out of 2
Realistic as PRA	U^a	${}_3C_3 \times P^3 = 10^{-6}$ 1 out of 3	${}_2C_2 \times P^2 = 10^{-4}$
	SC		1 out of 2

[a]Unreliability is calculated only for the first term
U unreliability; *SC* success criteria; *P* failure probability of each RHR train (10^{-2} is assumed)

A conventional BWR-5 has a 2 × 100% RHR configuration. The less reliable 3 × 50% RHR configuration is not acceptable for the ABWR. In the PSA, however, it can be assumed that the containment can remain intact up to about three times its design pressure. This in turn can reduce the minimum capacity requirement of RHR to <50%. Only in this situation can the 3 × 50% RHR configuration be more reliable than the 2 × 100% RHR configuration. This is because the success criteria for the 3 × 50% configuration changes to one-out-of-three from two-out-of-three success criteria in this situation. If this can be allowed in the RHR safety design, the reliability of the RHR system decreases in the order 4 × 50% > 3 × 50% > 2 × 100%. In this way, the 3 × 50% RHR configuration becomes acceptable from the standpoint of the safety design. This is, however, still a compromise. For the 4 × 50% RHR configuration, it is unnecessary to accept any compromise. Therefore, the three-division concept was just a backup for the four-division concept.

For reactor shutdown capability, the diverse functions of the FMCRD system were used. This system can insert control rods by using motors for normal operation, and it also has a hydraulic scram capability as a reactor shutdown system. By adding a safety-grade signal system to the FMCRD motor, which is independent of the protection system, a diverse control rod insertion system can be added to the original hydraulic scram system.

To improve the power source, the ABWR adopted three-division emergency diesel generators. A conventional RCIC system with a turbine-driven pump was also incorporated. It should be noted that the advantage of an RCIC system is its ability to deliver water directly into the reactor vessel during a station blackout. This means that the RCIC system can actually offer another power source for coolant injection. This RCIC capability lasts for ~8 h in a station blackout situation, which can provide recovery time for off-site power and failed emergency diesel generators. One additional emergency diesel generator can work longer than 8 h, but its ability to continue to run decreases considerably after 8 h. There is no major difference between the RCIC systems and having one additional diesel generator. Therefore, in BWRs, even a three-division emergency diesel generator configuration has a capability equivalent to a four-division diesel generator configuration.

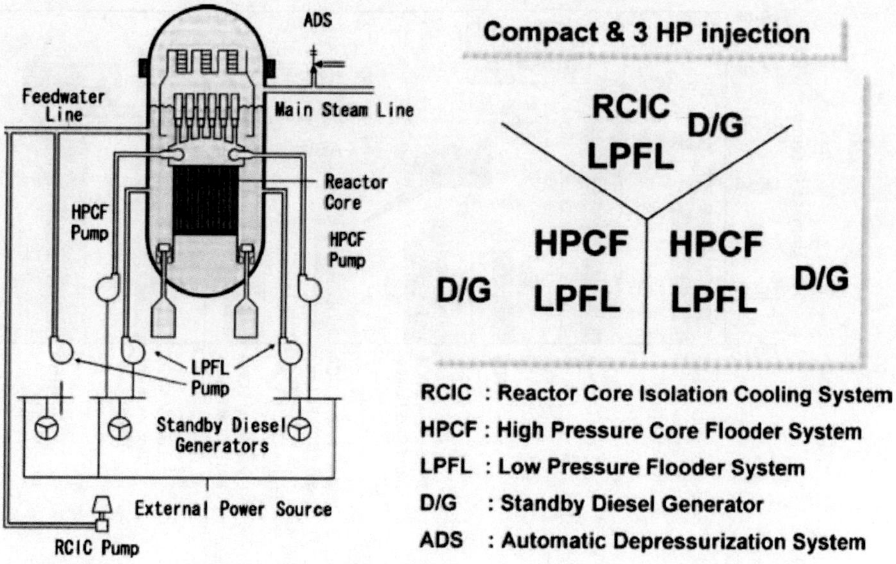

Fig. 1.13 ECCS configuration of the ABWR

On the other hand, when compared with a conventional BWR-5, the ABWR safety design has an RHR in each safety division. In the BWR-5, the HPCS system division does not include an RHR train. Therefore, a high suppression pool water temperature could damage the HPCS system during a station blackout, where only the HPCS system is operating with its dedicated diesel generator. This mechanism was one of the dominant sequences of the level 1 PSA. In the ABWR safety design, however, this mechanism hardly ever occurs because the RHR subsystems are distributed in each safety division, and the suppression pool water can be cooled when the HPCS system operates. Therefore, the three-division emergency diesel generator configuration of the ABWR has more capability than a conventional BWR-5 that has three diesel generators.

Finally, a complete three division concept was chosen as the ECCS configuration of the ABWR shown in Fig. 1.13.

1.2.3.2 PSA Performance of the ABWR

Figure 1.14 compares level 1 PSA results for internal events at power among Japanese BWR-4, BWR-5, and ABWR safety designs. The bases of the comparison, e.g., component failure rates, occurrence frequencies of transients, modeling of common-mode failures, and so on, are exactly the same among these plants. (This is a very important point so some additional discussions are provided as supplement to this chapter.)

Figure 1.14 clearly shows the safety improvement of the ABWR safety design, namely, approximately one order of magnitude reduction in the total core damage

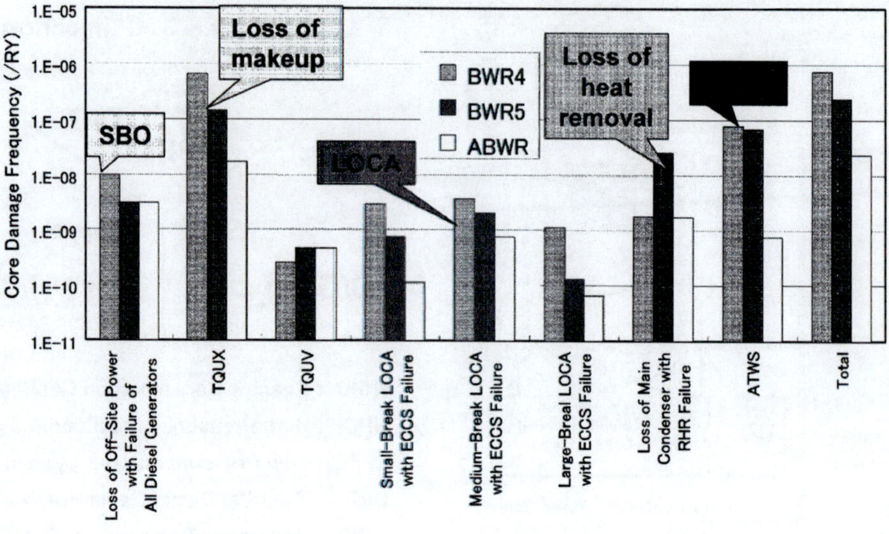

Fig. 1.14 Comparison of core damage frequency values for internal events at full power for Japanese BWRs. (Taken from [3] and used with permission from AESJ)

frequency. This is due to the reduction of three dominant sequence frequencies found in conventional BWRs, i.e., loss of feedwater with failure of high-pressure injection systems (TQUX), loss of main condenser with RHR failures and ATWS. Safety for these sequences is improved by redundancy enhancement of high-pressure core injection systems, redundancy enhancement of RHR systems and diversity enhancement of the scram system in the ABWR, respectively. Although dominant sequences of an ABWR are still transients followed by multiple failures, LOCA is overcome and not dominant in an ABWR, the same as in conventional BWRs.

It should be noted, however, that Fig. 1.14 shows only a relative comparison of the probabilistic safety performance of Japanese BWR plants. The absolute value of the core damage frequency is not so meaningful. This is because this level 1 PRA only covers internal events and full-power operation.

This PSA instead shows that the ABWR safety design reduced the risk of transients followed by multiple failures and that the core damage frequency caused by multiple failures in the mechanical portion of the plant is quite low.

1.2.3.3 ABWR Design Related to Safety Enhancement and/or Cost Reduction

ABWR design features related to safety enhancement and/or cost reduction safety are summarized in the following. In addition, various features to ensure safety but which are hard to quantify in a conceptual design stage are also summarized. The ABWR safety systems configurations are summarized in Table 1.5. A complete three-division safety system configuration is installed in the ABWR using part of the cost savings needed to cope with a DBA LOCA in conventional BWRs.

Table 1.5 BWR safety system innovations (taken from [3] and used with permission from AESJ)

Items	BWR4	BWR5	ABWR	Comments
ECCS/RHR	HPCI | (RCIC) CS | CS LPCI | LPCI LPCI | LPCI (RHR) | (RHR)	(RCIC) HPCS LPCS | LPCI LPCI | LPCI /RHR | /RHR	RCIC LPFL/RHR HPCF | HPCF LPFL | LPFL /RHR | /RHR	ABWR has 6×100% redundancy for core makeup at a LOCA
Division	2	3 (partial)	3 (full)	N−1 design
HP injection	2	2	3	Effect on TQUX[a]
RHR Hx	2×100%	2×100%	3×50%	Effect on TW[b]
D/G	2	3	3	Effect on SBO
Reactor shutdown	Hydraulic SCRAM	Hydraulic SCRAM	Hydraulic SCRAM + motor run-in	Effect on ATWS

[a] loss of feedwater with failure of high pressure injection systems
[b] loss of main condenser with RHR failures

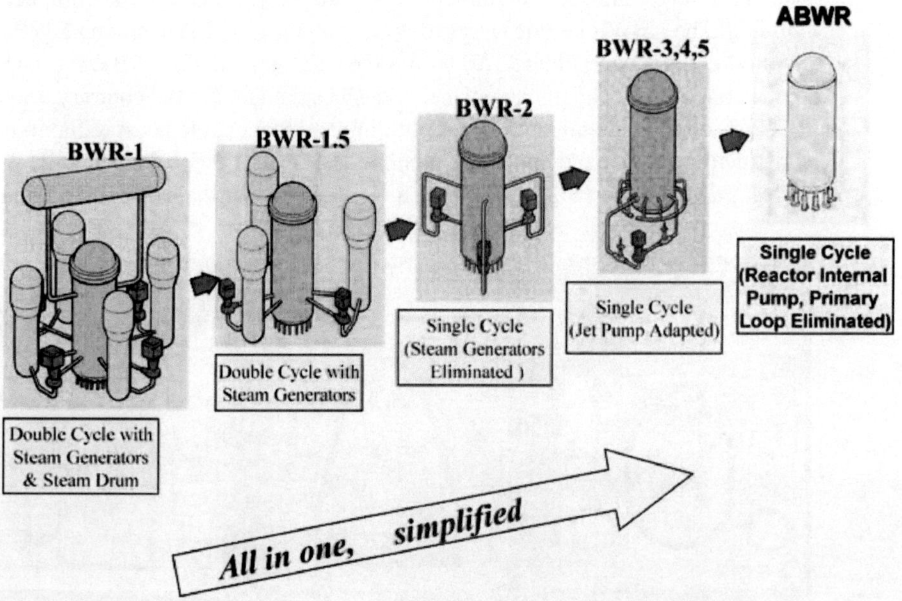

Fig. 1.15 Trends in primary system innovations of BWRs. (Taken from [3] and used with permission from AESJ)

Simplification of the Primary System

Figure 1.15 shows the trends in simplifications and innovations of the BWR primary system leading to the ABWR.

Adoption of internal pumps in the ABWR eliminates the external recirculation loops resulting in the most simplified primary system that has no large pipes connected below the core. Obtained merits are again listed below.

- Safety enhancement by achieving no core uncovering at DBA LOCA
- Reduction of ECCS capacity provided for DBA LOCA in conventional BWRs
- Cost reduction by eliminating the external recirculation loops needed in conventional BWRs
- Safety enhancement by LOCA frequency reduction because of total pipe length reduction inside the primary containment, although this was not included in PSA

Primary Containment Vessel Innovations

Due to the elimination of the external recirculation loops, the ABWR lowered the RPV into the pedestal. And the suppression chamber (S/C) was arranged very close to the RPV. With this closer arrangement, the ABWR reinforced concrete containment vessel (RCCV) could be very short. It is only 29.5 m high from the mat to the top slab. This very compact containment design also contributed to the compact reactor building. The ABWR has the largest output of about 1350 MWe among BWRs but the smallest containment. Figure 1.16 shows the PVC innovations of BWRs. The Mark I and II containments are made of steel and self-standing. On the contrary, the ABWR RCCV is combined with the reactor building and that enabled cost reduction, shorter construction period, and enhanced seismic design. The ABWR containment has the lowest gravity center; it is about 10 m lower than that of the Mark II containment.

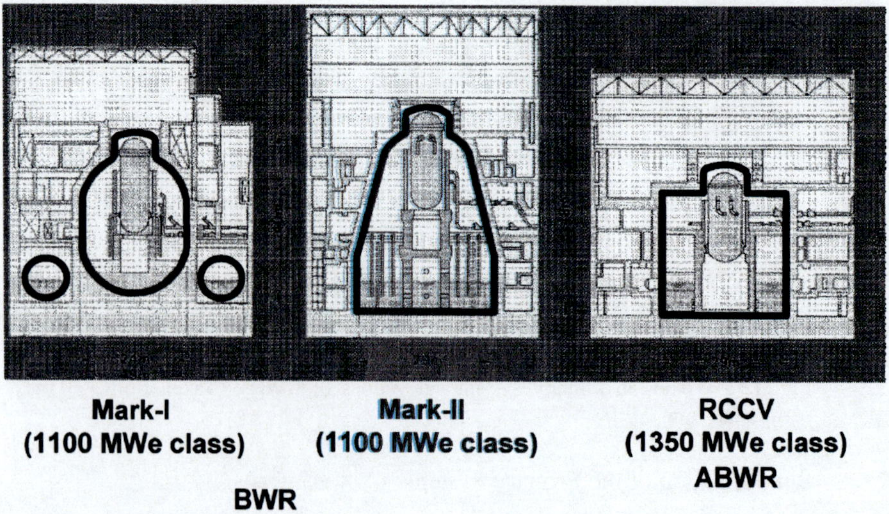

Fig. 1.16 Primary containment vessel innovation of BWRs. (Taken from [3] and used with permission from AESJ)

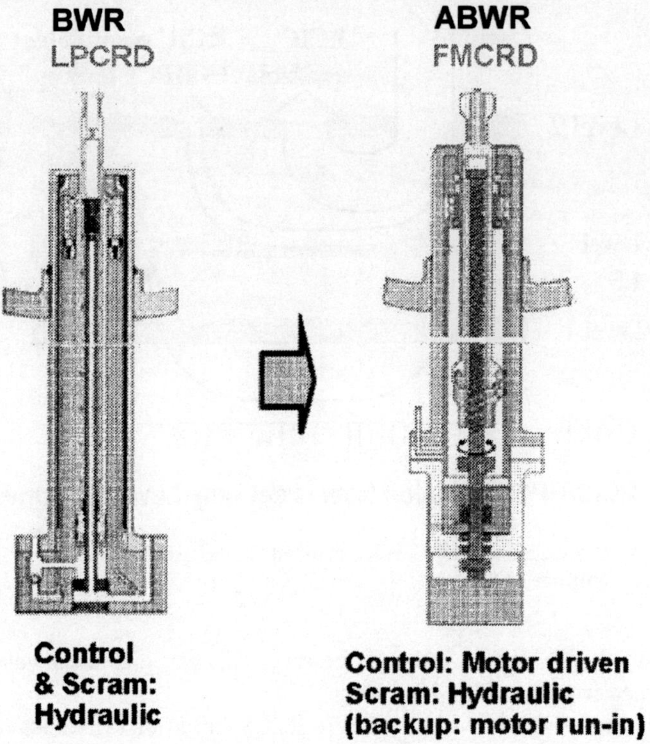

Fig. 1.17 Comparison between LPCRD and FMCRD. (Taken from [3] and used with permission from AESJ)

Adoption of Fine Motion Control Rod Drive

The ABWR improved reliability of the reactor shutdown system using the FMCRD. The FMCRD has a back-up motor run-in capability in addition to the hydraulic scram for complete diversity. Figure 1.17 compares the conventional locking piston CRD (LPCRD) and the FMCRD. Both have hydraulic scram but only the FMCRD has the motor run-in backup capability.

The FMCRD was adopted to improve normal operation. This, however, resulted in a large cost increase. On the other hand, the elimination of external recirculation greatly reduced the containment volume as well as the volume of the reactor building. These volume reductions brought about cost reductions that could compensate for the cost increase for the FMCRD. Therefore, the utilization of the FMCRD as an ATWS countermeasure did not cause any net cost increase.

ECCS Initiation Level Separation Between Transient and LOCA

The TMI-2 accident taught operators of nuclear power plants that once the ECCS is initiated an operator must not stop it. In order to facilitate this, the ECCS must be

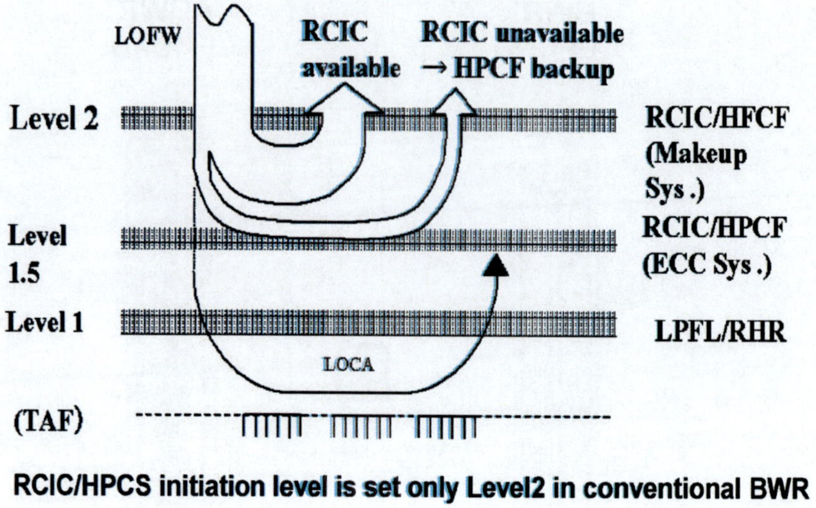

Fig. 1.18 ECCS initiation level separation in the ABWR. (Taken from [3] and used with permission from AESJ)

initiated only when it is truly necessary. This was one of the very important lessons learned from the TMI-2 accident.

Figure 1.18 shows the ABWR ECCS initiation water levels in the RPV that separate the HPCF initiation level from the RCIC initiation level. If a loss of feedwater transient occurs, the RCIC starts at the level 2 and the water level goes up. The two HPCFs are never initiated in the transient sequence and this moderates operator stress at the transient. However, if the RCIC fails to be initiated at level 2, then the water level goes down to level 1.5 and the two HPCFs are initiated to back up the RCIC. The HPCS of the conventional BWR also had the same function. The HPCS of the conventional BWR is initiated at level 2, namely, simultaneously with the RCIC, which is unnecessarily in a loss of feedwater transient.

This results in reducing human error probability (HEP) and enhances safety, although these are not modeled prudently in level 1 PSA.

Adoption of New Design Main Control Panel and Instrument and Control (I&C) Technologies

The ABWR adopts a newly designed main control panel using a state-of-the-art I&C system, named A-PODIA™ (Advanced Plant Operation by Display Information and Automation). This main control panel has various features among which the following features especially contribute to reduce HEP and to enhance safety, however these effects are hard to quantify.

- Information sharing by large display avoids miss-communication among operating crew resulting in reduced HEP.

- Compact main console using touch panel and flat display minimizes operator burden and avoids miss-selection of operation devices resulting in reduced HEP.
- Expansion of automated operation scope, especially automation for post-scram operation and control rod operation at startup resulting in reduced HEP. The former function reduces HEP directly, while the latter function minimizes operator burden and reduces transient occurrence frequency related to plant startup. Both contribute to safety enhancement.

The digital control system and the optical fiber network are employed throughout the ABWR for all plant systems, including safety-related systems, which realizes more reliability and greater performance than a conventional analog system. In the safety protection system, two-out-of-four logic is applied and that achieves more tolerant logics to both failure to initiate and spurious actuation.

Accident Management of the ABWR

Figure 1.19 shows accident management (AM) countermeasures of the ABWR. The Chernobyl 4 accident occurred after the ABWR safety design was set. Therefore, those AM countermeasures were added in exactly the same way as for conventional BWR plants. The AM countermeasures are not safety grade systems but still very effective to reduce the risk of severe accidents, although these effects are not included in the PSA.

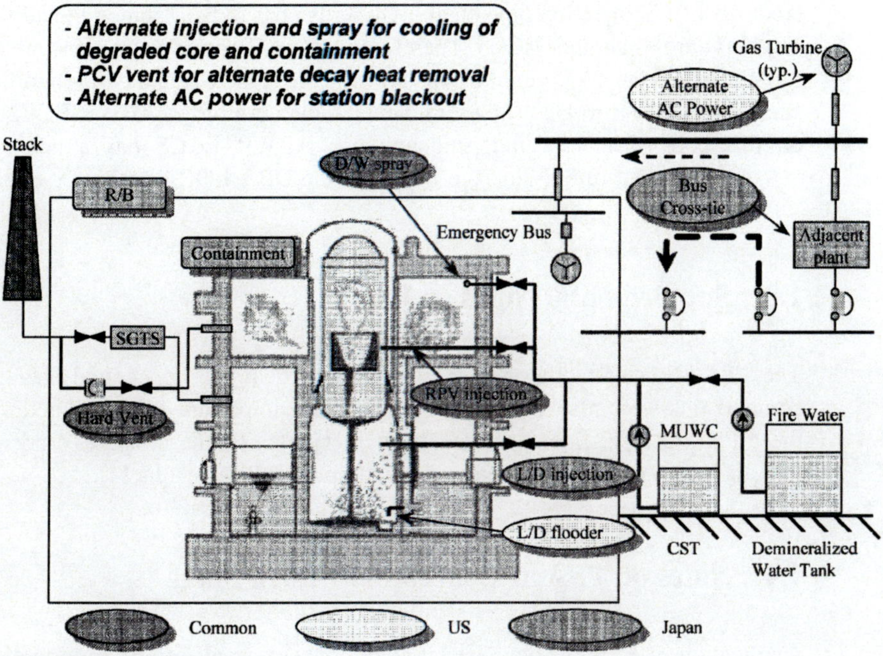

Fig. 1.19 Accident management countermeasures of the ABWR. (Taken from [3] and used with permission from AESJ)

1.2.4 Application in Maintenance

The major topic in the maintenance field concerning risk evaluation is online maintenance (scheduled maintenance during operation) to increase plant availability. For online maintenance, an (N-2) configuration is required deterministically, i.e., one system is ineffective due to maintenance and another maintains its ability to cope with the initiating event, allowing a third one to mitigate system failure. This means accident sequence frequency is limited below production of the initiating event frequency (IE) and mitigation system unavailability (P1, P2), i.e. IE $\times$ P1 $\times$ P2. But, even the (N-1) configuration may be allowed if the accident sequence frequency is extremely low and the duration of maintenance is very short.

The ECCS of the ABWR was designed to have more than N-2 reliability for most events except for a DBA LOCA. A single failure of an emergency diesel generator plus loss of off-site power is assumed at a DBA LOCA. For only this case the ABWR ECCS has single failure (N-l) reliability. Any one ECCS pump of the ABWR is intentionally designed to have enough capacity to compensate for any pipe break LOCA by itself independently in order to establish good PSA performance. This performance is called independency in the ECCS design requirements. This performance can be easily achieved because a DBA LOCA of the ABWR is not large and all the pipes are connected above the core. If the AC power source is available, ABWR ECCS has 6-pump independency for any pipe break LOCA. The turbine-driven RCIC, however, loses its safety function after the RPV is depressurized. An ECCS injection pipe break LOCA also has to be assumed in addition to a single failure of another ECCS. Therefore, if the AC power source is available, the ABWR ECCS has N-3 reliability for any pipe break LOCA and N-4 reliability for a small LOCA. Figure 1.20 shows a schematic diagram of the ABWR ECCS. Based on this performance of independency, the ABWR ECCS has a potential for enhancement to a full N-2 design including the DBA LOCA case very easily.

1.3 Supplemental Notes on PSA

The PSA shown in this chapter was performed at the conceptual stage. Some comparisons were also shown. There exist very important issues related to using these PSA results. This supplement provides notes on these issues, especially on PSA at conceptual design stage and on comparing PSA results [4].

1.4 Notes on PSA at Conceptual Design Stage

Application of PSA for new design plant concepts is also very important. In this case, however, there are some limitations. The first is the limitation of information. Precise plant design information cannot be provided for PSA engineers at a

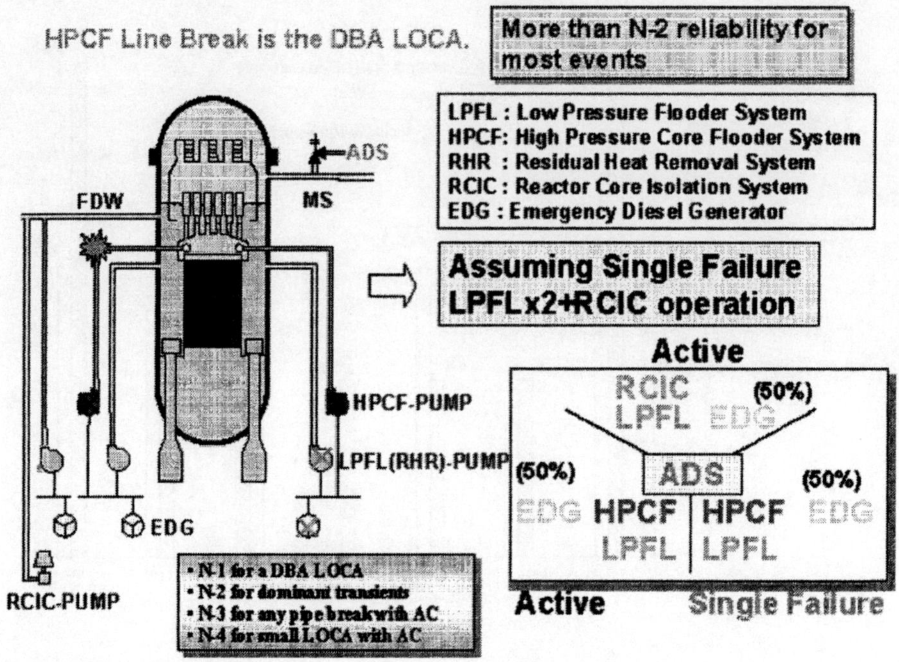

Fig. 1.20 Schematic drawing of ABWR ECCS with failure modes. (Taken from [3] and used with permission from AESJ)

conceptual design stage. The second is the limitation of time. Usually only a short time is allowed to conduct a PSA before deciding on the plant design. This is because plant design work itself requires much more time than PSA in the development of advanced reactors. PSA engineers are required to conduct PSAs for several plant concepts in order to choose the most favorable system and concept within a short period.

The ABWR level 1 PSA shown in this chapter was performed at a conceptual design stage about 10 years ago. It was by a conditional event tree method. No precise fault tree analysis was conducted because each safety system of the ABWR was almost the same as the conventional BWR safety systems. For the ECCS, the same amount of unreliability as in a conventional plant could be used. Only the network configuration was of interest for the ECCS in the PSA.

1.5 Notes on Comparing PSA Results

PSA results depend significantly on the analysis bases, such as scope, major premises, assumptions, data, and so on. Therefore, to compare PSA results in more detail, more careful attention should be paid to the analysis bases.

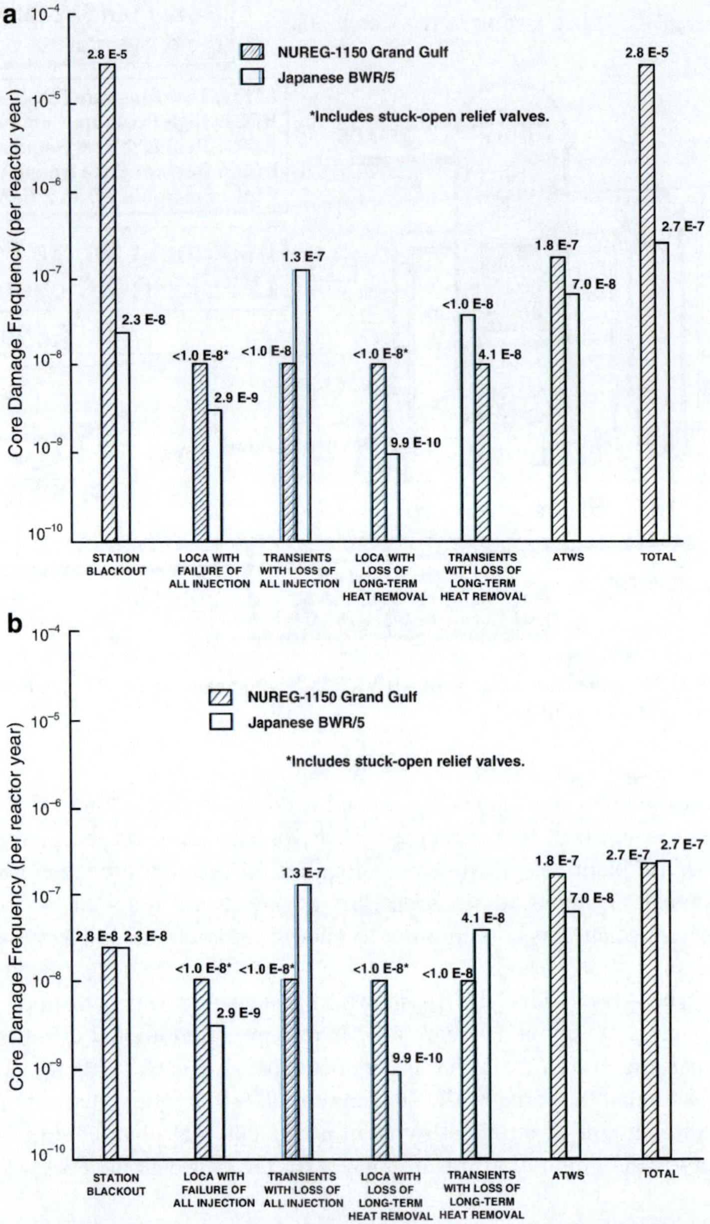

Fig. 1.21 Comparison of CDF between a Japanese BWR-5 and Grand Gulf results, (**a**) comparison of original results and (**b**) comparison with modification on station blackout sequence. (Taken from [4] and used with permission from ELSEVIER)

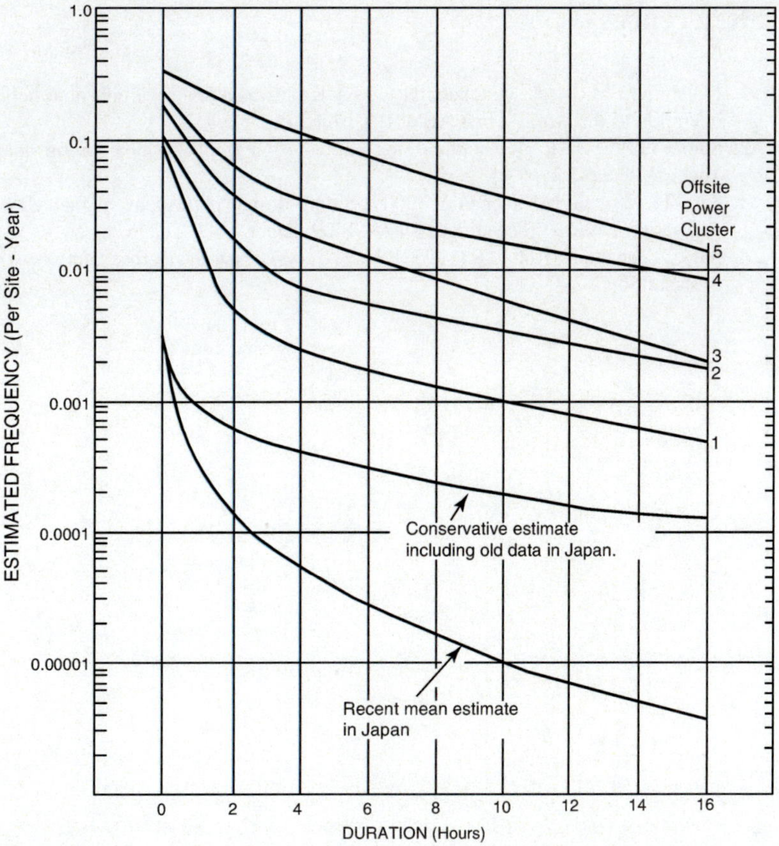

Fig. 1.22 Comparison of recovery curves of off-site power between Japan and the United States. (Taken from [4] and used with permission from ELSEVIER)

For example, Fig. 1.21a shows level 1 PSA results comparison between the Grand Gulf Nuclear Power Plant in NUREG-1150 and a Japanese BWR-5 shown in Ref. [4]. The only major difference comes from the station blackout column. For the other sequences there is little difference. Therefore, Japanese BWRs have a lower CDF because of higher reliability of power sources. This mostly comes from the high reliability of the off-site power. Figure 1.22 compares recovery characteristics of off-site power between Japan and the United States. There are about two orders of difference which can explain almost all the differences between Japanese level 1 PSA and that of the United States. If the difference in reliability of off-site power is corrected, level 1 PSA results between the Grand Gulf plant in NUREG-1150 and a Japanese BWR-5 are almost equivalent as shown in Fig. 1.21b.

References

1. Matsumura M, Ikeda M, Okabe N (1997) Kashiwazaki-Kariwa unit no.6 begins commercial operation as the world's first ABWR, Toshiba Rev 52(4):2–9
2. Sato T (1992) Basic philosophy of the safety design of the Toshiba boiling water reactor. Nucl Technol 99:22–35
3. Sato T, Akinaga M, Kojima Y (2009) Safety design philosophy of the ABWR for the next generation LWRs, Proceedings of ICAPP'09, Paper 9447, Tokyo, Japan
4. Sato T, Tanabe A, Kondo S (1995) PSA in design of passive/active safety reactors. Reliab Eng Syst Saf 50:17–32

Chapter 2
The Advanced Accumulator: A New Passive ECCS Component of the APWR

Tadashi Shiraishi

With the increased requirement for nuclear power generation as an effective countermeasure against global warming, Mitsubishi has developed the advanced pressurized water reactor (APWR) by adopting a new component of the emergency core cooling system (ECCS), a new instrumentation and control system, and other newfound improvements. The ECCS introduces a new passive component called the Advanced Accumulator which integrates both functions of the conventional accumulator and the low-pressure pump without any moving parts. The Advanced Accumulator uses a new fluidics device that automatically controls flow rates of injected water in case of a loss-of-coolant accident (LOCA). This fluidics device is referred to as a flow damper. In this chapter, the Advanced Accumulator is introduced from the background of its development to its principle, with some experimental results. Furthermore, the features of the flow damper are explained in detail.

2.1 Overview of the APWR

The development of the APWR was launched jointly by the Japanese government, five utility companies (Hokkaido Electric Power Co., Kansai Electric Power Co., Shikoku Electric Power Co., Kyushu Electric Power Co., and Japan Atomic Power Company) and suppliers, including Mitsubishi, in the early 1980s, under the Third Phase Improvement Standardization Program for Light Water Reactors [1]. The development was aimed at establishing an advanced standard light-water reactor with further enhanced reliability and safety, improved economy and more efficient usage of location with increased power, based on the results of the First and Second Phase Improvement Standardization Programs. The development also involved Mitsubishi's experience and technologies obtained through the design, construction,

T. Shiraishi (✉)
Mitsubishi Heavy Industries, Ltd, Tokyo, Japan
e-mail: tadashi_shiraishi@mhi.co.jp

T. Saito et al. (eds.), *Advances in Light Water Reactor Technologies*,
DOI 10.1007/978-1-4419-7101-2_2, © Springer Science+Business Media, LLC 2011

and operation of existing PWR plants. Even after the completion of the national programs, under the continuous support of the utilities, the APWR design has been improved through uprating by redesigning its core structure, optimizing its safety systems by introducing an advanced accumulator tank, and making other modifications. In March 2004, the Japan Atomic Power Company applied for the approval of changes in the reactor installation related to the addition of Tsuruga Units 3 and 4, the first two APWRs. As a result, the government has started the safety assessment of these units.

On the basis of the operating success of these units, Mitsubishi is making efforts to establish APWRs as a lineup of large-capacity standard PWR plants with excellent enhanced economy. This lineup includes 1,600 and 1,700 MWe class APWRs whose design certification in the US has already been applied for, with high-performance steam generators and steam turbines. The APWRs have realized a large capacity increase of about 30% or more, compared with the current 4-loop PWRs as shown in Fig. 2.1.

The main features of APWR plant components, as outlined in this section, are as follows:

1. Large reactor core and main components with large capacity (steam generator, primary coolant pump, pressurizer, and turbine)

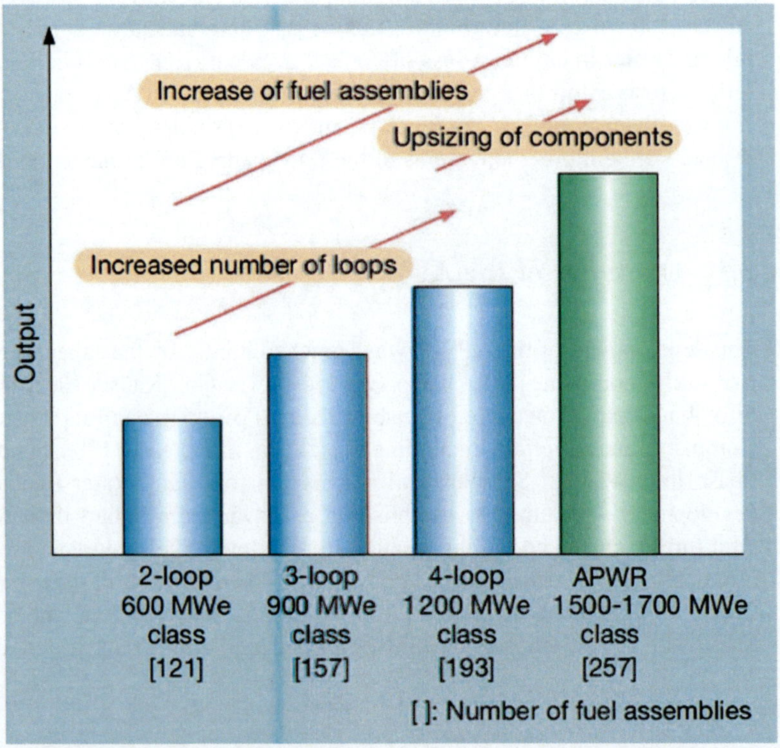

Fig. 2.1 The trend in output capacity of PWR plants

2. Advanced safety systems (four subsystems and refueling water storage pit installed in the reactor containment)
3. New instrumentation and control systems (advanced digital main control board)

2.1.1 Large Capacity Core

The APWR has adopted a large capacity core to achieve high thermal power and has increased the number of fuel assemblies from 193 in the current 4-loop type to 257 as shown in Fig. 2.2. An advanced 17 × 17 fuel assembly has been adopted as the fuel bundle, and also, a zircaloy grid that absorbs fewer neutrons, which is already used in current plants, has been incorporated for the effective use of uranium resources. Furthermore, the APWR allows the number of control rods to be set according to the quantity of loaded MOX fuel, so that the requirement for diverse operations, such as the use of a MOX core and high burnup, can be met flexibly.

2.1.2 Neutron Reflector

The APWR employs a neutron reflector as an internal component for effective use of uranium resources. The reflector has a simple structure that consists of stacked blocks of stainless steel rings without weld lines and with a few bolt connections, whereas the same internal in the current PWR has a baffle structure in which stainless steel plates are connected with many bolts as shown in Fig. 2.3. The new structure reduces neutron irradiation to the reactor vessel to about 1/3 so that the reliability of the vessel is improved.

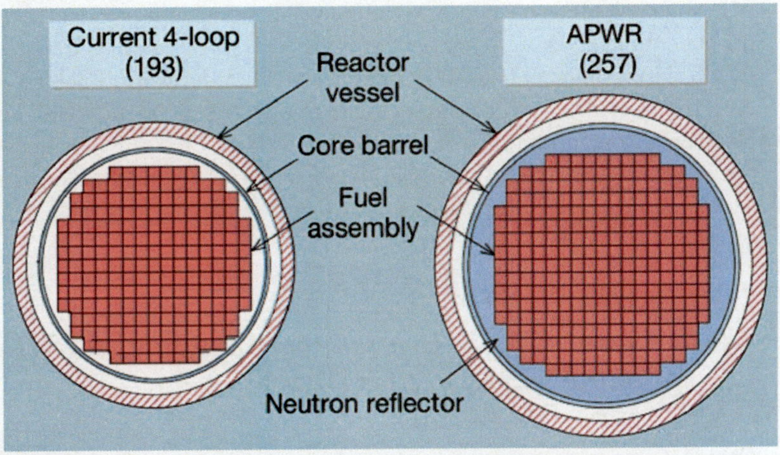

Fig. 2.2 The APWR adopts a larger core to increase the number of fuel assemblies from 193 to 257

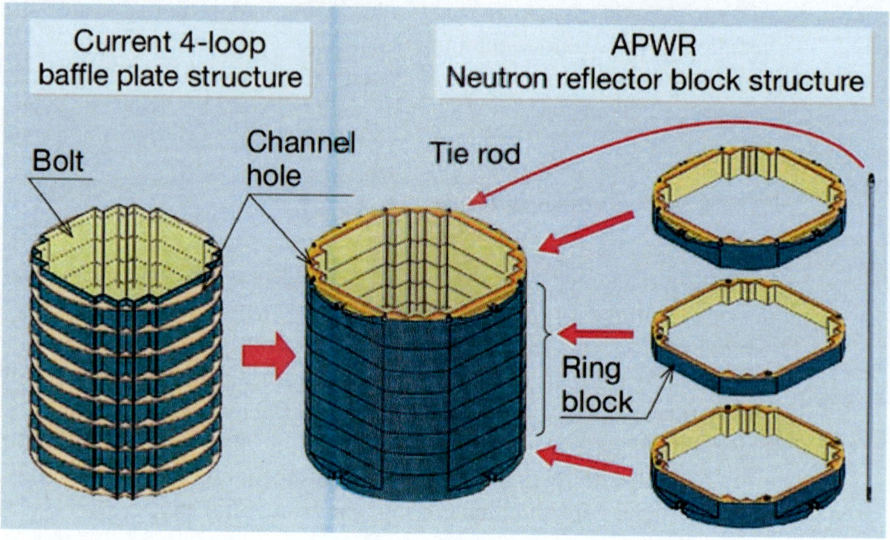

Fig. 2.3 The neutron reflector for the APWR consists of rings without weld lines and with only a few bolt connections

2.1.3 Advanced Safety Systems

In the APWR, the emergency core cooling system (ECCS) has a 4-train configuration (4 × 50% capacity) instead of the conventional 2-train configuration (2 × 100% capacity) to improve safety. The new configuration increases the reliability of equipment operation in the case of an accident as the best mix of active and passive safety systems. The systems of each train are installed near the corresponding loop to reduce the quantity of piping and enhance the separation and independence of each train. A refueling water storage pit is moved from outside of the containment vessel to the bottom of the vessel to serve as a water source for the ECCS during an accident. With this design, cooling water injected into the core during an accident can be automatically collected in the pit. This eliminates the changeover operation of the core cooling water source and enhances safety. Furthermore, an Advanced Accumulator with a passive concept has been adopted. It is explained in detail in this chapter.

2.1.4 Advanced Main Control Board and Integrated Digital Control and Protection System

Compact console panels are adopted in the advanced main control board to accomplish all monitoring and operations by touch screen displays. The operational

Fig. 2.4 Advanced main control board (prototype)

switches of the plant components and the necessary operation information are consolidated on the screen to improve the operators' performance as shown in Fig. 2.4. When an anomaly occurs in the plant, the panel's rich supporting functions automatically check the status of the plant and equipment operation and provide the necessary information. Compared with the conventional type, the advanced main control board is expected to reduce the operators' burden by about 30% and human error by about 50%.

2.1.5 Main Components with Increased Capacity

Main components with increased capacity have been developed to cope with the increased core output, and various technologies to improve performance and reliability are being adopted and verified. According to the lineup of plant electricity output, a compact, high-performance steam generator (SG) with a substantially larger heat transfer area than that of the conventional 4-loop type is employed as shown in Fig. 2.5. In order to minimize the increase of the outer dimensions of the increased capacity SG, the tube diameter has been decreased from 7/8 to 3/4 inches to reduce the diameter of the SG, and an improved moisture separator with a reduced number of stages has been adopted to reduce the SG height. This reduces the weight of the SG by about 10% or more compared with the larger SG based on the conventional design concept. To improve reliability, the number of antivibration bars installed at U-bends has been increased from six in the current plants to nine or more.

To provide high efficiency to the steam turbine, the last-stage blades of the low pressure turbine have been extended by adopting blades as long as 54–74 in. Ideal

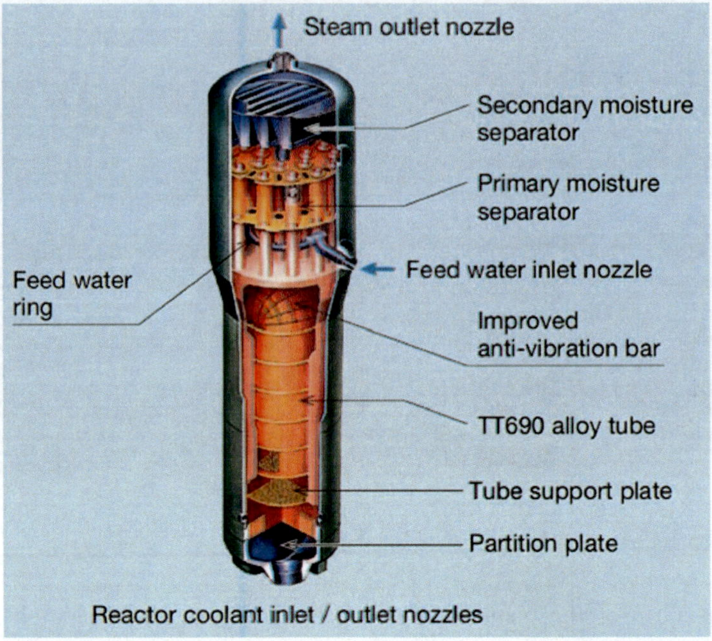

Fig. 2.5 The compact, high-performance steam generator for the APWR has a large heat transfer area in a minimized size

high-performance three-dimensional (3D) blades have been adopted to reduce blade loss by making a complete 3D flow design. Furthermore, integral shroud blades (ISBs) aimed at reducing vibration stress by forming an all-round stitch structure through contact with adjacent blades during revolution have also been adopted to enhance reliability. The developed blades were tested under actual steam conditions at Mitsubishi's own test facility, the world's largest class test facility, to verify their performance and reliability as shown in Fig. 2.6. These new technologies are also being applied in the replacement of turbines in existing domestic and overseas plants.

2.2 Development of the Advanced Accumulator

Globally speaking, development of passive safety systems for nuclear plants thrived in the 1980s against the backdrop of the Three-mile Island Accident in March of 1979 and the Chernobyl Accident in April of 1986. At the time, Mitsubishi Heavy Industries, Ltd. (MHI) had been developing a hybrid safety system, namely, orchestrating the merits of both passive and active safety systems [2]. The designers of the hybrid safety system requested the members of the R&D group to propose a device that changes the flow rate of the ECCS with high reliability and is maintenance-free for the plant life time. The solution we proposed was the Advanced Accumulator with a new fluidics device, called a flow damper which

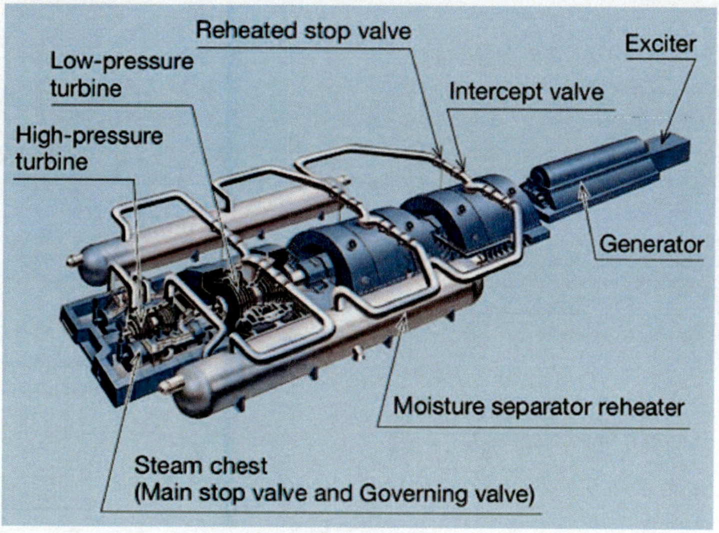

Fig. 2.6 Large-capacity, high-performance turbine generator system

has no moving parts. Since there are no moving parts, fluidic elements have extraordinary reliability and require no maintenance.

The Advanced Accumulator was invented using the science of fluidics in 1986 [3] and was developed for Mitsubishi's next generation PWR after the APWR from 1987 to 1994. The experimental results were reported in [4–10]. The results were successful in that the Advanced Accumulator was experimentally verified to have the basic functions that we expected, and the ratio of flow rates for large and small flow injections was confirmed to satisfy the requirement for the next generation PWR.

The development of the Advanced Accumulator for the APWR was then initiated in 1995 and completed in 1997. Since a fluidic element has no moving parts, its configuration had to be modified for different specifications. The Advanced Accumulator for the APWR was incorporated into the safety system design to provide the low-pressure injection function of the current ECCS using a conventional accumulator and a safety injection pump. This arrangement simplifies the configuration of ECCS and allows sufficient time to use gas-turbine generators for safety injection pumps.

2.3 ECCS of the APWR

A loss of coolant accident (LOCA) is the severest hypothetical accident. Figure 2.7 shows the scenario of a large break LOCA of PWR as follows:

1. One of the main coolant pipes is assumed to break with a large opening.
2. Pressure in the reactor vessel plummets towards atmospheric pressure.

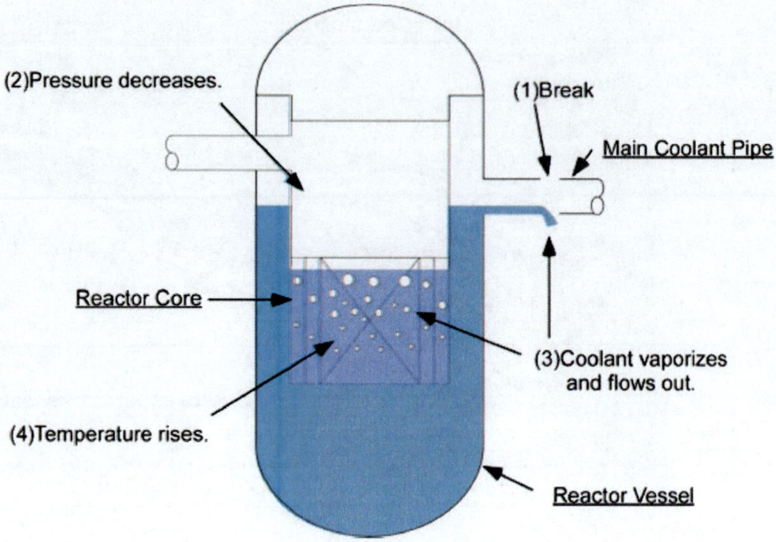

Fig. 2.7 At a large break LOCA, one of the main coolant pipes is assumed to break with a large opening. The coolant will then vaporize and flow out of the main coolant pipe

3. Coolant vaporizes in the rector vessel and flows out the opening of the main coolant pipe to expose the reactor core.
4. The fuel cladding temperature begins to rise and requires additional coolant to protect fuels from serious damage.

The ECCS supplies borated water into the reactor vessel to meet the requirement of sufficient cooling of the reactor core at LOCA. Figure 2.8 compares the ECCS configurations of the current 4-loop PWR and APWR. The current 4-loop PWR has two trains of injection systems, conventional accumulators, and a refueling water storage pit outside of the containment vessel. The APWR has some improvements, which include the following:

1. Four trains of the injection system composed of simplified pipe routing for higher reliability
2. Four Advanced Accumulators that allow for the elimination of low-head safety injection pumps
3. An in-containment refueling water storage pit located in the containment vessel for higher reliability

Thus, the current ECCS is composed of the accumulator injection system with conventional accumulators, a low-head injection subsystem, and a high-head injection subsystem. The new ECCS for APWR is composed of the accumulator injection system with Advanced Accumulators and a safety injection subsystem without low-head safety injection pumps.

Figure 2.9 shows a view of the safety system of APWR. There is a reactor vessel (2) at the center of the containment vessel (1), to which steam generators (3) and

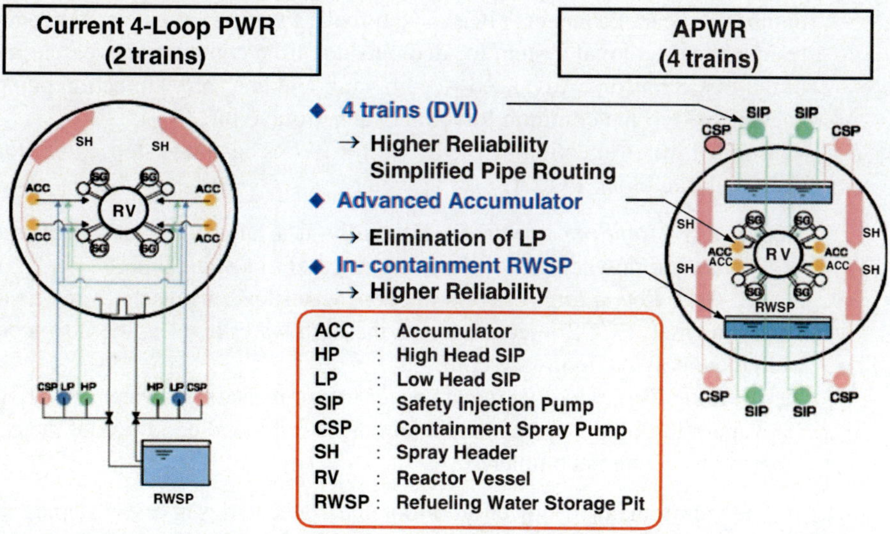

Fig. 2.8 Configurations of the ECCS for the current 4-loop PWR and APWR. The APWR has higher reliability and a simplified pipe routing without low-head injection pumps. The systems of each train are installed near the corresponding loop to reduce the quantity of piping and enhance the train separation and independence

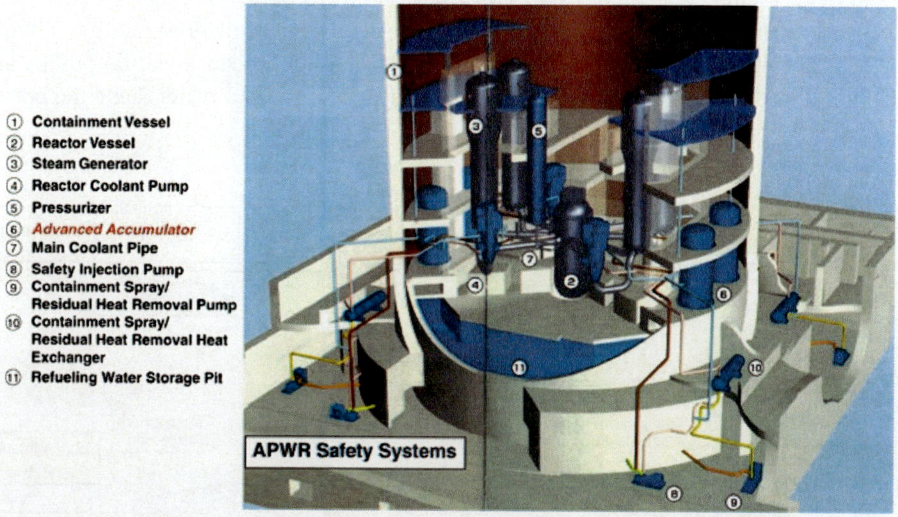

Fig. 2.9 Cut-away view of the safety system of the APWR shows the configuration of the components

reactor coolant pumps (4) are connected by the main coolant pipes (7) to form loops. A pressurizer (5) is connected to one of the loops. An Advanced Accumulator (6) is connected to every main coolant pipe (7). Safety injection pumps (8) are located outside of the containment vessel. Containment spray/residual heat removal pumps (9), heat exchangers (10), and a refueling water storage pit (11) compose the residual heat removal system located outside of the containment vessel. Each train of the ECCS has one Advanced Accumulator and one safety injection pump, and it is connected to the common refueling water storage pit.

Borated water injection by the ECCS for PWRs has three steps after blow down due to a large break LOCA.

Step 1: *Core Refilling* injects water rapidly at a large flow rate to fill the lower plenum and downcomer of the reactor vessel in a short time.

Step 2: *Core Reflooding* recovers the core water level using the water head in the downcomer. ECCS injection keeps the high water level in the downcomer and immediately re-floods the core.

Step 3: *Long-Term Cooling* injects water to compensate for water reduction due to evaporation by decay heat and maintains a reflooded condition of the core after core reflooding is completed.

The requirement for injection of borated water varies at every step as shown in Fig. 2.10. If a large break LOCA happens, a large flow rate of injected water is required at Step 1 for Core Refilling. The water head in the downcomer drives water into the reactor core at Step 2 for Core Reflooding. At this step, a relatively small flow rate is required because any excess water will flow out of the opening to no purpose. The reactor core will be covered with water at the end of the Core Reflooding and Step 3 starts for Long-Term Cooling.

The current 4-loop plants satisfy the requirement of flow rate by conventional accumulators at Step 1, and by low-head and high-head injection pumps at Steps 2 and 3. Each conventional accumulator injects borated water using the pressure of

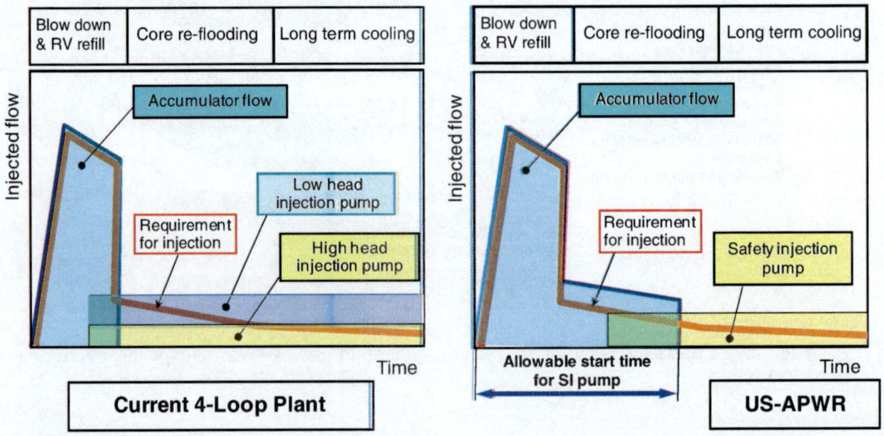

Fig. 2.10 The schematic drawings on the *left* and the *right* show the injection modes of the current 4-loop plant and of the APWR, respectively. The APWR is improved by using the Advanced Accumulators to eliminate the low-head injection pumps and to obtain a longer allowable start time for SI pumps

nitrogen gas stored in it, while low-head and high-head injection pumps are driven by diesel generators.

On the contrary, the APWR satisfies the requirement by the Advanced Accumulators at Steps 1 and 2, and by Safety Injection Pumps at Step 3. The advantages of the Advanced Accumulators are the automatic changeover of injected flow rate from Step 1 to Step 2 in the absence of moving parts, and eliminating the low-head injection subsystem. Each Advanced Accumulator injects borated water using the pressure of stored nitrogen gas, the same as the conventional accumulators do. Furthermore, longer time injection of the Advanced Accumulators allows a longer time to prepare for the start of the safety injection pumps. In other words, not only diesel generators but also gas-turbine generators can be selected for the safety injection pumps.

2.4 Characteristics of the Advanced Accumulator

The configurations of the current and new safety systems are shown in Fig. 2.11. The current safety system is composed of conventional accumulators, low-head injection pumps and high-head injection pumps, which are not shown within the figure. The new safety system is composed of Advanced Accumulators and safety injection pumps, which are also not represented within the figure. The Advanced Accumulator has a flow damper at the bottom of the tank. The Advanced Accumulator orchestrates large flow injection for refilling and small flow injection for

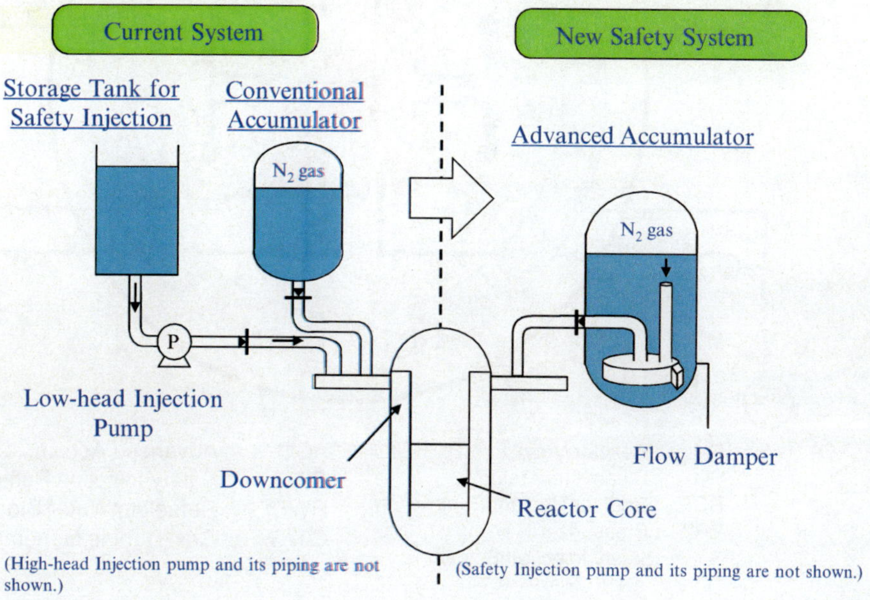

Fig. 2.11 The current safety system has a low-head injection system in addition to a conventional accumulator, while the new safety system has an Advanced Accumulator

reflooding. Since the low-head injection pumps are not needed, that allows a longer time to start the safety injection pumps.

Both conventional and Advanced Accumulators drive borated water with nitrogen gas filled at the tops of their tanks. There is a standpipe which detects water level in the tank of the Advanced Accumulator. The standpipe is connected to the large flow pipe of the flow damper to supply water only for large flow injection. Stopping the supply of flow from the standpipe changes the flow resistance of the flow damper without using moving parts.

Figure 2.12 shows the specific configuration of the ECCS for the APWR. Four Advanced Accumulators are installed and each Advanced Accumulator connects to a separate cold leg of the reactor coolant system (RCS). Four high-head safety injection subsystems are installed in order to inject water following injection by the Advanced Accumulators. There is no low-head injection subsystem.

The fundamental safety requirement for the ECCS is to limit the peak clad temperature (PCT) of fuel rods in the reactor core to 1,200°C during a large break

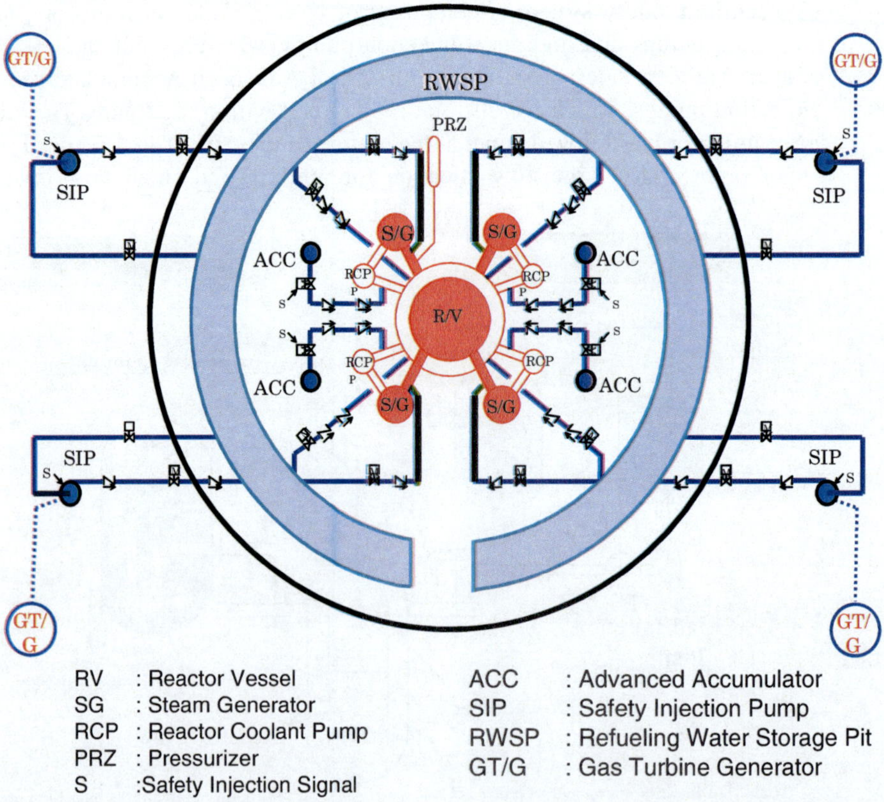

Fig. 2.12 The ECCS of the APWR has four pipelines for the safety injection pumps. Each Advanced Accumulator is connected to a separate cold leg. The refueling water storage pit is placed at the *bottom* of the containment vessel

LOCA. The functions of the Advanced Accumulators are encompassed within the following two steps:

1. To immediately refill the lower plenum and the downcomer of the reactor vessel during refilling period following blow down of reactor coolant
2. To establish a reflooding condition of the core by maintaining the water level in the downcomer after refilling the core

Hence, the performance requirements for the Advanced Accumulator design are the requirements for large flow injection, which comes from Step 1, and for small flow injection which comes from Step 2.

The requirements for large flow injection during the refilling period are that the water volume for large flow injection in the accumulator tank should cover the total volume of the lower plenum and downcomer regions of the reactor vessel, and that the lower plenum and the downcomer should be filled with borated water as rapidly as possible during the refilling period. The requirements for small flow injection are that the required small injected flow rate is determined by the performance requirements of the ECCS, along with the assumption that 3-out-of-4 sets of Advanced Accumulators are available. The large injected flow rate before flow-rate changeover is given by the expected flow rate at the end of large flow injection from calculated results. Consequently, the requirement for the flow-rate changeover ratio from large to small flow injection should be less than the maximum value required for small injected flow rate and be set with some margin.

2.5 Development of the Advanced Accumulator

The functions of the Advanced Accumulator are realized by a new fluidics device called a flow damper. It secures the large flow injection, the rapid flow changeover, the desired ratio of injected flow rates before and after the changeover, and the small flow injection. The flow damper is named after its function to restrict flow.

2.5.1 Structure of the Flow Damper

Figure 2.13 shows the structure of the flow damper and its installation at the bottom of the accumulator tank. There is a vortex chamber at the center of the flow damper. A small flow pipe is tangentially attached to the vortex chamber. A large flow pipe is radially attached to the chamber at one end and is connected to the standpipe at the other end. An outlet nozzle stands at the center of the vortex chamber, and is connected to the outlet pipe which leads to the injection pipe. An antivortex cap and an antivortex plate are set on the upper inlet of the standpipe and the lower inlet of the small flow pipe, respectively.

The antivortex cap installed on the inlet port of the standpipe prevents the formation of a vortex and gas entrainment in the flow damper just before the flow changeover. Additionally, it improves flow-rate changeover characteristics.

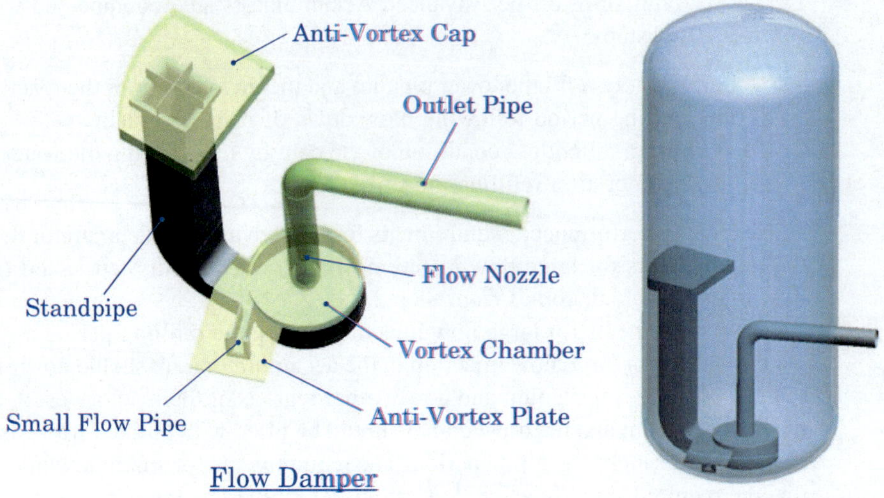

Fig. 2.13 Schematic views of the flow damper. Its detailed structure appears on the *left* and its installation in the accumulator tank is shown on the *right*

The inlet of the standpipe is set at the water level for switching flow from large to small flow injections. The small flow pipe is tangentially attached to the vortex chamber and has a configuration designed to reduce energy loss in it to make a strong vortex in the chamber. A throat is provided to increase flow resistance during small flow injection along with a diffuser to recover pressure during large flow injection in the outlet nozzle, which is smoothly connected to the injection pipe. The injection pipe is connected to a cold leg of the RCS. The antivortex plate on the small flow pipe prevents gas entrainment at the very last time of small flow injection.

2.5.2 Design of the Flow Damper

The mechanical configuration of the flow damper is shown in Fig. 2.14. The vortex chamber is chosen to be a cylindrical structure able to form a strong vortex in it. The small flow pipe is tangentially connected to the vortex chamber to yield a strong vortex for small flow injection. If energy loss is negligible in the chamber, the angular momentum is preserved to form a free vortex. The formation of a free vortex is very useful to get a large pressure drop across the vortex chamber. The pressure drop will be larger as the ratio of the radii of the chamber and the throat gets larger. But a larger ratio of the radii requires a larger space for the vortex chamber. Therefore, the optimum ratio for the radii will be achieved based on the required characteristics of the flow damper and the space in which the flow damper can be placed. The ratio of the radii is one of the key parameters of the flow damper. The vortex may be so strong that cavitation could appear at the center of the vortex

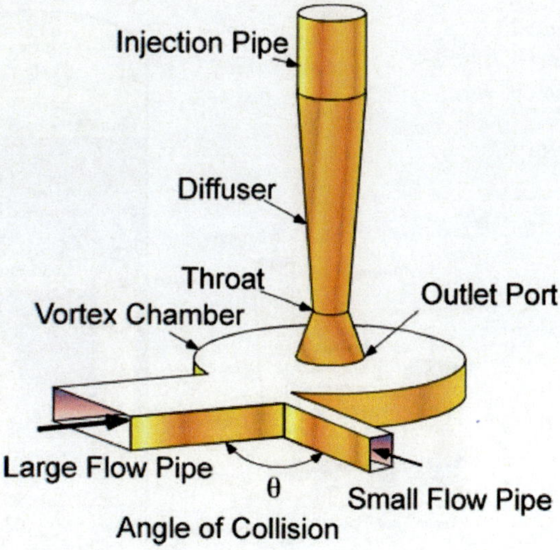

Fig. 2.14 Mechanical configuration of the flow damper. It consists of large and small flow pipes, a vortex chamber, and an outlet nozzle, which is composed of a reducer, a throat, and a diffuser. The reducer is between the outlet port and the throat

chamber. The reducer between the outlet port of the vortex chamber and the throat prevents the influence of cavitation on the flow and stabilizes the flow in it. Large energy loss occurs in the strong shear layer at the center of the vortex core and in the diffuser. Most of the energy is lost before the exit of the diffuser.

The large flow pipe is radially connected to the vortex chamber at a certain angle to cancel out the vortex formed by flow from the small flow pipe during large flow injection. The angle of collision of the large to small flows is one of the key parameters for the flow damper design. The width of the large flow pipe is set as large as possible to get a large ratio of flow rates. The detailed design of the configuration at the inlet ports of the large and small flow pipes will be one of the key items to get good characteristics of the flow damper. The reducer upstream from the throat prevents or minimizes the separation of flow at the exit of the outlet port and stabilizes the flow at the throat. The diffuser downstream from the throat will recover the static pressure during large flow injection. The size of the throat is determined by the required flow rate during large flow injection. The other dimensions can be determined by the size of the throat in a similar manner as the configuration of the flow damper model for which flow characteristics were investigated.

2.5.3 Principle of the Advanced Accumulator

Figure 2.15 shows the configurations and the principle of the Advanced Accumulator. The flow damper is installed at the bottom of the accumulator tank as shown in Fig. 2.15a, the center drawing. The lower inlet port of the small flow pipe is on the same level as the vortex chamber at the bottom of the tank. The standpipe is connected

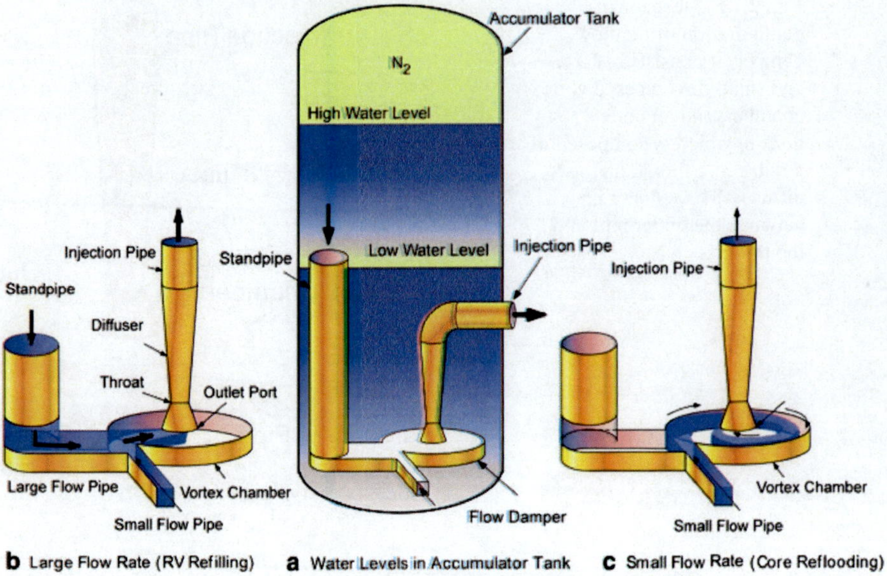

Fig. 2.15 The principle of the Advanced Accumulator showing flow patterns depending on the water level in the accumulator tank

to the large flow pipe. The inlet port of the standpipe is located at the middle level of the accumulator tank where there is a boundary between the large and small flow volumes. The outlet pipe is connected to the injection pipe at the boundary of the wall of the accumulator tank. The antivortex cap and plate are not illustrated in the figure for simplicity.

Initially, the water level in the accumulator tank is high above the inlet port of the standpipe. Once injection starts at a large break LOCA, water comes into both the upper and lower inlet ports, and the water level comes down. The flows from both of the inlets collide with each other and do not form a vortex in the vortex chamber as shown in Fig. 2.15b. Thus, the flow resistance of the flow damper comes from only the form resistance which is relatively small. A large flow rate is then obtained.

After the water level falls below the upper inlet port of the standpipe, water stops flowing into the standpipe. The other flow in the small flow pipe forms a strong and steady vortex in the vortex chamber as shown in Fig. 2.15c. Thus, the flow resistance of the flow damper comes from the strong and large vortex. The small flow rate is then achieved.

Consequently, the standpipe detects the water level at which flow rate must be changed, and formation of a strong vortex in the chamber reduces the flow rate in the absence of any moving parts. The accumulator tank should have the total capacity to accommodate the nitrogen gas, and large and small coolant flow volumes.

Figure 2.16 shows an example of flow rate transition of the Advanced Accumulator. As the injection starts, the flow rate goes to its maximum, and gradually falls due to

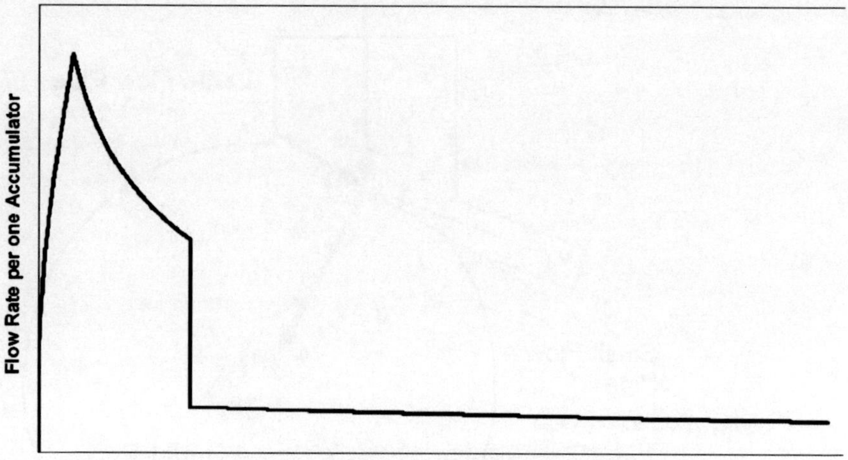

Fig. 2.16 An example of flow rate transition of the Advanced Accumulator. The flow rate is quickly changed from large flow injection to small flow injection

the pressure drop of nitrogen gas expansion. This is the large flow injection. At the time of flow changeover, the flow rate quickly switches and keeps small flow injection with a small decrease. This is the small flow injection.

2.5.4 Theoretical Consideration of the Flow Damper

For large flow injection, total angular momentum of flows from the large and small flow pipes must be zero. Furthermore, the resultant conflux must go straight to the outlet port of the vortex chamber in order not to form a vortex in the vortex chamber. Figure 2.17 shows the collision of flows from the large and small flow pipes and the resultant conflux in the vortex chamber. In other words, the tangential components of the momenta from the large and small flow pipes must have the same magnitude and opposite directions to each other, or

$$Q_S V_S \sin\varphi - Q_L V_L \sin(\varphi + \phi) = 0, \tag{2.1}$$

where Q is flow rate, V velocity, and ϕ and φ the angles defined in Fig. 2.17. The suffixes, S and L, indicate quantities for small and large flows respectively.

In addition to that, the sum of the radial components of the momenta from the large and small flow pipes must have the same magnitude as the momentum of the conflux and the same direction, or

$$-Q_S V_S \cos\varphi - Q_L V_L \cos(\varphi + \phi) = (Q_S + Q_L) V_O, \tag{2.2}$$

where V_o (<0) is the inward velocity of the resultant conflux.

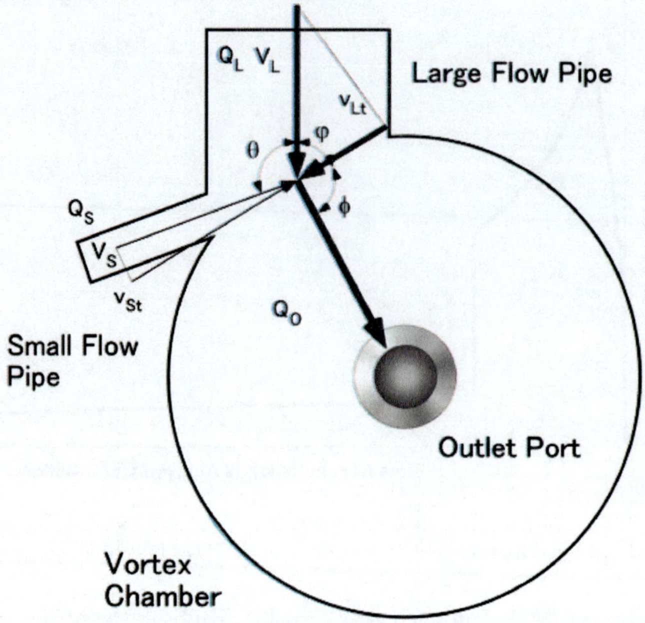

Fig. 2.17 Collision of large and small flows in the vortex chamber for large flow injection is considered. The angle is so designed that the angular momentum becomes zero. Additionally, radial momentum balance is also taken into account for stable formation of the resultant flow

If (2.2) is satisfied, there will be no pressure gradient at the collision point of the large and small flows so that the conflux can be stable. The collision angle, $\theta = \pi - \phi$, and the dimensions of the large and small flow pipes are selected to satisfy the condition mentioned above. The resultant conflux then goes straight to the outlet port of the vortex chamber without forming any vortex there.

The flow then enters the outlet nozzle. Given that the reducer produces a small energy loss, the flow rate is mainly controlled by the throat for large flow injection. Figure 2.18 shows the outlet nozzle and the control volume to examine the momentum balance of the flow in the diffuser. Applying the momentum balance to the control volume, the pressure of the throat, P_t, is determined by the following equation of the momentum balance;

$$P_{\mathrm{t}}\frac{\pi}{4}d_{\mathrm{p}}{}^{2} + \bar{C}_{\mathrm{p}}\frac{1}{2}\rho V_{\mathrm{t}}{}^{2}\frac{\pi}{4}\left(d_{\mathrm{p}}{}^{2} - d_{\mathrm{t}}^{2}\right) + \rho Q V_{\mathrm{t}} = P_{2}\frac{\pi}{4}d_{\mathrm{p}}{}^{2} + \rho Q V_{2}, \qquad (2.3)$$

where P is pressure, d diameter, V velocity, Q flow rate, and ρ the density of fluid. The suffixes, t, 2, p and w, indicate the quantities at the throat, the outlet section of the control volume, the outlet pipe and the diffuser wall, respectively. The mean pressure coefficient is the mean value of the pressure coefficient over the diffuser wall, and given by

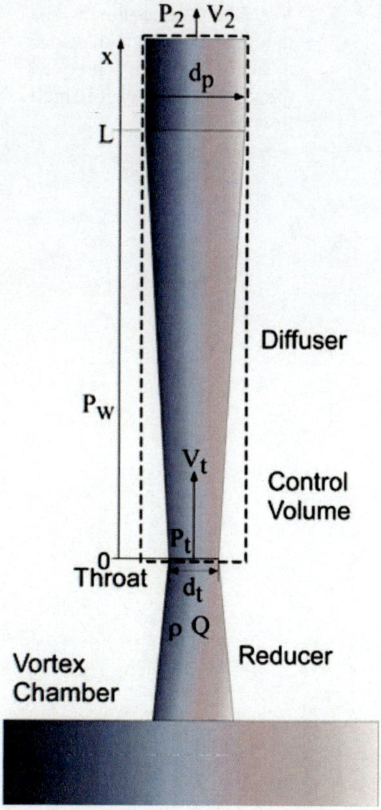

Fig. 2.18 The pressure at the throat is determined by the momentum balance of the flow in the diffuser. The *dotted line rectangle* is the control volume for which the momentum balance is considered

$$\bar{C}_p \equiv \frac{2}{L(d_2 + d_t)} \int_0^L \frac{P_w - P_t}{\rho {V_t}^2/2} \left(d_t + \frac{d_2 - d_t}{L} x \right) dx. \tag{2.4}$$

The first term on the left-hand side of (2.3) is the force acting on the upstream cross section of the control volume, the second term is the force acting on the wall of the diffuser, and the third term is the momentum flowing in the control volume. The first term on the right-hand side of (2.3) is the force acting on the downstream cross-section of the control volume and the second term is the momentum flowing out of the control volume. If cavitation occurs at the throat, pressure on the diffuser wall may vary and affect the flow rate of the damper.

For small flow injection, the small flow pipe is tangentially attached to the vortex chamber in order to make a strong vortex in the vortex chamber. Figure 2.19 shows the one-dimensional model of a vortex for small flow injection. The tangential velocity, v, at radius, r, is expressed as;

$$v = V \left(\frac{r}{R} \right)^n, \tag{2.5}$$

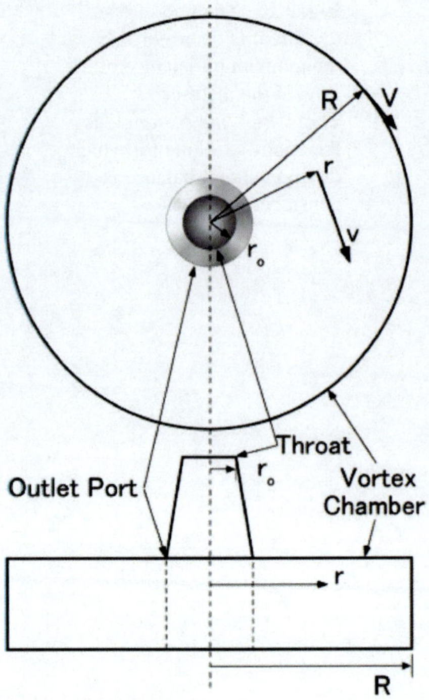

Fig. 2.19 A one-dimensional model of a vortex with respect to radius, *r*, is considered in the vortex chamber for small flow injection

where R is the radius of the vortex chamber, and V the velocity at $r = R$. If $n = 1$, (2.5) expresses a forced vortex, while if $n = -1$, (2.5) expresses a free vortex. Practically, n is between -1 and 1, and depends on the configuration and the size of the vortex chamber and the property of water.

Figure 2.20 shows the distributions of the dimensionless tangential velocity, v/V, with respect to the dimensionless radius, r/R, and with the parameter of the exponent, n, using (2.5). If a free vortex can be formed in the vortex chamber, a large tangential velocity will be formed near the center of the vortex for $n < 0$. A free vortex for $n = -1$ conserves its angular momentum for all values of the radius. Therefore, less energy loss will conserve a large amount of its angular momentum and make a strong vortex in the vortex chamber.

The equation of motion yields the pressure drop, Δp, from the radius, R, to an arbitrary radius, r, using (2.5) as:

$$\Delta p = \frac{1}{n}\frac{\rho}{2}V^2\left\{1 - \left(\frac{R}{r}\right)^{2n}\right\}. \tag{2.6}$$

From (2.6), the pressure drop coefficient from the radius, R, to the radius of the throat, r_o, is defined as

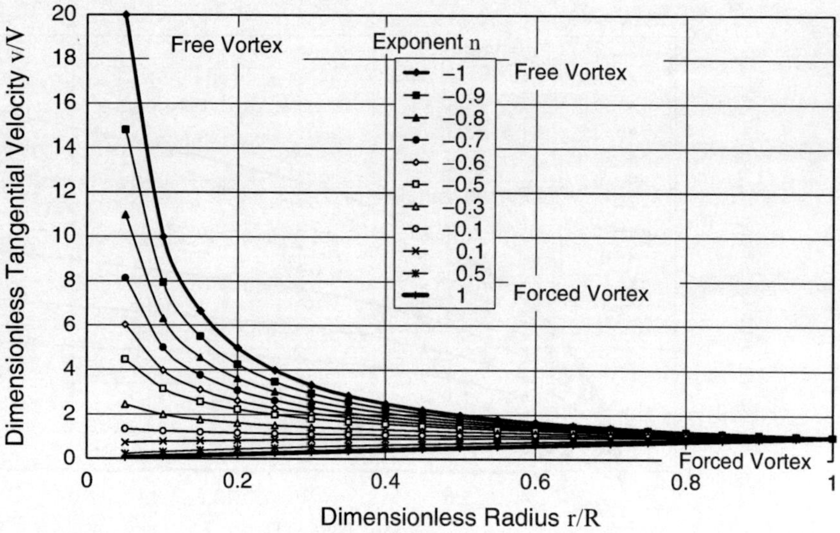

Fig. 2.20 This chart shows the distributions of the dimensionless tangential velocity, *v/V*, with respect to the dimensionless radius, *r/R*, for the parameter of the exponent, *n*, from the one-dimensional model of (2.5)

$$\zeta_s \equiv -\frac{\Delta p}{\rho V^2/2} = \frac{1}{n}\left\{\left(\frac{R}{r_o}\right)^{2n} - 1\right\}. \tag{2.7}$$

Figure 2.21 shows the pressure drop coefficient with respect to the vortex radius ratio. The larger the vortex radius ratio is, the larger the pressure drop coefficient of the vortex damper is. Moreover a free vortex, or $n = -1$, yields the largest pressure drop among the exponents $n = -1$ to 1. For a larger vortex radius ratio, a throat is inserted in the outlet nozzle. Furthermore, a diffuser is used to connect the throat to the outlet pipe to recover static pressure for large flow injection. The outlet pipe has the same diameter as that of the injection pipe and connects the outlet nozzle to the injection pipe.

The structure of a vortex in the vortex chamber is more complicated than that of the one-dimensional model. Velocity boundary layers practically develop on the upper- and lower-disk walls of the vortex chamber, while an inviscid swirl flow develops between them. Since viscosity reduces tangential velocity, the centrifugal force is so weak that the radial component of velocity becomes larger in the boundary layers than that in the inviscid vortex. This is the reason why the exponent, *n*, varies between -1 and 1. The swirl flow actually accelerates its tangential velocity component as it goes inward within the chamber. The accelerating flow and high Reynolds number generally restrain and suppress the development of the thickness of the boundary layers. Consequently, if the height of the vortex chamber is sufficiently larger than the thicknesses of the boundary layers, viscosity will negligibly affect the flow to yield a strong vortex in the chamber.

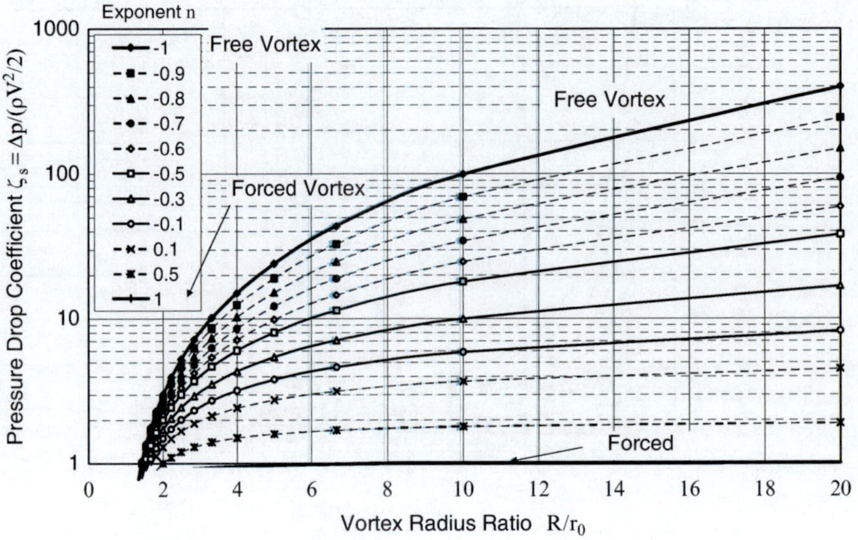

Fig. 2.21 Pressure drop coefficient, Δp, is shown here with respect to vortex radius ratio, R/r_o. The larger the vortex radius ratio is, the larger the pressure drop coefficient of the vortex damper is. The free vortex yields the largest pressure drop

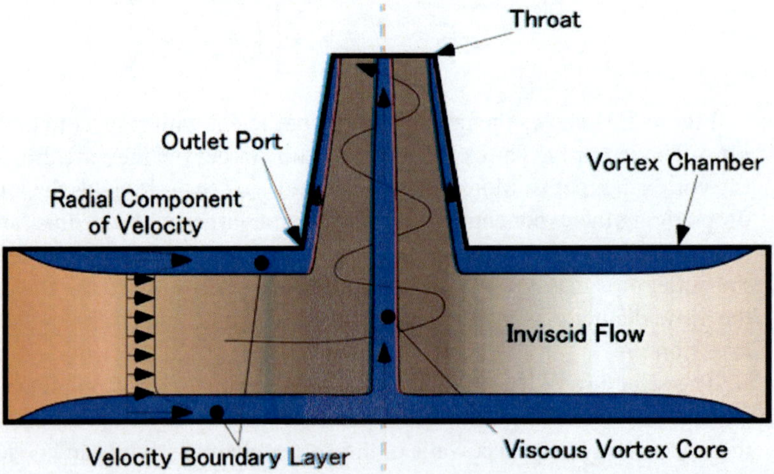

Fig. 2.22 This is the estimated structure of the vortex in the vortex chamber. There will be viscous boundary layers on the upper- and lower-disk walls. There will be a viscous vortex core at the center of the chamber. If the viscous boundary layers strongly affect the flow, velocity of the inviscid flow may be reduced

The estimated structure of the vortex in the chamber is shown in Fig. 2.22. A viscous vortex core will be formed at the center of the vortex. The flow in the boundary layer on the lower-disk wall will go into the viscous core, while the flow in the boundary layer on the upper-disk wall will go along the wall of the reducer to the throat. Inviscid flow will form a swirl flow between them.

If influence of viscosity in the boundary layers is not negligible, flow patterns in the chamber will be more complicated. High swirls with small rate of total flow may cause a reverse flow in the inviscid vortex and form a so-called doughnut pattern near the outlet port. This pattern may appear when the flow rate in the boundary layers due to large pressure drop induced by high swirls is larger than the supplied flow rate into the chamber.

Even if an ideal inviscid swirl with negligible boundary layers is formed in the vortex chamber, the vortex core cannot be neglected. In order to examine this influence of viscosity, a combined model of a nonstretched vortex with a stretched vortex is examined as shown in Fig. 2.23 [11]. The cylindrical coordinates are adopted. The radii of the vortex chamber and the outlet port are r_1 and r_0, respectively. The height of the vortex chamber is H. The tangential and radial components of velocity are v_θ and v_r, respectively. r is radius and p is pressure. Suffices 0 and 1 indicate quantities at r_0 and r_1, respectively.

The following assumptions are used.

1. The flow is axisymmetric and depends only on radius, r.
2. The flow is laminar incompressible steady viscous flow.
3. The flow is a stretched vortex in the central region with $r < r_0$, and a non-stretched vortex in all other regions with $r_0 < r < r_1$.

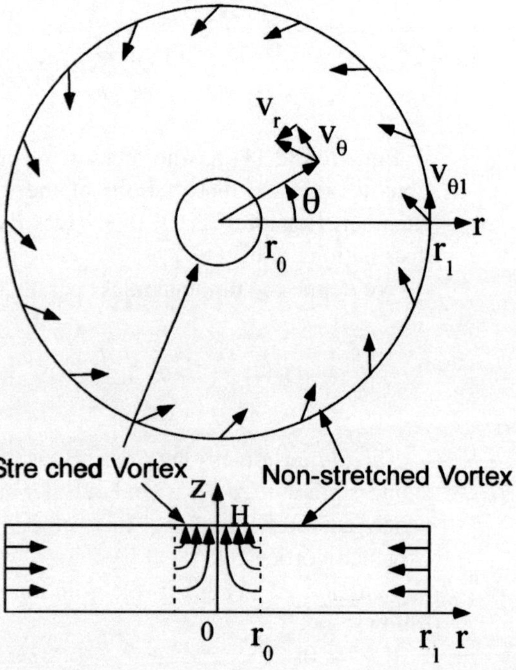

Fig. 2.23 A combined model of a nonstretched vortex with a stretched vortex is examined here. The flow is a stretched vortex for $r < r_0$, and a nonstretched vortex for $r_0 < r < r_1$. The boundary layers on the upper- and lower-disk walls are assumed to be negligible and small

4. The velocity and velocity gradient at the boundary of these regions at $r = r_0$ are continuous of their own.

The boundary conditions are $v_r = v_\theta = 0$ for $r = 0$, $v_\theta = v_{\theta 0}$, $p = p_0$ for $r = r_0$, and $v_\theta = v_{\theta 1}$ for $r = r_1$.

We let the flow rate be Q. The equation of continuity gives the characteristic flow rate, q as

$$q \equiv \frac{Q}{2\pi H} = -rv_r;\ for\ r_0 \le r \le r_1, \tag{2.8}$$

and the gradient of velocity component in the z-direction as

$$\frac{\partial v_z}{\partial z} = -\frac{1}{r}\frac{\partial}{\partial r}(rv_r) \equiv \frac{2q}{r_0{}^2} = \text{cons tan t; for } r \le r_0 \tag{2.9}$$

Equation (2.9) represents a stretched vortex, while (2.8) is a nonstretched vortex. The r-component of the velocity is given by these equations for nonstretched and stretched vortices.

The equation of motion yields

$$\rho\left(v_r\frac{\partial v_r}{\partial r} - \frac{v_\theta}{r} + v_z\frac{\partial v_r}{\partial z}\right) = -\frac{\partial p}{\partial r}, \tag{2.10}$$

and

$$v_r\frac{1}{r}\frac{\partial}{\partial r}(rv_\theta) = \nu\frac{\partial}{\partial r}\left[\frac{1}{r}\frac{\partial}{\partial r}(rv_\theta)\right]. \tag{2.11}$$

Equation (2.11) has no pressure term and can be solved under boundary conditions to give the distributions of the θ-component of the velocity in the vortex chamber. Equation (2.10) then gives the pressure distributions using the velocity distributions in it.

We define the dimensionless parameters as:

$$r^* \equiv \frac{r}{r_0},\quad r_1{}^* \equiv \frac{r_1}{r_0},\quad q^* \equiv \frac{q}{\nu},\quad v_\theta{}^* \equiv \frac{v_\theta}{v_{\theta 1}},\quad p^* \equiv \frac{p - p_0}{\rho v_{\theta 1}{}^2},\quad \text{and}\quad v_{\theta 0}{}^* \equiv \frac{v_{\theta 0}}{v_{\theta 1}}.$$

The dimensionless flow rate, q^*, is a kind of Reynolds number, and the dimensionless pressure, p^*, is a kind of Euler number. $v_{\theta 0}{}^*$ is the inner boundary condition which smoothly connects the velocity distributions of the stretched and nonstretched vortices.

Equation (2.11) gives the solution of the tangential velocity component as follows:

If $q^* \ne 0,\ 2,$

$$v_\theta{}^* = \frac{r_1{}^*}{r^*}\frac{2-q^*-\left(2-q^*r^{*2-q^*}\right)\exp(-q^*/2)}{2-q^*-\left(2-q^*r_1{}^{*2-q^*}\right)\exp(-q^*/2)}\ \text{for}\ 1\le r^*\le r_1{}^*, \tag{2.12a}$$

$$v_\theta{}^* = \frac{r_1{}^*}{r^*}\frac{(2-q^*)\left[1-\exp\left(-q^*r^{*2}/2\right)\right]}{2-q^*-\left(2-q^*r_1{}^{*2-q^*}\right)\exp(-q^*/2)};\ \text{for}\ r^*\le 1. \tag{2.12b}$$

If $q^* = 2$,

$$v_\theta{}^* = \frac{r_1{}^*}{r^*}\frac{1-(1-2\ln r^*)\exp(-1)}{1-(1-2\ln r_1{}^*)\exp(-1)};\ \text{for}\ 1\le r^*\le r_1{}^*, \tag{2.13a}$$

$$v_\theta{}^* = \frac{r_1{}^*}{r^*}\frac{1-\exp\left(-r^{*2}\right)}{1-(1-2\ln r_1{}^*)\exp(-1)};\ \text{for}\ r^*\le 1. \tag{2.13b}$$

And, if $q^* = 0$,

$$v_\theta{}^* = \frac{r^*}{r_1{}^*};\ \text{for}\ 0\le r^*\le r_1{}^*. \tag{2.14}$$

These equations are continuous to each other both for $q^* = 2$ and 0. Furthermore, these equations show that the dimensionless physical number used to determine the tangential velocity, v_θ, is only the Reynolds number, q^*. A forced vortex is given by (2.14) for $q^* = 0$, and a free vortex, by (2.12) for $q^* \to \infty$. In other words, a forced vortex is given only for a vortex with zero flow rate, and a free vortex is given only for inviscid flow. These solutions are more complicated than the one-dimensional vortex given by (2.5), but do not need to assume the exponent, n.

Some example solutions of these equations are shown in Fig. 2.24. The calculation conditions are $r_1{}^* = 1$ for the stretched vortices, and $r_1{}^* = 10$ for the combined vortices. The graphs compare the tangential velocity distributions of stretched vortices, or $r_0/r_1 = 1$, and combined vortices with the ratio of the radii, $r_0/r_1 = 0.1$. The tangential velocity in the region of $r^* = 0.1$–1.0 comes close to the free vortex when $q^* = 500$ for the stretched vortex, and $q^* = 7$ for the combined vortex. Thus, the nonstretched vortex has an advantage to make a strong vortex. That is because a stretched vortex loses its angular momentum rapidly, while a nonstretched vortex conserves its angular momentum better than a stretched vortex does. It is seen that the maximum tangential velocity of the combined vortex is formed in the stretched vortex. It implies that a nonstretched vortex preserves its angular momentum well and forms the maximum tangential velocity at the innermost location so that the stretched vortex can make the maximum tangential velocity inside itself.

Hereafter, the equations representing pressure distributions are examined. The inner boundary condition, $v_{\theta 0}{}^*$, is given for $r^* = 1$ by (2.12), (2.13) and (2.14).

Equation (2.10) gives the solution of the pressure as follows:

If $q^* \neq 0, 1, 2$

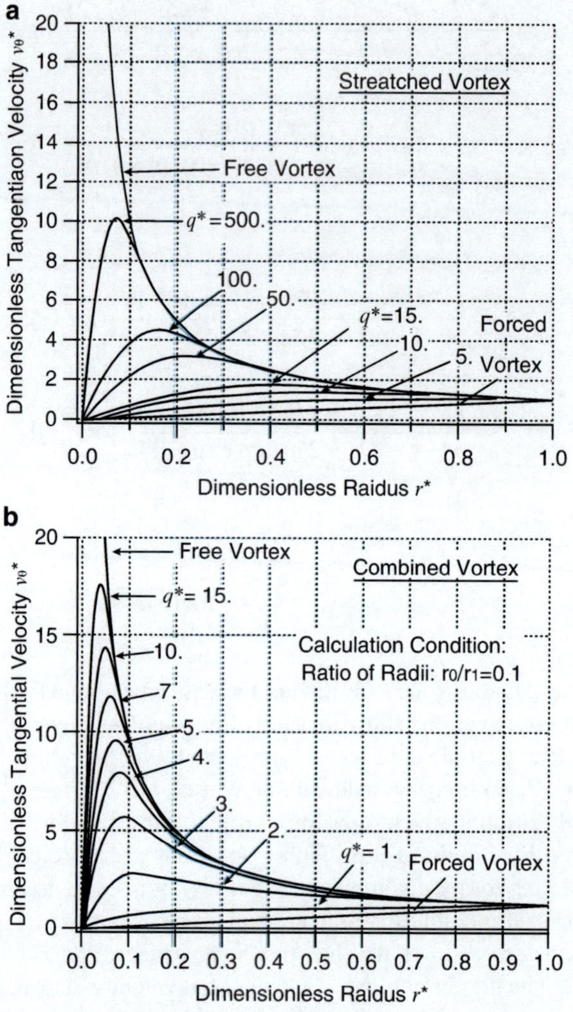

Fig. 2.24 Tangential velocity distributions of stretched and combined vortices. The stretched vortex yields a tangential velocity distribution close to that of a free vortex at the dimensionless flow rate of $q^* = 500$, while the combined vortex yields a tangential velocity distribution collapsed on that of a free vortex at the dimensionless flow rate of $q^* = 15$

$$p^* = \frac{({r_1}^* - {v_{\theta 0}}^*)^2 \left({r^*}^{2-q^*} - 1\right)}{2(1-q^*)\left({r_1}^{*2-q^*} - 1\right)^2} - \frac{2({r_1}^* - {v_{\theta 0}}^*)}{\frac{q}{\nu}\left({r_1}^{*2-q^*} - 1\right)} \left({v_{\theta 0}}^* - \frac{{r_1}^* - {v_{\theta 0}}^*}{{r_1}^{*2-q^*} - 1}\right)$$
$$\times \left({r^*}^{-q^*} - 1\right) - \frac{1}{2}\left[\left({v_{\theta 0}}^* - \frac{{r_1}^* - {v_{\theta 0}}^*}{{r_1}^{*2-q^*} - 1}\right)^2 + \left(\frac{q}{r_0 v_{\theta 1}}\right)^2\right]$$
$$\times \left(\frac{1}{r^{*2}} - 1\right); \text{ for } 1 \le r^* \le {r_1}^*, \tag{2.15a}$$

$$p^* = \frac{(2-q^*)^2 {r_1}^{*2} q^* \sum_{n=1}^{\infty} \frac{1}{nn!}\left(-\frac{q^*}{2}\right)^n (1-2^n)\left(r^{*2n}-1\right)}{2[2-q^*+(q^* {r_1}^{*2-q^*}-2)\exp(-q^*/2)]^2}$$
$$-\frac{(2-q^*)^2 {r_1}^{*2}\left\{\frac{1}{r^{*2}}\left[\exp\left(-\frac{q^* r^{*2}}{2}\right)-1\right]^2-\left[\exp\left(-\frac{pq^*}{2}\right)-1\right]^2\right\}}{2[2-q^*+(q^* {r_1}^{*2-q^*}-2)\exp(-q^*/2)]^2}$$
$$-\frac{1}{2}\left(\frac{q}{v_{\theta 1} r_0}\right)^2\left[\frac{1}{r^{*2}}-1+\left(\frac{2H}{r_0}\right)^2\left(1-\frac{z}{H}\right)^2\right]-\frac{g}{{v_{\theta 1}}^2}$$
$$\times (z-H);\ \text{for}\ 0<r^*\leq 1, \tag{2.15b}$$

$$p^* = \frac{(2-q^*)^2 {r_1}^{*2}\left\{q^* \sum_{n=1}^{\infty} \frac{1}{nn!}\left(-\frac{q^*}{2}\right)^n (2^n-1)+\left[\exp\left(-\frac{q^*}{2}\right)-1\right]^2\right\}}{2[2-q^*+(q^* {r_1}^{*2-q^*}-2)\exp(-q^*/2)]^2}$$
$$-\frac{1}{2}\left(\frac{q}{v_{\theta 1} r_0}\right)^2\left[-1+\left(\frac{2H}{r_0}\right)^2\left(1-\frac{z}{H}\right)^2\right]-\frac{g}{{v_{\theta 1}}^2}(z-H); \tag{2.15c}$$
$$\text{for}\ r^*=0.$$

If $q^* = 2$,

$$p^* = \frac{({r_1}^* - {v_{\theta 0}}^*)^2}{4\ln {r_1}^*}\left[\frac{1}{r^{*2}}(2\ln^2 r^* + 2\ln r^* + 1) - 1\right] - \frac{{r_1}^* - {v_{\theta 0}}^*}{2\ln {r_1}^*}$$
$$\times \left[\frac{1}{r^{*2}}(2\ln r^* + 1) - 1\right] - \frac{1}{2}\left\{{v_{\theta 0}}^{*2} + \left(\frac{q}{r_0 v_{\theta 1}}\right)^2\right\}$$
$$\times \left(\frac{1}{r^{*2}} - 1\right);\ \text{for}\ 1\leq r^*\leq {r_1}^*, \tag{2.16a}$$

$$p^* = \frac{2{r_1}^{*2}\sum_{n=1}^{\infty}\frac{1}{nn!}(-1)^n(1-2^n)\left[r^{*2n}-1\right]}{2[1+(2\ln {r_1}^*-1)\exp(-1)]^2}$$
$$-\frac{{r_1}^{*2}\left\{\frac{1}{r^{*2}}\left[\exp(-r^{*2})-1\right]^2-\left[\exp(-1)-1\right]^2\right\}}{2[1+(2\ln {r_1}^*-1)\exp(-1)]^2}$$
$$-\frac{1}{2}\left(\frac{q}{v_{\theta 1} r_0}\right)^2\left[\frac{1}{r^{*2}}-1+\left(\frac{2H}{r_0}\right)^2\left(1-\frac{z}{H}\right)^2\right]-\frac{g}{{v_{\theta 1}}^2}$$
$$\times (z-H);\ \text{for}\ 0<r^*\leq 1, \tag{2.16b}$$

$$p^* = \frac{r_1{}^{*2}\left\{2\sum_{n=1}^{\infty}\frac{1}{nn!}(-1)^n(2^n-1)+[\exp(-1)-1]^2\right\}}{2[1+(2\ln r_1{}^*-1)\exp(-1)]^2} - \frac{1}{2}\left(\frac{q}{v_{\theta 1} r_0}\right)^2\left[-1+\left(\frac{2H}{r_0}\right)^2\left(1-\frac{z}{H}\right)^2\right] - \frac{g}{v_{\theta 1}{}^2}(z-H);\ \text{for}\ r^* = 0. \tag{2.16c}$$

If $q^* = 1$,

$$p^* = \frac{(r_1{}^* - v_{\theta 0}{}^*)^2}{(r_1{}^*-1)^2}\ln r^* - \frac{2(r_1{}^* - v_{\theta 0}{}^*)}{r_1{}^*-1}\left(\frac{v_{\theta 0}}{v_{\theta 1}} - \frac{r_1{}^* - v_{\theta 0}{}^*}{r_1{}^*-1}\right)(r^*-1) - \frac{1}{2} \times\left[\left(v_{\theta 0}{}^* - \frac{r_1{}^* - v_{\theta 0}{}^*}{r_1{}^*-1}\right)^2 + \left(\frac{q}{r_0 v_{\theta 1}}\right)^2\right]\left(\frac{1}{r^{*2}}-1\right);\ \text{for}\ 1 \le r^* \le r_1{}^*, \tag{2.17a}$$

$$p^* = \frac{r_1{}^{*2}\sum_{n=1}^{\infty}\frac{1}{nn!}\left(-\frac{1}{2}\right)^n(1-2^n)\left(r^{*2n}-1\right)}{2[1+(r_1{}^{*2-q^*}-2)\exp(-1/2)]^2} - \frac{r_1{}^{*2}\left\{\frac{1}{r^{*2}}\left[\exp\left(-\frac{r^{*2}}{2}\right)-1\right]^2 - \left[\exp\left(-\frac{1}{2}\right)-1\right]^2\right\}}{2[1+(r_1{}^{*2-q^*}-2)\exp(-1/2)]^2} - \frac{1}{2}\left(\frac{q}{v_{\theta 1} r_0}\right)^2\left[\frac{1}{r^{*2}}-1+\left(\frac{2H}{r_0}\right)^2\left(1-\frac{z}{H}\right)^2\right] - \frac{g}{v_{\theta 1}{}^2} \times (z-H);\ \text{for}\ 0 < r^* \le 1, \tag{2.17b}$$

$$p^* = \frac{r_1{}^{*2}\left\{q^*\sum_{n=1}^{\infty}\frac{1}{nn!}\left(-\frac{1}{2}\right)^n(2^n-1)+\left[\exp\left(-\frac{1}{2}\right)-1\right]^2\right\}}{2[1+(r_1{}^*-2)\exp(-1/2)]^2};\ \text{for}\ r^*=0. - \frac{1}{2}\left(\frac{q}{v_{\theta 1} r_0}\right)^2\left[-1+\left(\frac{2H}{r_0}\right)^2\left(1-\frac{z}{H}\right)^2\right] - \frac{g}{v_{\theta 1}{}^2}(z-H) \tag{2.17c}$$

If $q^* = 0$,

$$p^* = \frac{(r_1{}^* - v_{\theta 0}{}^*)^2(r^{*2} - 1)}{2(r_1{}^{*2} - 1)^2} - \frac{2(r_1{}^* - v_{\theta 0}{}^*)}{r_1{}^{*2} - 1}\left(v_{\theta 0}{}^* - \frac{r_1{}^* - v_{\theta 0}{}^*}{r_1{}^{*2} - 1}\right)\ln r^*$$

$$-\frac{1}{2}\left(v_{\theta 0}{}^* - \frac{r_1{}^* - v_{\theta 0}{}^*}{r_1{}^{*2} - 1}\right)^2\left(\frac{1}{r^{*2}} - 1\right);\ \text{for } 1 \le r^* \le r_1{}^*, \quad (2.18a)$$

$$p^* = \frac{1}{2}r_1{}^*2\left[r^{*2} - 1\right] - \frac{g}{v_{\theta 1}{}^2}(z - H);\ \text{for } 0 \le r^* \le 1. \quad (2.18b)$$

These equations are continuous to each other for $q^* = 2$, 1 and 0. The pressure drop is a function of not only a Reynolds number but also the radial inlet velocity, v_{r1}, and the gravity. Fig. 2.25 shows some pressure distributions of stretched and combined vortices for $q^* = 1$ to 15. The calculation conditions are $r_1{}^* = 1$ for the stretched vortices, and $r_1{}^* = 10$ for the combined vortices the same as for the tangential velocities in Fig. 2.24. The inlet radial velocity is given as $v_{r1}/v_{\theta 1} = 1$ for both vortices and $z = 0$ in common. The pressure drop of the combined vortex for $q^* = 15$ is about 100 times that of the stretched vortex for $q^* = 15$. This is the reason why we need to select the form of the vortex chamber.

2.5.5 Transition of Water Level in the Standpipe

The important role of the standpipe is to prevent gas leakage at the changeover of flow rate and during the small flow injection in addition to detection of water level in the accumulator tank. The size of the standpipe is determined by the prevention of gas leakage at the changeover.

The behavior of the water level during the flow changeover is shown in Fig. 2.26. There are three steps after the formation of the water level in the standpipe as follows:

Step 0: Water in the standpipe is still flowing at the end of large flow injection. Then, the water level in the accumulator tank drops across the inlet port of the standpipe, and water stops flowing in. This forms a water level in the standpipe.

Step 1: The water level still descends due to the inertia of the water column in the standpipe. The water level then has to stop to form a stationary water level so that gas leakage is prevented. The static pressure at the exit of the large flow pipe is smaller than the pressure in the accumulator tank at the amount of the dynamic pressure in the small flow pipe. The water level comes down once below the balanced level corresponding to the static pressure to stop its movement due to its inertia.

Step 2: Then, the static pressure at the exit of the standpipe pushes the water level back to the balanced level.

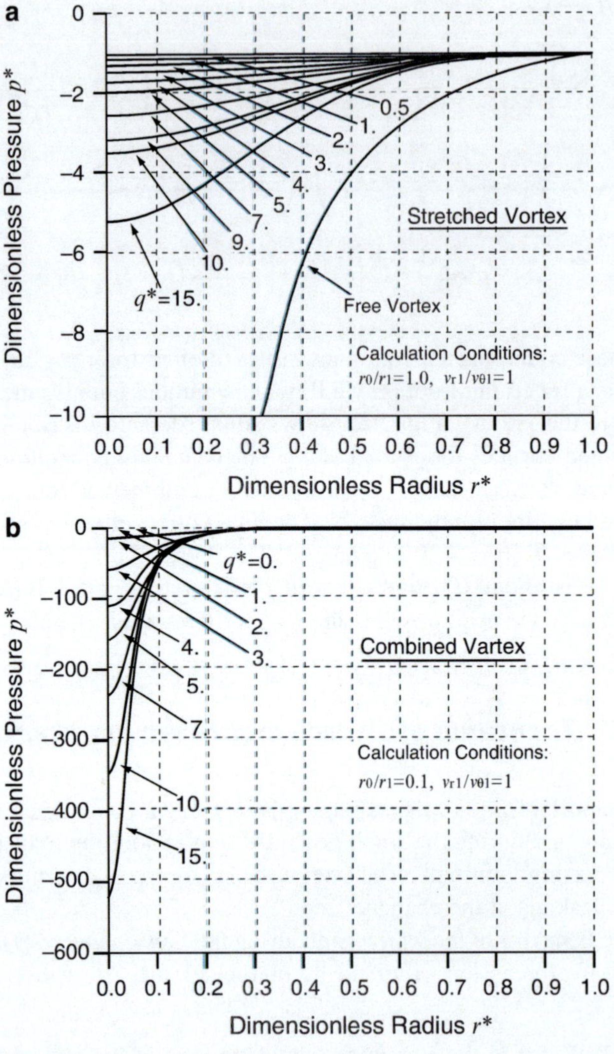

Fig. 2.25 Pressure distributions of stretched and combined vortices are shown for $q^* = 1$ to 15. A combined vortex yields much larger pressure drop than a stretched vortex does

Step 3: Thereafter, it stops at the balanced level and gradually drops along with the water level in the accumulator tank.

The schematic chart in Fig. 2.27 shows the behavior of the water level in the standpipe as time passes. For large flow injection, the inlet port of the standpipe is underwater. When the water level in the accumulator tank drops to the inlet port of the standpipe and the flow changeover is initiated, a water level in the standpipe appears and plummets due to the inertia of the water column and causes an undershoot. This undershoot produces a force to stop the movement of the water

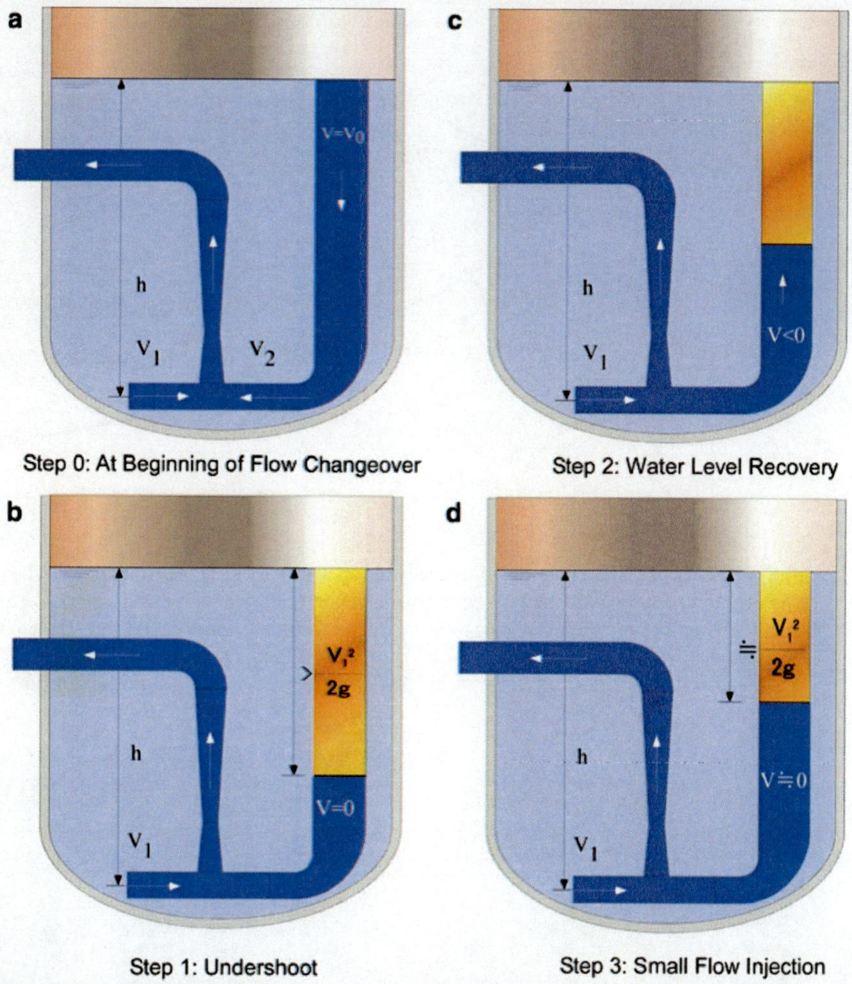

Fig. 2.26 At the flow changeover, the water level appears in the standpipe and undergoes the transition. The transition of water level is controlled by the momentum balance of the water column

column and recovers the water level to a balanced level with the static pressure at the exit of the standpipe which is equivalent to that of the large flow pipe. The water level then gradually decreases as the water level in the accumulator tank decreases for small flow injection.

The transition of the water column is controlled by the momentum balance. Figure 2.28 shows a one-dimensional model of the water column in the standpipe. The one-dimensional momentum equation applied to the water column is expressed as:

$$\frac{\mathrm{d}}{\mathrm{d}t}(hv) = -v^2 + gh - \zeta\frac{1}{2}|v|v - \left(gH - \frac{1}{2}{v_s}^2\right). \quad (2.19)$$

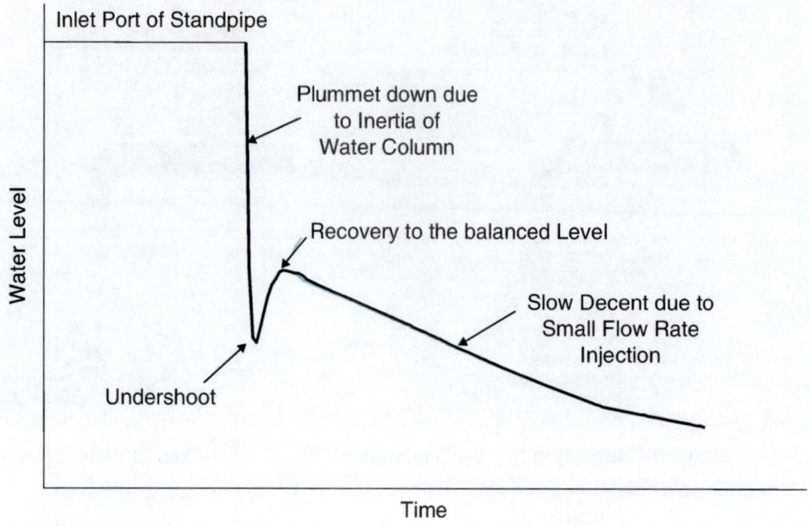

Fig. 2.27 Schematic chart showing transition of water level in the standpipe. When the water level in the accumulator tank falls below the inlet port of the standpipe, the water level also appears in the standpipe, and plummets due to its inertia to stop at the undershoot level. Then, it recovers to the balanced level with the static pressure at the exit of the large flow pipe. The water level retains its balance with the static pressure at the exit of the large flow pipe during small flow injection

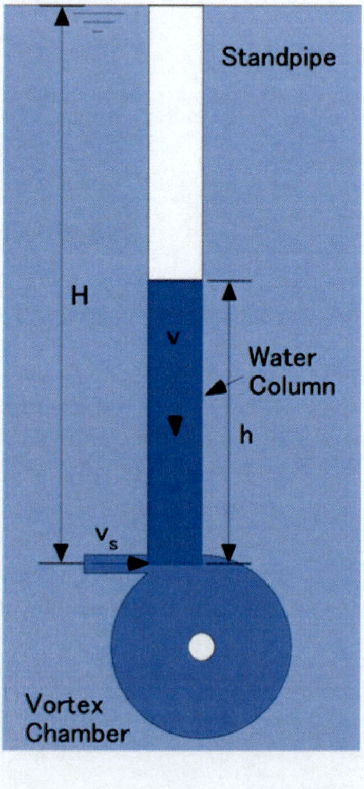

Fig. 2.28 The one-dimensional momentum model can explain the transition of the water column in the standpipe at the flow changeover. A control volume is applied to the water column to get the momentum balance to yield the one-dimensional momentum (2.19). The static pressure at the exit of the large flow pipe equals to the pressure in the accumulator tank minus dynamic pressure in the small flow

The notations are shown in Fig. 2.28 h and v are the length and the velocity of the water column, respectively. t is time and g is the gravitational acceleration. ζ is the loss coefficient of the standpipe. H is the water level in the accumulator tank. v_s is the velocity on the small flow pipe.

The term on the left-hand side of (2.19) is the momentum change rate of the water column. On the right-hand side: the first term is the outgoing momentum; the second term is the gravitational force; the third term is the resistance of the standpipe; and the fourth term is the static pressure at the exit of the standpipe. This equation is solved step by step with a small time interval to provide a maximum drop in the water level. The balanced level is given by (2.19) for $v = 0$ as:

$$h = H - \frac{1}{2g} v_s^{\,2}. \tag{2.20}$$

The comparison of the calculated maximum drops of water level in the standpipe by the one-dimensional momentum (2.19) to the measured values taken by the full-height 1/2-scale model is shown in Fig. 2.29 for the test conditions shown at Table 2.1. They are in good agreement with each other. The water levels are sustained in the standpipe, and they prevent gas leakage throughout, even for the severest Case 3. The initial velocity of the water column at the switchover is the main factor to determine the maximum drop of the water level in the standpipe. If the initial velocity is small, the maximum drop will be also small and preserve the water level in the standpipe to prevent gas leakage. Alternatively, if the initial velocity is too large, the maximum drop will so large that gas leakage may occur.

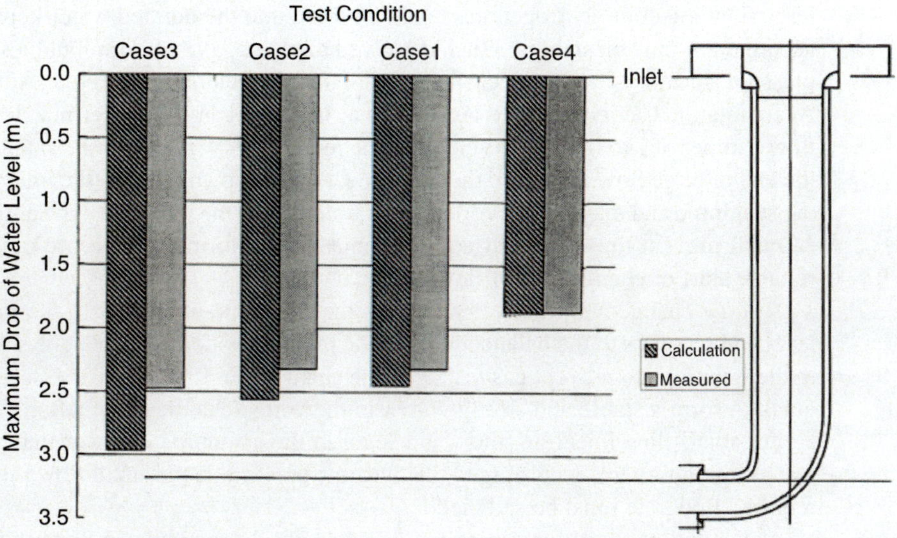

Fig. 2.29 The comparison of the calculated maximum drops of water level in the standpipe to the measured values shows good agreement between them

Table 2.1 Test conditions of injection under the actual pressure with the full-height 1/2-scale model apparatus at Takasago R&D Center, MHI

Test case	Pressure in test tank (MPaG)	Pressure in exhaust tank (MPaG)	Objective
Case 1	4.04	0.098	To obtain data for evaluation of ECCS performance during large LOCA
Case 2	4.53	0.098	To obtain data for high pressure design
Case 3	5.22	0.49	To obtain data of large differential pressure
Case 4	4.04	0.49	To obtain data of small differential pressure

Therefore, the cross-sectional area of the standpipe has to be determined by this calculation. The standpipe can be either a circular or square cylinder in shape. There should be a bell mouth at the inlet of the standpipe to rectify flow in it. The antivortex cap should give enough clearance on the inlet port of the standpipe so that it does not block the flow.

2.6 Confirmation Tests of the Advanced Accumulator

If a large break LOCA happens during operation of the Advanced Accumulator, the pressure in the accumulator tank decreases from about 4 MPa to about 1 MPa for large flow injection. Nitrogen gas could dissolve into the borated water kept in the accumulator tank on standby. Such dissolved nitrogen gas may form bubbles at low pressure locations and affect the flow rate characteristics of the Advanced Accumulator. Cavitation may also occur at the throat in the outlet nozzle of the flow damper due to high flow velocity. The required flow rate must be satisfied. At the end of large flow injection, the water level comes down close to the inlet port of the standpipe and may form a vortex and a water fall at the inlet port. The antivortex cap will prevent these free surface phenomena from forming to secure quick and reliable start of changeover of flow rate.

At flow changeover, the water level in the standpipe plummets to the minimal level and recovers to the balanced level. The minimal water level must be sustained in the standpipe to prevent gas leakage. The changeover of flow must be secure and quick to form a stable and strong vortex in the vortex chamber thereafter.

For small flow injection, the water level in the standpipe is lower than that in the accumulator tank according to the dynamic pressure in the small flow pipe. The required flow rate must be satisfied.

At the end of small flow injection, the Advanced Accumulator finishes its role and leaves water only in the dead-water region at the bottom of the accumulator tank.

The characteristics of the Advanced Accumulator are determined by the following factors:

- For large flow injection
  - Cavitation at the throat of the outlet nozzle may affect flow rates.
  - Cavitation factors and Reynolds numbers will be the key parameters to determine flow rates.
- For flow changeover
  - Flow injection is securely changed over at the determined level of water in the accumulator tank.
  - The transition of flow changeover is preferably finished in a short time.
- For small flow injection
  - A strong and steady vortex is formed in the vortex chamber.
  - A large pressure drop exists along the radius of the vortex chamber.

The confirmation tests were carried out to examine whether the expected performance was achieved by the operational principle of the Advanced Accumulator and to discuss the performance requirements. The confirmatory testing program was conducted as a joint study among the five Japanese utilities and MHI.

2.6.1 Purpose of Scale Testing

For the development of the Advanced Accumulator, the following items should be confirmed:

1. The principle of the flow damper
2. The performance of the flow damper during large and small flow injections
3. The influence of dissolved nitrogen gas on the performance of the flow damper
4. The dimensionless numbers (cavitation factor and flow coefficient) to represent flow characteristics
5. The independency of flow characteristics from scales of flow damper
6. The transition of water level in the standpipe at flow changeover
7. The water level in the accumulator tank at flow changeover with respect to the inlet port of the standpipe
8. The prevention of vortex formation by the antivortex cap at the end of large flow injection

Four kinds of scale models were made to confirm these items.

1. The 1/8.4-scale visualization tests were carried out to demonstrate the principle of switching flow to confirm Item 1.
2. The 1/3.5-sclae visualization tests were carried out for demonstration of quick shutoff of flow into the standpipe to confirm Item 8.

3. The 1/5-scale visualization tests at low-pressure injections were carried out for acquisition of the flow rate characteristics of the flow damper to confirm Items 4 and 5.
4. The full-height 1/2-scale tests at the actual pressure injections were carried out to demonstrate the total characteristics of the actual Advanced Accumulator to confirm Items 2–7.

2.6.2 Principle of the Advanced Accumulator

The overall functions of the Advanced Accumulator were demonstrated by visualization tests of flow with the 1/8.4-scale model. The test apparatus consisted of an accumulator tank, a flow damper, an exhaust tank, and an injection pipe as shown in Fig. 2.30. The scale of the flow damper was 1/8.4, which was selected so that it could be moved anywhere for the tests. The vortex chamber was in an upright position. The front panel was made of transparent acrylic resin to observe water levels in the accumulator tank and the standpipe, and flow in the flow damper.

The objectives of the tests were:

1. Confirmation of the operational principle of the flow damper
2. Confirmation of behavior of water level in the standpipe at flow changeover and during small flow injection.

The test apparatus can visualize flows and water levels in the accumulator tank, the flow damper and the standpipe during large flow injection, changeover from large to small flow rates, and small flow injection. It can also be used to observe formation of a vortex in the vortex chamber. Finally, the motion of water level in the standpipe at flow changeover can be observed as well.

Since there was no special requirement for the test conditions, pressure in the Advanced Accumulator was set slightly lower than 0.1 MPaG, and the exhaust tank was opened to the atmosphere. A compressor supplied air to pressurize the tank in place of nitrogen gas. In addition to flow visualization, flow rate was measured and displayed on a screen. Pressure in the accumulator was monitored with a pressure gauge.

The basic performance of the accumulator was confirmed as follows.

1. When water level was higher than the top of the standpipe, or during large flow injection:

   (a) Flows from the standpipe and the small flow pipe collided with each other, and the conflux directly went to the outlet port in the vortex chamber.
   (b) A vortex was not formed in the vortex chamber.
   (c) A large flow rate appeared.
   (d) There was no gas-entraining vortex formed at the inlet port of the standpipe.

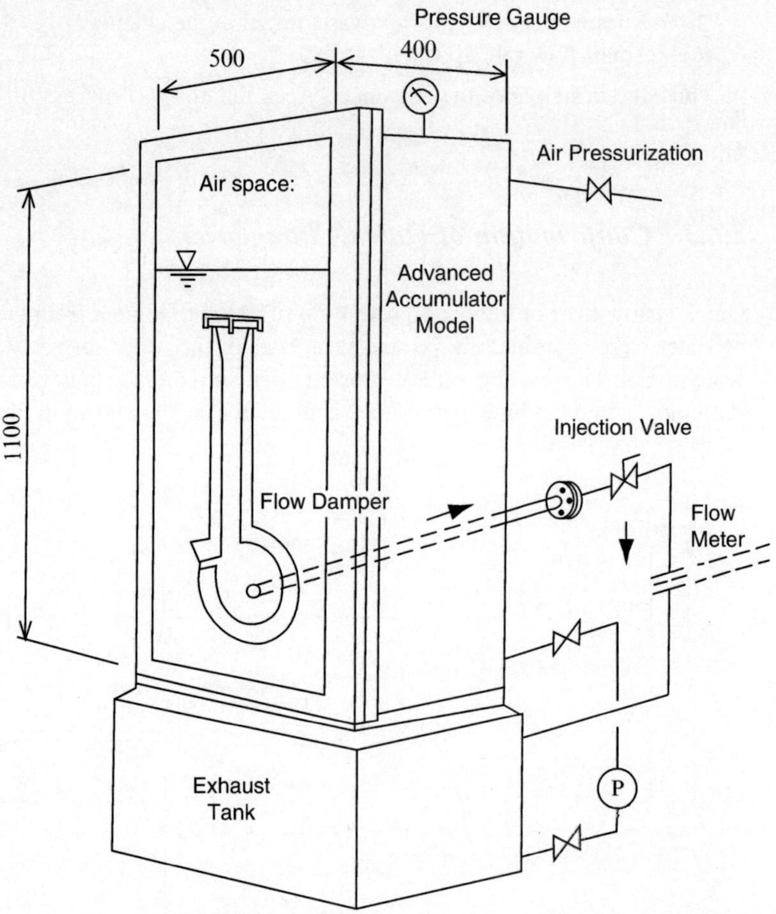

Fig. 2.30 1/8.4-scale model of the Advanced Accumulator used to demonstrate the principle. The flow damper was set at the upright position and the front wall was made of transparent acrylic resin to observe the water levels and the flow in it

2. During flow changeover:
   (a) The water level that appeared in the standpipe plummeted to the minimal level and quickly recovered to the balanced level.
   (b) The flow in the standpipe immediately stopped, and the water column was sustained in it.
   (c) A vortex was quickly formed in the vortex chamber.
   (d) Gas did not enter through the standpipe to the vortex chamber.
3. When water level was lower than the top of the standpipe, or during small flow injection:
   (a) The flow from the standpipe stopped and only the flow from the small flow pipe came into the vortex chamber.

(b) A strong and steady vortex was formed in the chamber.
(c) A small flow rate appeared.

Thus, the principle of the Advanced Accumulator was successfully confirmed.

2.6.3 Confirmation of Quick Changeover

Quick changeover of flow and prevention of gas entrainment from the free surface of water were examined by visualization tests of the water surface with the 1/3.5-scale model. The test apparatus consisted of an antivortex cap, the upper part of the standpipe, and the middle part of the accumulator tank as shown in Fig. 2.31. There

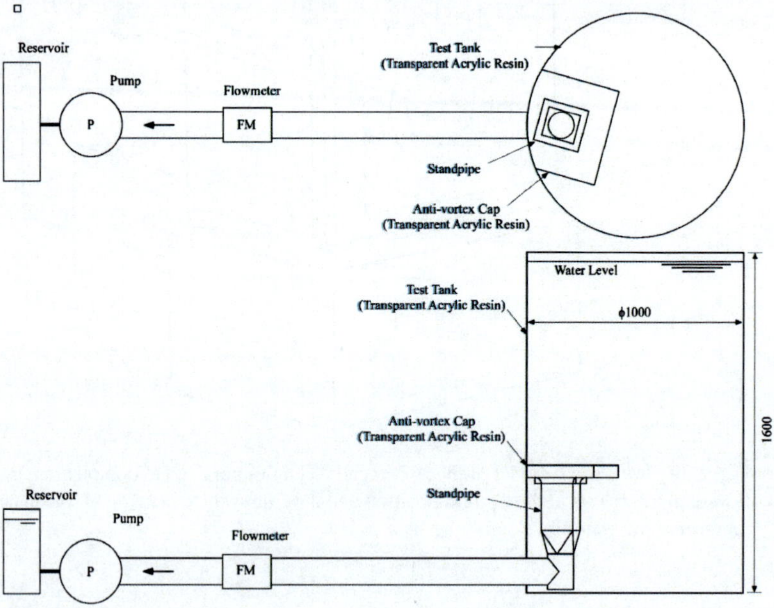

Fig. 2.31 1/3.5-scale model of the upper part of the standpipe and the anti-vortex cap in the accumulator tank used to confirm quick shutoff of flow into the standpipe

were also a pump and a flow meter in the pipe which are not shown in the figure. The accumulator tank and the antivortex cap were made of transparent acrylic resin to observe the shape and the behavior of water surfaces both in the tank and the antivortex cap. The scale of the model was 1/3.5, which was selected for easy observation of water surfaces.

The objectives of the tests were the certification of quick flow changeover and the prevention of vortex formation by the antivortex cap. The antivortex cap is known to prevent a vortex and a water fall causing gas entrainment at the inlet port of the standpipe. Without the antivortex cap, it was foreseen that gas entrainment due to a vortex and a water fall may affect the flow rate.

Because the tests were conducted to investigate the phenomena of water surfaces, a Froude number was adopted to determine the test conditions. The transition of flow rate at the tests was manually simulated as that of large flow injection of the Advanced Accumulator. Hence, the transition of the water level in the accumulator tank was simulated. The flow rate was measured by an ultrasonic flow meter in the injection pipe. The water level was measured by a ruler attached on the sidewall of the test tank. The transition and the phenomena of water levels were recorded on video.

Figure 2.32 shows examples of transitions of flow rate in the standpipe with and without the antivortex cap. In the case without the anti-vortex cap, gas entrainment started at 26 s. Fluctuation of the flow rate was generated for several seconds due to gas entrainment. It resulted from the formation of a vortex and a water fall at the inlet port of the standpipe. It took about 5 s for the flow rate to decrease to zero.

In the case with the antivortex cap, it took approximately 1 s for the flow rate to become zero after 25.5 s. There was no vortex or water fall formed on the inlet port of the standpipe. The flow rate thus switched more quickly than that without the antivortex cap.

Consequently, it was confirmed that the antivortex cap prevented formation of a vortex and a water fall causing gas entrainment, and it assured quick changeover of the flow rate.

2.6.4 Performance of the Flow Damper

Acquisition of the flow-rate characteristics of the flow damper was implemented by the low-pressure injection tests with the 1/5-scale model. The test facility consisted of a test tank, a flow damper model with a standpipe, an injection pipe and an exhaust tank as shown in Fig. 2.33. The configurations of the flow damper and the standpipe were similar to those of the actual accumulator for measurement of quantitative data of the flow damper. The flow damper was placed outside of the test tank. The lower-disk wall of the vortex chamber was made of transparent acrylic resin in order to observe and record flow in the vortex chamber on video. The 1/5-scale was selected so that several flow dampers could be installed and tested to acquire accurate data for evaluation of the characteristics of the flow damper. A resistance control valve was installed in the injection pipe to simulate the

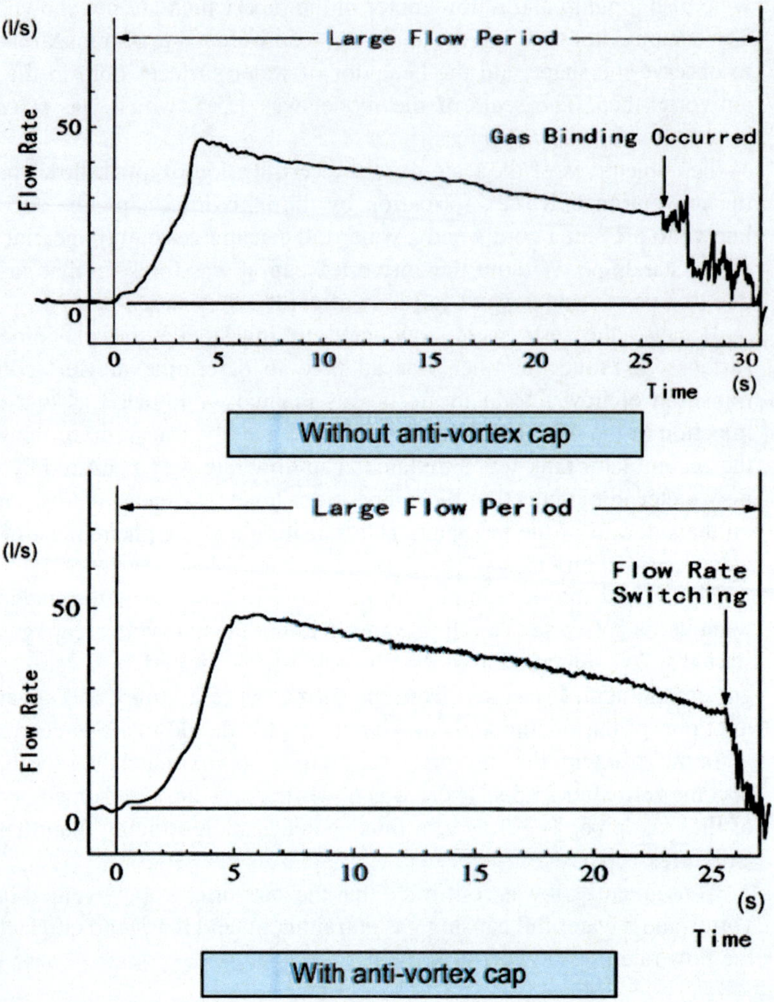

Fig. 2.32 Examples of flow rate transitions with and without the antivortex cap show that changeover of flow rate with the antivortex cap was much faster than that without the antivortex cap

flow resistance of the injection pipe of actual plants. There was an isolation valve at the end of the injection pipe. It was quickly opened to start injection of flow. The exhaust tank corresponded to the main coolant system. The test tank was supplied with and pressurized by nitrogen gas before every test.

The objectives of the tests were:

1. Confirmation of the operational principle of the flow damper
2. Acquisition of performance data during large and small flow injections

The dimensionless parameters of the flow damper are Reynolds number and cavitation factor. The flow rate of the flow damper is represented by the flow rate coefficient.

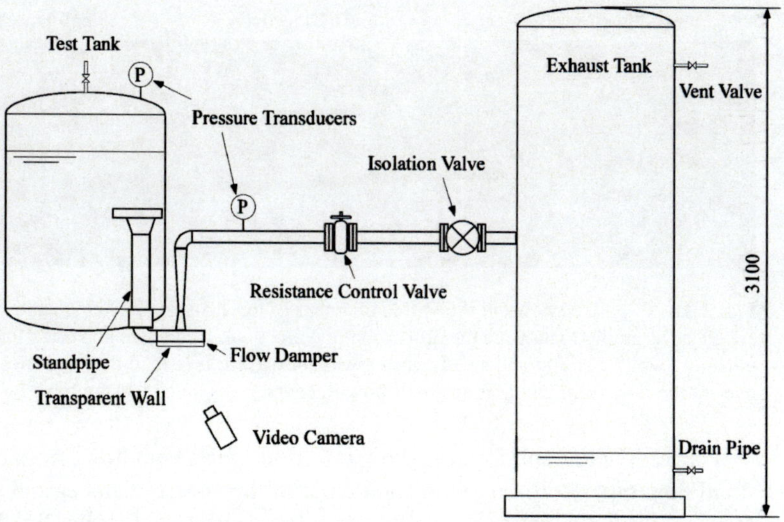

Fig. 2.33 1/5-scale test facility for acquisition of the characteristics of the flow damper. The flow damper was placed outside of the test tank for observation and recording of flow in the vortex chamber on video

If influence of viscosity on the characteristics of the flow damper is negligibly small, a Reynolds number can be excluded from consideration. This means that the characteristics of the flow damper taken by the 1/5-scale model collapse to those taken by the 1/2-scale model. The dependency on a Reynolds number is discussed later in this chapter. On the other hand, if cavitation occurs in the flow damper and affects the flow rate, the characteristics of the flow damper will depend on the cavitation factor. Consequently, the flow rate coefficient of the flow damper should be examined with respect to the cavitation factor. The test conditions were selected to get data in the widest range of cavitation factors possible in the test facility.

To derive the acquisition of flow rate characteristics, pressures in the test tank, the injection pipe, and the exhaust tank were measured with pressure transducers. Water levels in the test tank and the standpipe were measured by a differential pressure transducer and an electrocapacitance level meter, respectively. These data were recorded by a personal computer. The flow rates were calculated from the water level in the test tank. Moreover, flow rate coefficients and cavitation factors were calculated from these data.

The flow in the vortex chamber was observed through the transparent lower-disk wall and recorded on video to examine the vortex formation in it at changeover of flow rates. Figure 2.34 shows a set of flow images observed through the lower-disk wall of the vortex chamber. Ink was injected as a tracer from the small flow pipe on the bottom right side to indicate small flow, and from a small hole on the upper-disk wall to indicate large flow. The large flow pipe was located on the right top side of each photograph, and the outlet port was at the center.

Fig. 2.34 Visualized flows in the vortex chamber of the 1/5-scale model are shown. White lines are added as guides to indicate the observed flow directions. There was no vortex for the large flow injection, while a strong and steady vortex was formed for the small flow injection. The changeover of the flow from the large to small flow injections was quick and continuous

The left photograph for large flow injection shows both flows from the large and small flow pipes colliding with each other in the vortex chamber and the resultant conflux going straight to the outlet port. The central photograph for flow changeover shows transition of flow from the large to small flow injections, and how a vortex promptly began to form. The right photograph for small flow injection shows how flow from the small flow pipe tangentially entered in the vortex chamber and formed a strong and steady vortex in it.

Flow rate coefficient, C_v, and cavitation factor, σ_v, are defined as:

$$C_{\mathrm{v}} = \frac{Q}{A_{\mathrm{d}}\sqrt{2(p_{\mathrm{a}} - p_{\mathrm{dT}})/\rho}}, \tag{2.21}$$

and

$$\sigma_{\mathrm{v}} = \frac{p_{\mathrm{d}} - p_{\mathrm{v}}}{p_{\mathrm{a}} - p_{\mathrm{dT}}}, \tag{2.22}$$

where Q is flow rate, A_{d} is the area of the injection pipe, p_{a} is pressure in the accumulator tank, p_{d} and p_{dT} are static and total pressures in the injection pipe, respectively. Finally, ρ and p_{v} are density and vapor pressure of fluid, respectively.

An example of the flow rate characteristics of the flow damper obtained by the 1/5-scale model is shown in Fig. 2.35. The data were divided into two groups for large and small flow injections. The flow rate coefficient depended on the cavitation factor for large flow injection, while it was independent of the cavitation factor for small flow injection. The data for cavitation factor near $\sigma_{\mathrm{v}} = 9$ obtained at the last stage of small flow injection had relatively larger errors than the other data.

The ratio of the flow rate coefficient of the flow damper for large to small flow injection at cavitation factor, $\sigma_{\mathrm{v}} = 4.5$, was about 10. It was confirmed that the flow damper could yield a large flow rate ratio. The flow rate characteristics were determined by the flow at the throat of the outlet nozzle for large flow injection. If cavitation occurred at the throat, the flow rate coefficient might be reduced.

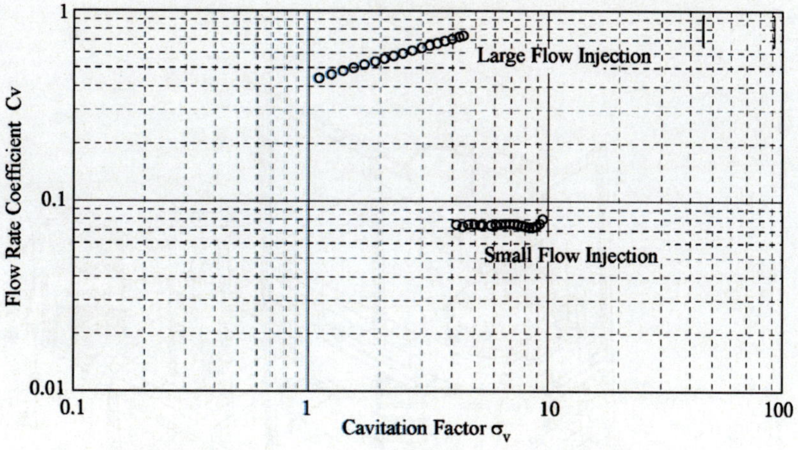

Fig. 2.35 Example flow rate characteristics of the flow damper obtained by the 1/5-scale model. The flow rate coefficients were clearly divided into two groups for large and small flow injections

The smaller the cavitation factor was, the bigger the reduction of the flow rate coefficient because of larger occurrence of cavitation.

For small flow injection, the flow rate characteristics were determined by a strong vortex in the vortex chamber. There was no cavitation at the throat of the outlet nozzle. This was why the flow rate coefficient was independent of the cavitation factor. There may be cavitation at the center of a strong vortex in the vortex chamber. But the reducer at the outlet port confined the cavitation within the vortex chamber so that it did not reach the throat.

2.6.5 *Total Performance of the Advanced Accumulator*

Acquisition of the data of the total performance was carried out with the full-height 1/2-scale model of the Advanced Accumulator under the actual pressure conditions. The test facility consisted of the accumulator model, which was the test tank of about 9 m in height, and contained the flow damper along with the standpipe, the injection pipe, and the exhaust tank as shown in Fig. 2.36. The auxiliary devices were composed of a liquid nitrogen tank and an evaporator to supply nitrogen gas to the accumulator tank and the exhaust tank. The exhaust tank corresponding to the main coolant system was kept at an arbitrary pressure from atmospheric pressure to 0.5 MPaG. The accumulator tank was initially supplied with water and then nitrogen gas to a given pressure before every test.

Since the pressure in the accumulator tank was the same as the actual one, the velocity of flow in the flow damper was the same as that in the actual damper under the corresponding conditions. The flow damper and outlet pipe were half-scale. The diameters of the accumulator tank and the standpipe were also half-scale, but their heights were full-scale so that the transitions of water levels were the same as the

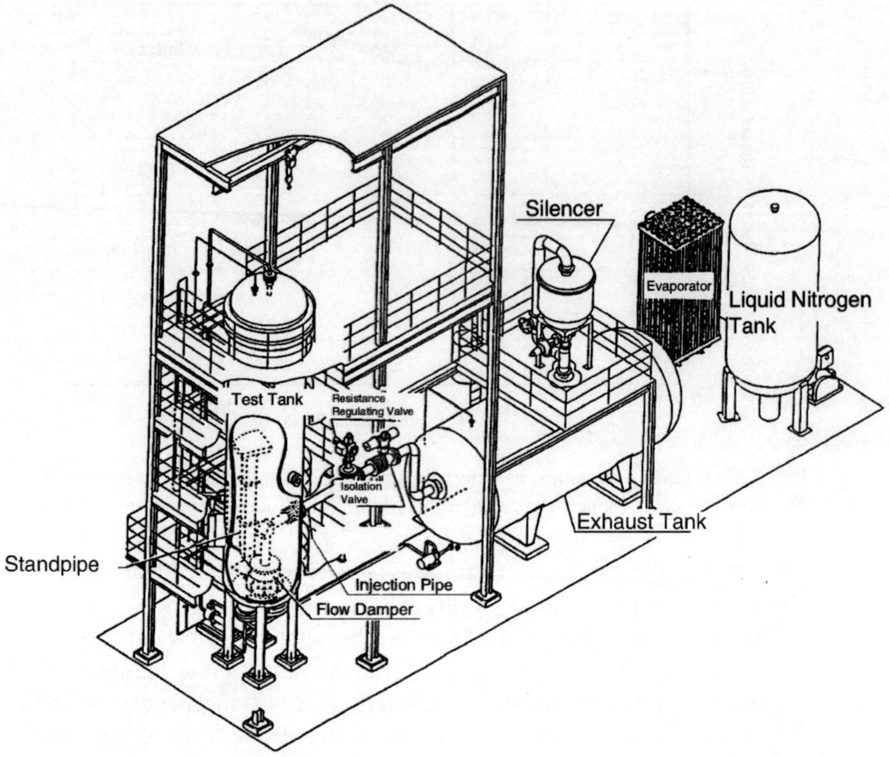

Fig. 2.36 A cut-away view of the test facility of the full-height 1/2-scale model of the Advanced Accumulator under the actual pressure conditions. The flow damper was installed at the bottom of the test tank. It could simulate the entire integrated functions of the Advanced Accumulator with the real time injection of flow

actual ones under the actual pressure conditions. Consequently, the time scale was also the same as the actual one.

The objectives of the tests were:

1. Confirmation of the expected performance
2. Confirmation of the flow rate characteristics
3. Assessment of the water level at flow changeover
4. Assessment of the influence of dissolved nitrogen gas.

It was expected that the principle confirmed by the 1/8.4-scale model would be implemented without any unexpected problem in the large-scale accumulator under the actual pressure conditions in the sequence of the large flow injection, changeover of flow rates, and small flow injection.

The flow rate characteristics, which were obtained by the half-scale model under the actual pressure conditions, should be compared with those obtained by the 1/5-scale model. If they agree with each other, it can be concluded that the Reynolds numbers will negligibly affect the characteristics of the flow damper.

Then, the characteristics of the half-scale model of the Advanced Accumulator will be adoptable to the full-scale accumulator.

The water level at flow changeover is important to secure the total volume and the duration of large flow injection. Since the descending velocity of the water level in the model tank of the accumulator is the same as that in the actual Advanced Accumulator, assessment of the accuracy of the water level at the flow changeover will be reliable.

For the evaluation of the influence of dissolved nitrogen gas, the water in the test tank was saturated with nitrogen gas before the tests. There were a ring header at the bottom of the test tank to emit nitrogen bubbles underwater and a spray nozzle at the top of the test tank to spray water into the gas space. The water for the spray was pumped up by a circulation pump as shown in Fig. 2.37.

The pressure in the test tank was monitored while bubbling gas and spraying water. The pressure first decreased due to dissolution of nitrogen gas into water, and later approached a constant value when water became saturated with nitrogen. Nitrogen gas was supplied to the test tank until the pressure became unchanged at the given value. It took about 3 or 4 h to get saturation in this test tank. After confirming the pressure became stable, the bubbling and the spraying were stopped. Hence, the water would not only be saturated with nitrogen but would also contain many cavitation nuclei. It must be much easier for cavitation to occur for this condition of the water than the condition of borated water in the actual accumulator tank.

Pressures in the test tank, the injection pipe and the exhaust tank were measured with pressure transducers. Water levels in the test tank and the standpipe were

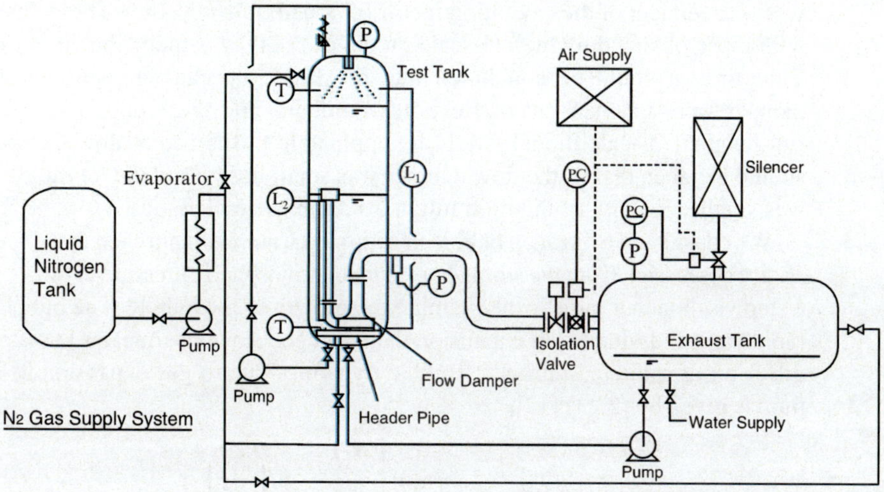

Fig. 2.37 Schematic of the test facility. Nitrogen gas was supplied from the liquid nitrogen tank through the evaporator not only for pressurization of the test facility, but also for dissolution of nitrogen gas into water. The ring header provided bubbling at the bottom of the test tank and the spray nozzle provided a water spray from the circulation pump. Quantities measured were pressures in the test tank, the injection pipe, and the exhaust tank, water levels in the test tank and the standpipe, and temperatures in the test tank

measured with a differential pressure transducer and an electrocapacitance level meter, respectively. The flow rates were calculated from the water level in the test tank. Temperatures in the test tank were measured with thermocouples.

Figure 2.38 shows an example of the test results in Case 1 obtained by the full-height 1/2-scale model. Figure 2.38a shows the transitions of water levels in the test tank and the standpipe, which were identical to those of the actual accumulator. The water level in the test tank rapidly decreased to the top of the standpipe for large flow injection, and slowly decreased for small flow injection thereafter. When the water level in the test tank reached the top of the standpipe, the water level appeared in the standpipe and plummeted to form the undershoot. The water level then recovered to the balanced level and slowly decreased as the water level in the test tank slowly descended.

Figure 2.38b shows the transitions of flow rate of the test calculated from the gradient of the transition of the water level in the test tank. The actual flow rate expected was four times this flow rate. Quick changeover of the flow rate was clearly seen. The decrease for large flow injection was caused by the pressure drop in the test tank due to the expansion of nitrogen gas.

Figure 2.38c shows the transitions of pressures in the test tank and the injection pipe, which were identical to those of the actual accumulator. Pressure rapidly fell once the injection started. The pressure in the injection pipe quickly changed at the changeover of flow rate, and became the same as that in the exhaust tank.

The flow-rate characteristics of the flow damper in Case 1 are shown in Fig. 2.39. The data were divided into two groups for large and small flow injections. The flow rate coefficient depended on the cavitation factor for large flow injection, while it was independent of the cavitation factor for small flow injection. These data agreed with those obtained by the 1/5-scale model of the flow damper shown in Fig. 2.35. Therefore, it could be concluded that the flow rate characteristics of the flow damper were independent of Reynolds numbers, and these data obtained by the full-height 1/2-scale model would be applicable to the actual flow damper. Care should be taken that, if the flow damper was small in size and the Reynolds number was small, viscosity of the fluid might affect its characteristics.

We considered a nitrogen bubble in saturated water with nitrogen that experiences abrupt depression from the storage pressure to atmospheric pressure at time $t = 0$ s in a stepwise manner in order to examine the growth of the bubble. The bubble at first rapidly expands due to gas expansion then nitrogen slowly permeates in water due to diffusion of saturated nitrogen. Bubble dynamics due to gas expansion in inviscid fluid is given by (2.23) [12]

$$P(t) = p_0 + \frac{1}{2}\rho\left[\frac{\mathrm{d}^2(R^2)}{\mathrm{d}t^2} + \left(\frac{\mathrm{d}R}{\mathrm{d}t}\right)^2\right], \tag{2.23}$$

where R is the radius of the bubble; $P(t)$ is pressure on the surface of the bubble; t is time; ρ is the density of the fluid; and p_0 is the ambient pressure.

The distension of a spherical bubble with an initial radius of 0.05 mm due to gas expansion is shown in Fig. 2.40. The surface tension on the bubble was taken into

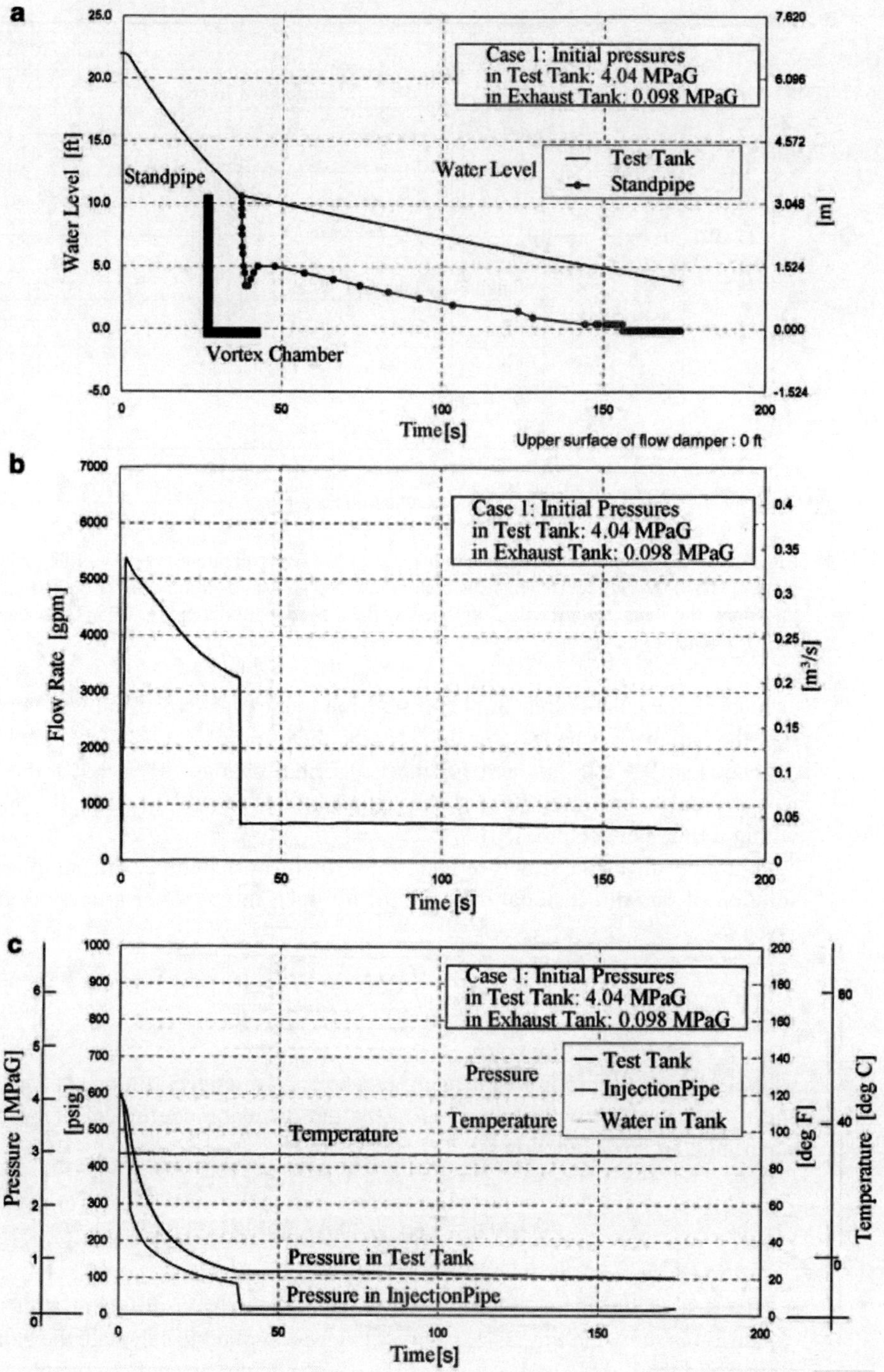

Fig. 2.38 An example of transitions of water levels in the test tank and the standpipe

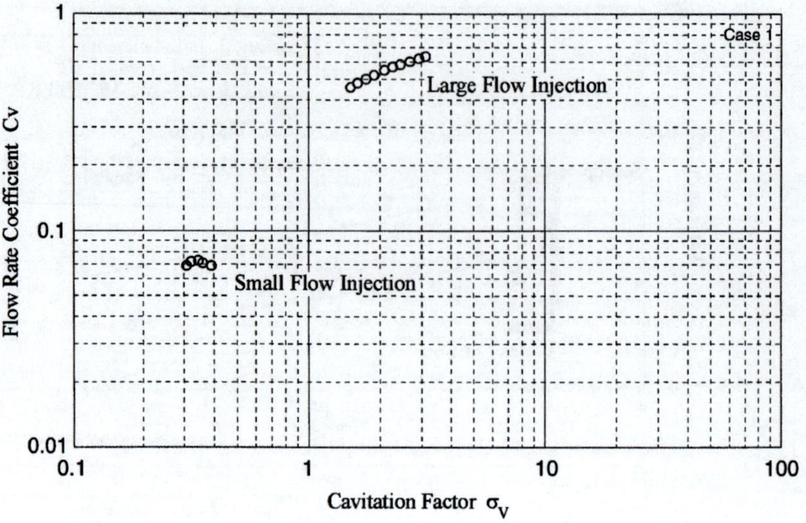

Fig. 2.39 An example of flow rate characteristics in Case 1 obtained by the full-height 1/2-scale model. The flow rate coefficient was clearly divided into two groups for large and small flow injections, the same as for the data obtained by the 1/5-scale model in Fig. 2.35. These data agreed well with each other

account in the calculation. Bubble distension due to gas expansion was very rapid. For the adiabatic change, or the specific heat ratio $\gamma = 1.4$, the bubble was distended in 9×10^{-6} s, and for the isothermal change, or $\gamma = 1.0$, the bubble was distended in 1.6×10^{-5} s. An actual bubble would be abruptly distended within a time between these times.

Nitrogen diffusion slowly affects the growth of a bubble. For simplicity, the solution of one-dimensional diffusion of nitrogen in the water around a bubble is given as:

$$c = c_R + (c_\infty - c_R)\mathrm{erf}\frac{x}{2\sqrt{Dt}}, \tag{2.24}$$

where c is concentration of nitrogen, c_R and c_∞ are concentrations at radii $r = R$ and $r \to \infty$, respectively. $x = r - R$. D is the diffusion coefficient of nitrogen in water, and an error function is

$$\mathrm{erf}\eta = \frac{2}{\sqrt{\pi}}\int_0^\eta \exp(-\xi^2)\mathrm{d}\xi. \tag{2.25}$$

The distension of the spherical bubble due to nitrogen diffusion after the gas expansion is shown in Fig. 2.41. Henry's law was applied to calculate the balance of pressure in the bubble and density of nitrogen in water around the bubble.

It was shown that the distension due to the diffusion of nitrogen was very slow, and the growth of the bubble was small for about 0.15 s which is the time for a bubble

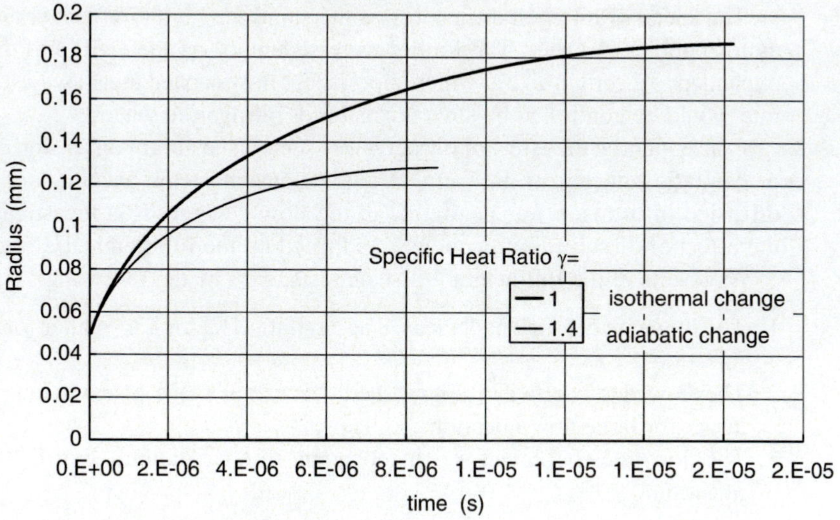

Fig. 2.40 Examples of distensions of a nitrogen bubble due to gas expansion are shown for specific heat ratios $\gamma = 1$ and 1.4, namely isothermal and adiabatic changes, respectively. Both distensions finished in 2×10^{-5} s

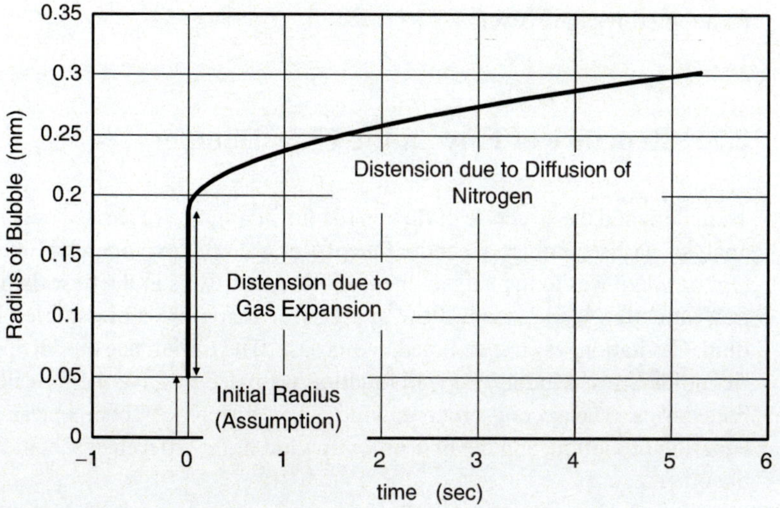

Fig. 2.41 An example of distension of a nitrogen bubble due to diffusion after the adiabatic gas expansion. This distension was the order of 1 s and much slower than that due to the gas expansion shown in Fig. 2.40

in water to pass through the vortex chamber and the throat. The diffusion around a bubble depends on its radius and is nonlinear. But the diffusion of nitrogen in this case was also very slow.

The speed of nitrogen diffusion was the same for cavitation nuclei on the walls as for bubbles in water. The former were sedentary on the walls and the bubbles expanded to a certain size at which superjacent flow carried them away. The growth rate would be controlled by slow diffusion of nitrogen in water.

Consequently, the effect of nitrogen was seen in the abrupt distension of bubbles, or cavitation nuclei, in the form of gas expansion in the accumulator tank, and diffusion of nitrogen was negligible in the flow damper. This investigation about the effect of dissolved nitrogen was confirmed by the full-height 1/2-scale tests.

As results of the full-height 1/2-scale tests, we saw the following:

1. The flow rate coefficient decreased as cavitation factor got smaller for large flow injection.
2. The flow rate coefficient approached a constant value as cavitation factor got larger for large flow injection.
3. The flow rate coefficient was independent of a cavitation factor for small flow injection.

Finding 1 was reasonable because cavitation was stronger for a smaller cavitation factor. Finding 2 was reasonable because cavitation vanished for a large cavitation factor. Finding 3 was reasonable because flow rate was small, and the pressure at the throat was almost the same as that in the exhaust tank which was larger than the vapor pressure.

Consequently, the Advanced Accumulator has been developed, and is going to be adopted for the APWR.

2.7 Structure of Flow in the Flow Damper

To understand the structure of flow in the flow damper, we carried out computational analysis with a commercial code, Fluent Ver.6.2.16 (developed by Fluent Inc.) [13]. Our problem was to investigate the flow characteristics in the flow damper for small flow injection using steady flow analysis of incompressible single-phase viscous fluid. Cavitation was not included in this case. The turbulence model applied was the Reynolds Stress Model. The wall function was used to solve the flow near solid wall boundaries. The second-order upwind finite difference scheme was used for the equation of motion, and the first-order upwind finite difference scheme was used for the others.

Figure 2.42 shows an example of the flow pattern visualized by tracers for small flow injection at the nominal condition. A combined vortex of a free vortex with a forced vortex at the center of it was formed in the vortex chamber and rapidly decreased in the reducer. The maximum velocity reaches 40 m/s in the combined vortex. There was a weak circulating flow induced in the stand pipe. The vortex became a swirl in the reducer and entered the diffuser. The swirl further decreases in the diffuser. Since the swirl flowed along the wall of the diffuser and a backflow was induced on the axis of the diffuser, it transported pressure from the injection

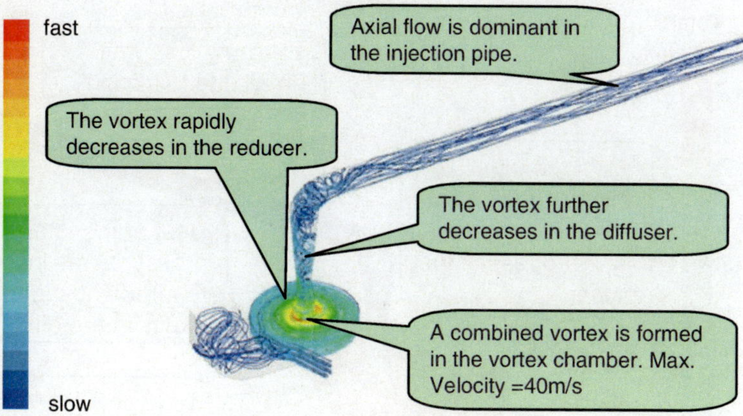

Fig. 2.42 The flow pattern for small flow injection. The vortex formed in the chamber rapidly decreases in the outlet nozzle, and a small swirl remains in the injection pipe

pipe to the throat. The backflow kept the pressure at the throat close to that in the injection pipe. The axial flow was dominant in the injection pipe, and the swirl flow vanished. The structure of the flow was just what we expected.

Figure 2.43 shows an example of the total pressure distribution for small flow injection at the nominal condition. Total pressure was conserved in the free vortex region and 90% of the total pressure was lost by shear stress at the vortex core in the chamber and the reducer. The rest of the total pressure was lost by turbulent shear stress in the diffuser. The pressure loss was slight in the injection pipe.

Now, we were able to see the mechanism of the flow damper. In the vortex chamber, static pressure was converted to dynamic pressure to form a strong swirl with very high velocity. But, the total pressure was conserved. The high velocity of the swirl yielded high shear stress at the center of the swirl so that the most of the total pressure was lost in the reducer. The remaining weak swirl was also lost by turbulent shear stress in the diffuser to form an axial flow in the injection pipe. Consequently, the vortex chamber converted static pressure to dynamic pressure and energy loss was produced at the center of the vortex chamber and in the outlet nozzle.

Figure 2.44 shows the distributions of the total pressure and the turbulent energy in the vertical cross section of the flow damper. Figure 2.44a indicates a free vortex was formed in the vortex chamber at the center of which there was minimum total pressure. The boundary layers on the upper- and lower-disk walls of the vortex chamber were very thin, and the inviscid flow prevailed in the chamber, except at its center near the outlet port. It ensured two dimensionality of the vortex flow in the chamber and independency of the flow rate characteristics of the flow damper from Reynolds numbers. The location of the minimum total pressure corresponded to the location of the minimum static pressure. If the velocity was very high, cavitation might occur and be confined there by the reducer. The total pressure plummeted in

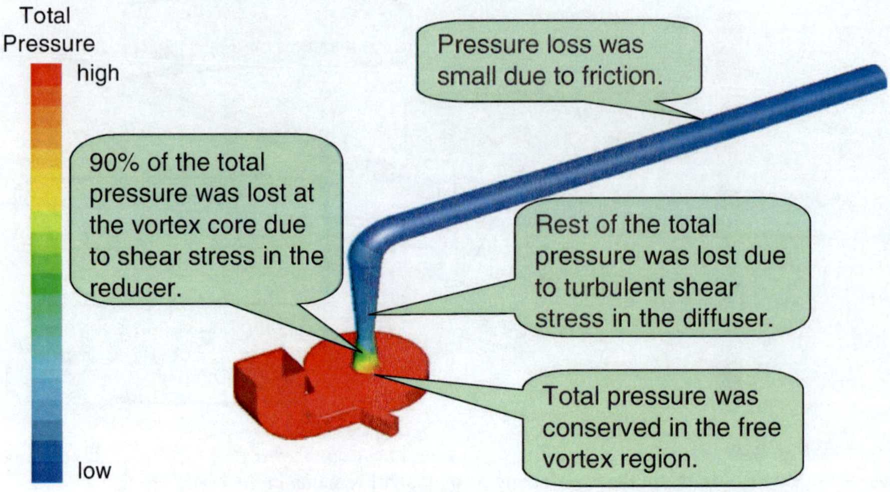

Fig. 2.43 The total pressure distribution for small flow injection. A free vortex is formed in the vortex chamber except in the vortex core and 90% of the total pressure is lost by the shear stress at the vortex core in the chamber and the reducer

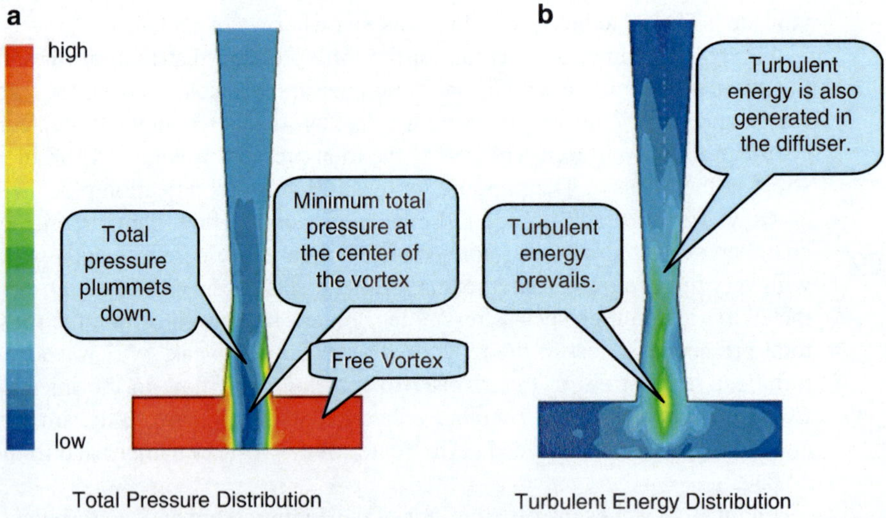

Fig. 2.44 Distributions of the total pressure and the turbulent energy in the vertical cross section of the flow damper. It was clear that total pressure was conserved in the vortex chamber to form a free vortex except in the vortex core. Turbulent energy prevailed at the vortex core

the reducer. But, the backflow on the axis of the diffuser recovered the pressure at the throat close to that in the injection pipe.

Figure 2.44b indicates the turbulent energy prevailed at the core of the swirl on the centerline of the vortex chamber and in the reducer. It was produced by shear stress. Turbulent energy was also generated in the diffuser just after the throat due to reduced velocity of flow. The pattern of the turbulent energy in the diffuser was made by the swirl near the wall and the backflow at the center.

2.8 Conclusion

Mitsubishi has developed a new passive safety component for the APWR called the Advanced Accumulator, which uses a fluidics device called the flow damper. It can control flow rate without any moving parts so that the reliability of the Advanced Accumulator is very high. The background of the development and the features of the Advanced Accumulator are explained in this chapter. The characteristics of the flow damper are investigated in detail and some results are introduced here. The structure of flow in the flow damper is also explained.

Acknowledgments We acknowledge the five Japanese utilities, Hokkaido Electric Power Co., Kansai Electric Power Co., Shikoku Electric Power Co., Kyushu Electric Power Co., and Japan Atomic Power Company, for their understanding and encouragement to develop the Advanced Accumulator. We thank many supporters and cooperators, especially Mr. Hisato Watakabe for his excellent skills for carrying out all the experiments to develop the Advanced Accumulator, and Mr. Takayoshi Sugizaki for his distinguished management of the development.

References

1. Suzuki S et al (2008) Global development of Mitsubishi standard APWR as an effective countermeasure against global warming. Mitsubishi Heavy Industries Technical Review 45(3)
2. Makihara Y et al (1993) Study of the PWR hybrid safety system. Nucl Eng Des 144:247–256
3. Shiraishi T (1994) Emergency water supply system for nuclear reactor. Japanese Patent H6-44060
4. Shiraishi T et al (1991) On flow controlled accumulator for Mitsubishi's simplified PWR (MS-300/600). Proc JSME B (in Japanese)
5. Shiraishi T et al (1992) Development of the flow controlled accumulator. ANP'92, Tokyo
6. Sugizaki T et al (1992) Design Studies for a passive safeguards system. NURETH-5
7. Shiraishi T et al (1994) Assessment of the performance of the flow controlled accumulator for next generation PWR. Proc JSME B (in Japanese)
8. Shiraishi T et al (1994) Characteristics of the flow-controlled accumulator. Nucl Technol 108:181–190
9. Shiraishi T et al (1994) Development of the advanced accumulator for next generation PWR. Therm Nucl Power Eng 45(6):43–49 (in Japanese)
10. Shiraishi T et al (1994) Development of the advanced accumulator. Mitsubishi Heavy Industries Technical Review 31(1)
11. Shiraishi T (1997) Flow control by a vortex (flow damper). Turbo Mach 25(9):54–61 (in Japanese)
12. Landau LD, Lifshitz EM (1975) Fluid mechanics. Pergamon Press, Oxford
13. Takata T et al (2009) CFD on small flow injection of advanced accumulator in APWR. Mitsubishi Heavy Industries Technical Review, 46(2)

Chapter 3
Severe Accident Mitigation Features of APR1400

Sang-Baik Kim and Seung-Jong Oh

The APR1400 (advanced power reactor, 1,400 MWe) is a standard advanced evolutionary light water reactor (ALWR) in the Republic of Korea. It is now under construction as Shin-Kori units 3 and 4. The APR1400 is designed with an additional safety margin to improve the protection of the public health, mainly focusing on typical initiators such as transients and small-break loss-of-coolant accidents (LOCAs) as well as safety against severe accidents. This section outlines the major mitigating design features of severe accidents, summarizes the results of the full-scope PSA, and presents the main results of a deterministic evaluation of severe accident issues. The APR1400 design is robust and capable of mitigating the consequences of a wide spectrum of severe accident scenarios while maintaining containment integrity and minimizing radiation release to the general public.

3.1 Introduction

The Advanced Power Reactor 1400 (APR1400), a standard ALWR, was developed by Korea Hydro and Nuclear Power Co. (KHNP) in 1992. The design is based on the experience that has accumulated through the development of the Korean Standard Nuclear Power Plant (KSNP, OPR1000) design, a 1,000 MWe pressurized water reactor (PWR). APR1400 incorporates a number of design modifications and improvements to meet the needs of utility companies as they pertain to enhanced safety and economic goals and to address new licensing issues related to the mitigation of severe accidents. APR1400 was developed in three phases. The first phase was the conceptual design phase. After surveying the candidate reactor types,

S.-B. Kim (✉)
Korea Atomic Energy Research Institute, Daejeon, Korea
e-mail: sbkim2@kaeri.re.kr

S.-J. Oh
Korea Hydro & Nuclear Power Co., Daejeon, Korea

T. Saito et al. (eds.), *Advances in Light Water Reactor Technologies*,
DOI 10.1007/978-1-4419-7101-2_3,

KHNP chose to develop an evolutionary PWR and set top-tier requirements. The second phase is the basic design phase, which started in March of 1995 and continued till February of 1999. As the basic design of APR1400 was completed, Standard Safety Analysis Report (SSAR) [1] and specifications for major NSSS equipment was also completed. The third phase was started in March of 1999. In this phase, design optimization was performed to improve the economic competitiveness, operability, and maintainability while maintaining the overall safety goal of the design. APR1400 completed the third phase as scheduled in 2001 and received its design certification from Korean regulatory agency in the May of 2002. APR1400 was built as the next in line of all nuclear power plant in the Republic of Korea following the 12 standard 1,000 MWe plants either under construction or in operation at that time. The site for the APR1400 design is close to the Kori Nuclear Power Plant (NPP) site. The construction project for the twin

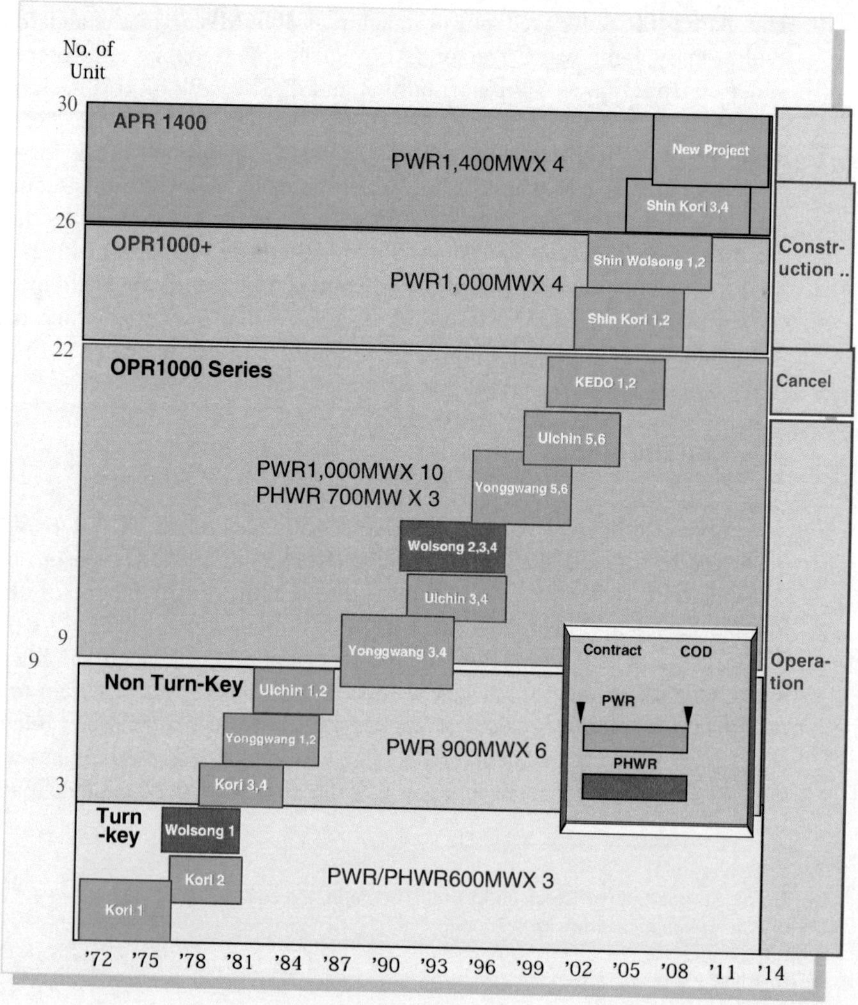

Fig. 3.1 Overview of Korean nuclear power plant program

units, Shin-Kori units 3 and 4 is in progress with the goal of commercial operation in 2013. Figure 3.1 shows the overall NPP program and its history in Korea.

Safety is a requirement of paramount importance in the operation of nuclear power plants. One of the APR1400 development policies is to increase the level of safety significantly. To implement this policy, APR1400 was designed with an additional safety margin to improve the protection of the investments as well as to protect public health. To improve the safety of the plant even further, it is important to focus on more likely initiators such as transients as well as small-break LOCA and SGTR events. Moreover, considering the TMI-2 incident, design features against severe accidents are also necessary. In order to implement this safety objective, quantitative safety goals for the design were established via a probabilistic approach [2]. These are outlined below.

The total core damage frequency (CDF) should not exceed 10E-5 per year, considering both internal and external initiating events. In addition, the frequency of core damage with reactor coolant pressure remaining high should not exceed 10E-6 per year.

The whole body dose for a person at the site boundary should not exceed 0.01 Sv (1 rem) during 24 h after the initiation of core damage, even in the event of containment failure. The frequency of exceeding such a limit should be less than 10E-6 per year.

To achieve these quantitative goals, the defense-in-depth concept remains as the fundamental principle of safety, requiring a balance between accident prevention and mitigation. With respect to accident prevention, the increased design margin and system simplification represent a major design improvement. The consideration of accident mitigation calls for the incorporation of design features to cope with severe accident as well as design basis accidents.

The design certification process in Korea is similar to that in the U.S. The certification rule similar to 10CFR52, Subpart B, Standard Design Certification, has been finalized. Regarding the technical requirements, most of the current licensing requirements are set. Safety requirements against design basis accidents are identical to those in currently operating plants [3]. The difference is in the area of severe accident mitigation features. To address PSA and severe accidents in new plant licensing, NRC has previously issued guidance, including the following documents:

1. The NRC policy statement on severe reactor accidents regarding future designs and existing plants (50 FR 32138)
2. The NRC policy statement on safety goals for the operation of nuclear power plants (51 FR 28044)
3. The NRC policy statement on nuclear power plant standardization (52FR 34844)
4. 10 CFR Part 52, "early site permits, standard design certification, and combined licenses for nuclear power plants"
5. SECY-90-016, "evolutionary light water reactor (LWR) certification issues and their relationship to current regulatory requirements"
6. SECY-93-087, "policy, technical, and licensing issues pertaining to evolutionary and advanced light water reactor (ALWR) designs"

Whereas the first three documents provide guidance as to the appropriate course for addressing severe accidents, 10 CFR Part 52 contains general requirements for addressing severe accidents. The Staff Requirement Memorandums (SRMs) of SECY-90-016 and SECY-93-087 define Commission-approved position implementing features for preventing severe accidents and mitigating their effects. 10 CFR Part 52 requires compliance with the TMI requirement in 10 CFR 50.34(f), the resolution of unresolved safety issues and generic safety issues, and the completion of a design-specific probabilistic risk assessment. SECY-90-016 and SECY-93-087 form the basis for a deterministic evaluation of severe accident performance for APR1400.

In this section, the APR1400 design and its safety features are introduced briefly. The paper then outlines the major mitigating design features of severe accidents, summarizes the results of the Level 1 and 2 PSA, and reviews the results of the deterministic evaluation of severe accident issues based on the description in the Standard Safety Analysis Report (SSAR) of APR1400. National and international test programs and computational analyses are still ongoing as a means of resolving the particular safety issues associated with the unique design features of APR1400. These results can be reflected in the final safety analysis report for the approval of the Shin-Kori 3 and 4 operating license and for the further improvement of the APR1400 design as part of the APR + development project in Korea.

3.2 Description of Nuclear Systems and Safety Systems

The primary loop configuration of APR1400 is similar to that of the currently operating Korean Standard Nuclear Power Plant, OPR1000. The nuclear steam supply system (NSSS) is designed to operate at a rated thermal output of 4000 MWt to produce approximately 1,450 MWe of electric power. The major components of the primary circuit are the reactor vessel, two reactor coolant loops (each containing one hot leg, two cold legs, one steam generator (SG), and two reactor coolant pumps) and a pressurizer connected to one of the hot legs. The SGs are located at a higher elevation than the reactor vessel to promote natural circulation. For the vent and drain, the elevation of the pressurizer and the surge line is higher than that of reactor coolant piping. A schematic diagram of the arrangements and locations of the primary components and safety-related systems is shown in Fig. 3.2. The APR1400 core consists of 241 fuel assemblies. Each fuel assembly consists of 236 fuel rods (16 × 16 array) and 5 guide tubes. The core is designed for an operating cycle of 18 or more months with discharge burnup up to 60,000MWD/MTU. The thermal margin is approximately 13%. The capacities of the pressurizer and SGs are greater than that of current design. The increased capacities of the pressurizer accommodate the plant transient without power operated relief valves. Conventional spring-loaded safety valves mounted onto the top of the pressurizer are replaced by pilot-operated safety relief valves (POSRVs). The POSRVs perform reactor coolant system (RCS) overpressure protection and safety depressurization functions. The major design objectives of the APR1400 are given in Table 3.1 [2].

The safety systems consist of the safety injection system (SIS), safety depressurization system, the in-containment refueling water storage system, and the

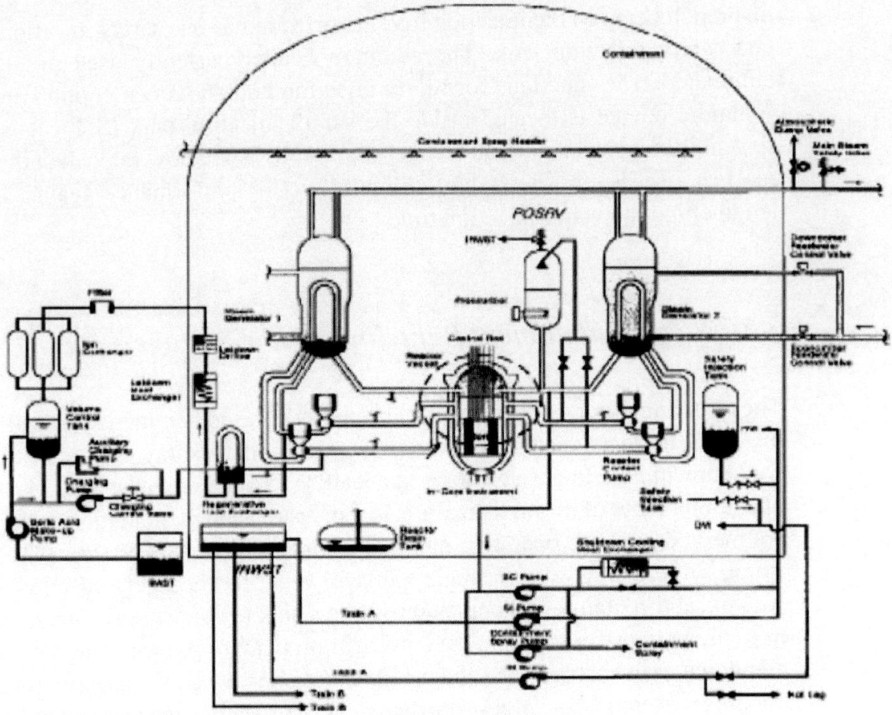

Fig. 3.2 Schematic diagram of primary components and safety system

Table 3.1 Major design objectives for APR1400

General requirement	Performance requirement and economic goal
Type and capacity: PWR, 4,000 MWt	Plant availability: greater than 90%
(NSSS system thermal power)	Unplanned trips: less than 0.8 per year
Plant lifetime: 60 years	Refueling interval: 18 months
Seismic design: SSE 0.3g	Construction period: 48 months (*N*th plant)
Safety goals	
Core damage frequency <1.0E-5/RY	
Frequency of radiation release <1.0E-6/RY	
Occup. radiation exposure <1 man Sv per RY	

containment spray system (CSS). A schematic diagram of the arrangements and locations of the safety system is also shown in Fig. 3.2.

3.2.1 Safety Injection System

The SIS is designed to inject water into the upper downcomer directly. The safety injection lines are mechanically four trains and electrically two divisions without a tie branch between the injection lines. Each train has one safety injection pump and one safety injection tank. The common header currently used in SIS trains is eliminated. The functions for safety injection and shutdown cooling are separate. A fluidic device is located inside the safety injection tank (SIT). It is a passive system that injects the borated water into the RCS at a low rate when the SIT level reaches a set level. The system will enhance the performance against LOCAs by lengthening the water injection time.

3.2.2 In-Containment Refueling Water Storage Tank

The refueling water storage tank is located inside the containment. The spillover from the RCS through the break as well as containment spray will return to the in-containment refueling water storage tank (IRWST). Through the IRWST, the current operation modes of the high pressure, low pressure, and recirculation during a LOCA are merged into one operation mode. The functions of IRWST are as follows: the storage of refueling water, a single source of water for the safety injection, shutdown cooling, and containment spray pumps, a heat sink to condense the steam discharged from the pressurizer for rapid depressurization (RD) to prevent a high-pressure core meltdown or to enable a feed and bleed operation and a coolant supply for the cavity flooding system in case of a severe accident to protect against a core melt.

3.2.3 Auxiliary Feed Water System

The auxiliary feed water system (AFWS) is designed to supply feedwater to the SGs for RCS heat removal in the event of a loss of the main/startup feedwater system. In addition, the AFWS refills the SGs following a LOCA to minimize leakage through preexisting tube leaks. The AFWS has divisions and four trains. The reliability of the AFWS is increased through the use of two 100% motor-driven pumps, two 100% turbine-driven pumps and two independent safety-related emergency feed-water storage tanks as a water source in place of a condensate storage tank.

3.2.4 Containment Spray System

The CSS is a safety grade system designed to reduce the containment pressure and temperature in the event of a main steam line break or a LOCA. It is also designed to remove fission product from the containment atmosphere following a LOCA. The CSS uses the IRWST, and it has two independent trains. The CSS provides a spray of borated water to the containment atmosphere from the upper regions of the containment. The spray flow is provided by the containment spray pumps which take suction from the IRWST. The CS pumps are designed to be functionally interchangeable with the shutdown cooling system (SCS) pumps. The CS pumps and CS heat exchangers can be used as a backup for the SCS pumps and heat exchanger to provide residual heat removal or to provide cooling of the IRWST.

3.3 Design Features against Severe Accident

The design approach differentiating the APR1400 from operating nuclear power plants such as Korean Standard Nuclear Plants (OPR1000) is related to how the former takes into account a severe accident in the design phase. Measures to cope with a severe accident are divided into the two categories of prevention and mitigation so as to minimize the possibility and consequences of a severe accident. The severe accident prevention features can be summarized as follows [4]:

Increased design margin such as a larger pressurizer, larger SGs, and an increased thermal margin
Reliable engineering safety features (ESF) system such as SIS, AFWS and CSS
Extended ESF system such as SDS with IRWST and an alternate AC source
Containment bypass prevention

The key differences from the current nuclear power plant design may be the consideration of severe accident mitigation in the design. The APR1400 as currently developed, incorporates design features that generally address severe accident issues, as follows:

For phenomena likely to cause an Early Containment Failure (ECF), for instance, within 24 h after an accident, a mitigation system shall be provided or the design should address the phenomena even if the probability of such an accident is low.
For phenomena which may potentially lead to a late containment failure if not properly prevented, the mitigation system or design measures should be considered in conjunction with the probabilistic safety goal and the cost of incorporating such features to address the phenomena.

This approach is intended to enhance the effectiveness of the investment in safety by avoiding undue over-investment in highly improbable accidents. In addition, a realistic assessment is recommended for severe accident analyses. The major design features for the mitigation of severe accidents are addressed below, based on the APR1400 SSAR[1].

3.3.1 Robust Containment

The containment vessel and parts associated with its penetration is a low-leakage cylindrical concrete shell designed to withstand a postulated LOCA or a MSLB (Main Steam Line Break). Additionally, the containment vessel provides a barrier against the release of radioactive materials which may be present in the containment atmosphere following an accident.

The cylindrical containment is 150 ft. (45.72 m) in diameter, and the nominal value of the net free volume is 3.2 million cubic feet (90,614 m^3). It is constructed of prestressed concrete and is designed to protect the inner containment from missile threats, to promote mixing throughout the containment atmosphere, and to accommodate condensable and noncondensable gas releases considering the design basis and the potential for severe accidents. The internal structures, which consist of reinforced concrete, enclose the reactor vessel and other primary system components as a form of providing a biological shielding. In severe accident scenarios, it is of paramount importance to provide a strong containment design to meet severe accident internal pressurization challenges. To this end, several structural analyses have been conducted to characterize the containment strength of the APR1400. An evaluation using the ABAQUS computer program indicated that the pressure limit in accordance with the ASME Factored Load Category liner plate allowable strain criteria is 115 psig (9.12 kg/cm^2) at a temperature of 350°F (176.7°C).

3.3.2 *Safety Depressurization and Vent System*

The safety depressurization and vent system (SDVS) is a multipurpose dedicated system specially designed to serve important roles in severe accident prevention and mitigation. In the context of severe accident prevention, the SDVS performs the following functions:

- Venting of the reactor coolant system
  The reactor coolant gas vent (RCGV) function of the SDVS provides a means of venting noncondensable gases from the pressurizer and the reactor vessel upper head to the reactor drain tank during post-accident conditions. In addition, the RCGV provides: (1) A safety-grade means to depressurize the RCS in the event that the pressurizer main spray and auxiliary spray systems are unavailable, (2) A means of venting the pressurizer and reactor vessel upper head during pre-refueling and post-refueling operations.
- Feed and bleed operation

  The rapid depressurization (RD) function, or bleed function of the SDVS, provides a manual means of depressurizing the RCS quickly when normal and auxiliary feedwater (AFW) is unavailable to remove core decay heat through the SGs. This function is achieved via a remote manual operator control. Whenever an event, e.g., a total loss of feedwater (TLOFW), results in a high RCS pressure with a loss of RCS inventory, the SDVS valves may be opened by the operator, causing a controlled depressurization of the RCS. As the RCS pressure decreases, the Safety Injection (SI) pumps start, initiating feed flow to the RCS and restoring the RCS liquid inventory. The RD function allows for both short- and long-term decay heat removal procedures.
- RD during a severe accident
  The RD feature of the SDVS also serves an important role in severe accident mitigation. In the event that a high-pressure meltdown scenario develops and the

feed portion of the feed-and-bleed operation cannot be established due to the unavailability of the SI pumps, the SDVS can be used to depressurize the RCS to ensure that a high-pressure melt ejection (HPME) event does not occur, thereby minimizing the potential for direct containment heating (DCH) following a vessel breach (VB).

The SDVS valve size is selected to meet both feed-and-bleed and DCH severe core damage depressurization goals. For a worst-case TLOFW event, the valve size ensures adequate feed-and-bleed capability. The severe core damage depressurization goal is to ensure that the SDVS can depressurize the RCS from 2,000 to 250 psia (13.6–1.7 MPa) prior to a reactor vessel melt-through.

3.3.3 Evaluation of Hydrogen Mitigation System

During a degraded core accident, hydrogen is produced at a greater rate than that of the design basis LOCA. The hydrogen mitigation system (HMS) is designed to accommodate the hydrogen production from 100% fuel clad metal–water reaction and limit the average hydrogen concentration in containment to 10% in accordance with 10 CFR 50.34(f). These limits are imposed to preclude detonations in the containment that may jeopadize the containment integrity or damage essential equipment. The HMS consists of a system of passive auto-catalytic re-combiners (PARs) complemented by glow plug igniters installed within the containment. The PARs serve all but accident sequences in which mild and slow hydrogen release rates are expected; they are located all over the containment area. In contrast, the igniters supplement PARs under a very-low-probability accident in which a very rapid release of hydrogen is expected; they are placed near source locations to promote the combustion of hydrogen in a controlled manner such that containment integrity is maintained. The APR1400 HMS consisting of 26 PARs and ten-igniters is distributed throughout the containment area such that the overall average concentration goal of 10 CFR50.34 (f) may be met. Figure 3.3 shows the location of the PARs in the APR1400 containment area. The details of this design will be finalized during the course of the Shin-Kori 3 and 4 project. The igniters are powered from Class IE buses which receive power from Preferred Offsite Power I or Preferred Offsite Power II (two distinct and separate sources of offsite power). In the event of a loss of off-site power, the igniters are powered from an emergency diesel generator. During a station blackout (SBO), the igniters are powered from the Alternate AC facility.

3.3.4 Cavity Flooding System

The function of the cavity flooding system (CFS) is to provide a means of flooding the reactor cavity during a severe accident for the purpose of cooling the core debris in the reactor cavity and scrubbing fission product releases. Flooding of the reactor cavity is an EPRI URD evolutionary plant design requirement and serves several

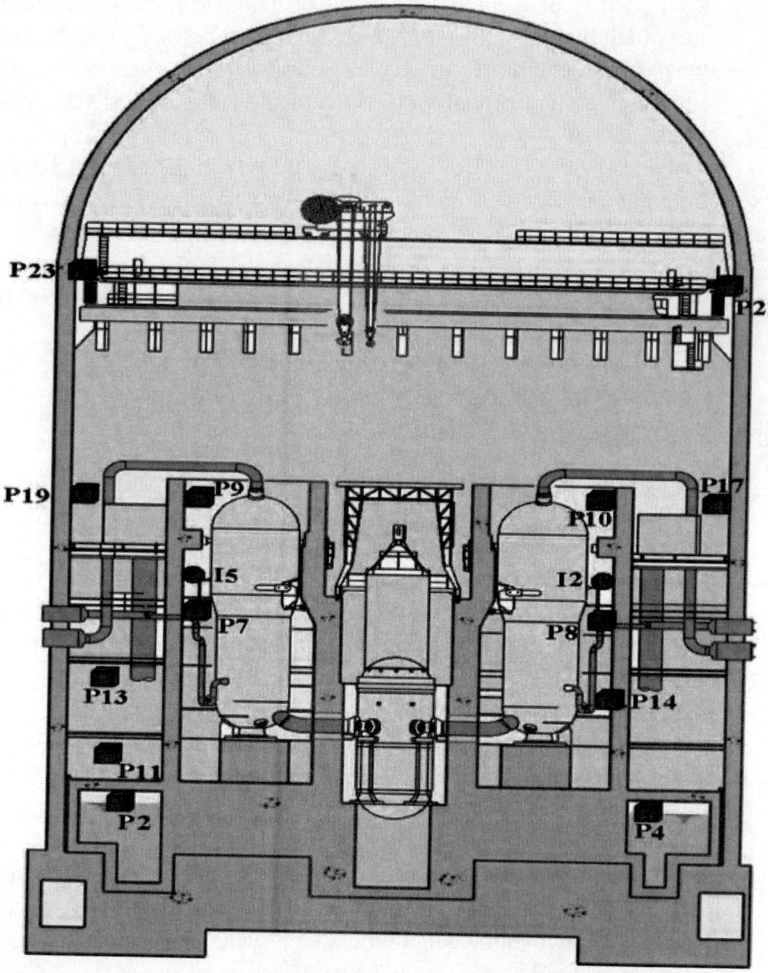

Fig. 3.3 Locations of PARs in APR1400 hydrogen mitigation system (HMS)

purposes in the overall strategy to mitigate the consequences of a severe accident. These include:

Minimizing or eliminating corium-concrete attack
Minimizing or eliminating the generation of combustible gas (hydrogen and carbon monoxide)
Reducing fission products released due to corium-concrete interaction
Scrubbing fission products released from trapped core debris

The components of the CFS include the IRWST, the holdup volume tank (HVT), the reactor cavity, connecting pipes, valves, and associated power supplies. This system is used in conjunction with the CSS to form a closed or recirculating water cooling system by providing a continuous cooling water supply to the corium debris. The quenching of the corium produces steam which is condensed by the containment spray flow. The CFS takes water from the IRWST and directs it to the reactor cavity. The water flows first into the HVT by way of two 14-in. (diameter) HVT spillways and then into the reactor cavity by way of two 10-in. reactor cavity spillways. A schematic drawing of the cavity flooding system is presented in Fig. 3.4. Once actuated for the opening of the spillway valves by means of the active attribute, the movement of the water from the IRWST source to the cavity occurs passively due to the natural hydraulic driving heads of the system.

3.3.5 *Reactor Cavity Design*

The APR1400 reactor cavity is configured to promote retention of, and heat removal from, the postulated core debris during a severe accident. Thus, it plays several roles in accident mitigation. Corium retention in the core debris chamber through a tortuous flow path eliminates the potential for significant DCH (Direct Containment Heating)-induced containment loadings. The large-cavity floor area allows for spreading of the core debris, enhancing its coolability within the reactor cavity region. Figure 3.4 also shows the configuration of the APR1400 reactor cavity design. The important features of the reactor cavity include

A large cavity volume

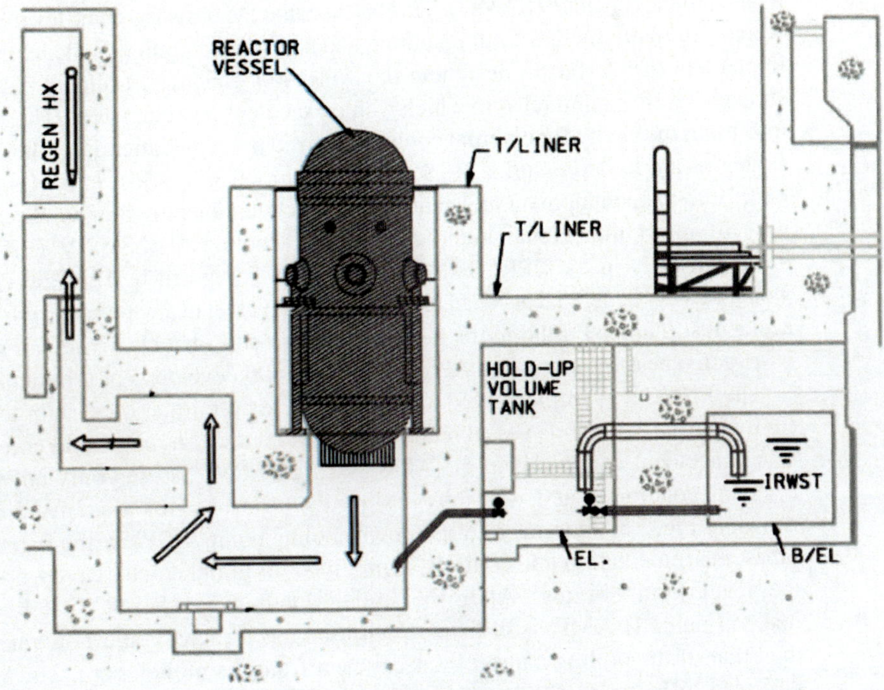

Fig. 3.4 Reactor cavity and cavity flooding system

- A closed vertical instrument shaft
- A convoluted gas vent
- A large recessed corium debris chamber
- A large-cavity floor area
- A minimum concrete thickness of 3 ft. from the cavity floor to the containment embedded shell
- Robust cavity strength

3.4 Severe Accident Management and External Reactor Vessel Cooling

In the USA, as a part of the integration plan for the closure of severe accident issues (SECY88-147, May 1988), the accident management program was implemented at each nuclear plant [5]. The focus was on what could be done once a severe accident occurs while at the same time recognizing the possible adverse consequence and inherent uncertainty [6]. As with TMI-2, severe accidents were assumed to be a result of multiple failures; hence, predicting a scenario a priori was not clear-cut. EPRI developed a technical basis for severe accident management action. Each owner group further developed what became known as the owner's group severe accident management guidance (SAMG). This has become the basis for plant-level SAMG. In Korea, similar to the USA, all operating and new plants require the development and implement of SAMG. In designing the APR1400, candidate accident management strategies were examined as to which would execute this strategy best. The strategy of inject into the cavity for external cooling received particular attention. The high-level strategies are to depressurize RCS, to inject into RCS, to inject into SGs, to spray/inject into the containment area, and to reduce containment hydrogen. As a subcategory of inject into containment, the external cooling of the RPV (ERVC) was examined early in the EPRI SAMG technical basis report [7]. The ERVC strategy would be very beneficial as it can retain corium in-vessel in a wide spectrum of severe accident scenarios. Furthermore, there have been a considerable number of studies of its effectiveness as part of the AP600 and APR1400 development programs.

The ERVC was implemented as a severe accident mitigation system to be used for the purpose of the in-vessel retention of corium under hypothetical core melting severe accident conditions in the APR1400. The ERVC is used only under severe accident conditions and was thus designed on the basis of a safety margin. As shown in Fig. 3.5, one train of a shutdown cooling pump (SCP), with related valves, pipes, instrumentation and controls, is provided for initial reactor cavity flooding to the level of a hot leg. After the initial flooding by the SCP, a boric acid makeup pump (BAMP) is utilized to refill the reactor cavity at a flow rate greater than that of the boiling caused by decay heat from the molten core. The ERVC is designed to be manually operable only when the core exit temperature reaches a certain temperature following a severe accident. The operating procedure for the

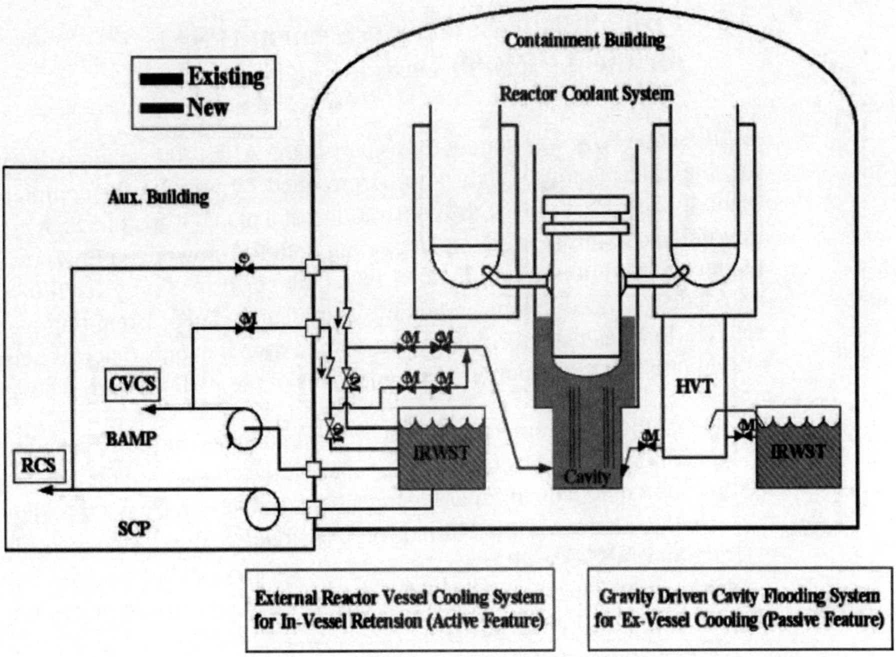

Fig. 3.5 External reactor cavity flood flooding system

ERVC was developed through severe accident analyses and probabilistic safety assessments.

The in-vessel retention of corium through the ERVC is pursued as a key severe accident management strategy. Probabilistic safety assessments of various PWRs, including the APR1400, clearly show that the threat to containment integrity is reduced if the reactor pressure vessel (RPV) does not fail. For ERVC, the RCS pressure should be sufficiently low and the RPV bottom head should be submerged before molten core debris relocation into reactor vessel lower plenum, to prevent reactor vessel creep ruptures and thermal shock. For in-vessel retention in the APR1400 design, two major uncertainty issues have been raised: the first is the effectiveness of the heat removal capacity by the water flowing through the annular space between the RPV and the thermal insulation. The second is the integrity of the in-core instrument nozzles that are welded at the RPV bottom head under the expected core debris thermal load. To resolve these issues, an integral effort involving experiments and analysis has been done for the APR1400 design. The evaluation shows that this strategy is very effective for most of depressurized core damage scenarios. The evaluation found that the margin is small for one hypothetical limiting scenario: a full core melt with a large LOCA. However, the likelihood of this scenario is insignificant. From a severe accident management perspective, the best strategy would be combining ERVC strategy with the strategy of injecting into the RCS.

3.5 Probabilistic Safety Assessment (PSA) for the APR1400 Design

A full-scope PSA was conducted as part of the APR1400 standard design excludin the site information, which is to be provided by utilities for completeness of th plant-specific PSA during the construction of a plant [1, 8]. The PSA addresses bot internal and external initiators, including both full-power and low-power/shutdow modes of operation. Level II & III PSAs are performed only for full-power opera tion. These procedures provide more significant results. Bounding plant site char acterization was used for the evaluation of external events (such as seismic events to evaluate risk to the public. The objectives of the APR1400 PSA are given below

- To satisfy the requirement of the Korean regulatory authority in which a design- specific PSA shall be conducted as a part of the application for design certificatior
- To provide a mechanism for assessing a balanced design from a risk standpoint (i.e., such that there are no outliers or individual features that contribute a large fraction of the overall risk)
- To demonstrate that the detailed plant design will be capable of meeting the safety goals imposed by the utility
- To serve as a tool that can be used interactively with the design process to aid in improving the design in an efficient and cost-effective manner

The methodology employed in the Level I portion of the APR1400 PSA is consistent with methodology outlined in USNRC's PSA Procedure Guideline, i.e., NUREG/CR-2300 and 2815 [9, 10]. The PSA-based seismic margin assessment (SMA) is used to evaluate seismic events. The simplified PSA is used to evaluate internal fire and flooding events. Other external events are evaluated qualitatively. The methodology used for the Level II PSA is consistent with that used in NUREG-1150[11]; it is described in the PRA Procedure Guideline. The Level III PSA also uses the methodology described in the PRA Procedure Guide.

The CDF (Core Damage Frequency) for internal events was estimated to be 2.3E-6 per year. The LOCA categories of initiating events dominate (43%) the CDF profile. Of the LOCA categories, small LOCA (17.2%) and SG tube rupture (10.3%) dominate the CDF. In the transient categories, the loss of feed water (20.9%), SBO (14.9%), and anticipated transient without scram (14.9%) events dominate the CDF. Table 3.2 presents the CDF contributions by initiating events. The CDF of the APR1400 design is less than that of the conventional Korean NPP by a factor of 4. The external event analysis for APR1400 was conducted using the bounding site characteristics. The CDF due to external events is 4.4E-7/RY considering fire- and flood-induced events.

Table 3.3 shows the fractions of each containment failure mode for all events, given that core damage exists. For internal events, a late containment failure (LCF), isolation failure (NOISOL), and containment bypass (BYPASS) are most dominant modes. Eventually, the containment fails if the containment heat removal fails and is not recovered. However, the late failure mode occurs several days after accident initiation. The emergency containment spray backup system (ECSBS) plays an important role in

Table 3.2 Core damage frequency contributions by internal events for full power operation

Initiating events	CDF contributions by initiating events(%)
Large loss-of-coolant accident (LLOCA)	5.11
Medium loss-of-coolant accident (MLOCA)	7.29
Small loss-of-coolant accident (SLOCA)	16.44
Steam generator tube rupture (SGTR)	10.31
Interfacing system loss-of-coolant accident (ISLOCA)	0.18
Reactor vessel rupture (RVR)	3.94
Large secondary side break (LSSB)	1.82
Loss of main feedwater (LOFW)	20.93
General transient (GTRN)	3.76
Loss of condenser vacuum (LOCV)	0.05
Loss of 4.16 KV (LOKV)	0.00
Loss of 125 V DC (LODC)	2.74
Loss of component cooling water (LOCCW)	0.89
Loss of offsite power (LOOP)	4.05
Station blackout (SBO)	14.89
Anticipated transient without scram (ATWS)	7.69
Total	100.00

Adapted from Proceedings of ICAPP '03, May 2003 [8] and used with permission from American Nuclear Society

preventing late containment failure. The burn of combustible gases can result in a late containment failure. A containment isolation failure is mainly caused by a containment failure before a VB (Vessel Breach). A containment bypass is mainly caused by a SG tube rupture with an unisolable leak from ruptured SG tubes. ECF is mainly caused by a hydrogen burn and/or the DCH; these events are dependent on the maximum pressure load that is produced and the ultimate strength of the containment area. The basemat melt-through (BMT) contribution is low due to the cavity flooding system (CFS). The CFS is designed so that the core debris in the cavity is submerged. A dry cavity allows core-concrete interaction and results in an eventual BMT if a containment failure by over-pressure and over-temperature does not occur.

External events are initiated mainly by transients, which have no RCS breaks. The severe accident mitigation features are assumed to be less reliable compared to those of internal events. Even when a loss of active features is caused by external events, the passive features to mitigate containment failure, i.e., robust containment, cavity design, and PARs are effective for external events.

The CDF safety goal meets the design goal of 1.0E-5/RY. The containment failure frequency for all events is expected to be 2.8E-7/RY, which is less than the design goal of 1.0E-6/RY. The results of the containment performance analysis in terms of the conditional probability of a containment failure indicate that the APR1400 design does not have any particular vulnerability to core melt and containment failure. This assessment result is based on the standard design information and will be updated in the detailed design stage considering the site information and detailed design information.

Table 3.3 Conditional containment failure probability for full power operation

	Internal events		External events			
			Fire		Flood	
Mode	Frequency (/ry)	Fraction (%)	Frequency (/ry)	Fraction (%)	Frequency (/ry)	Fraction (%)
NO CF	2E-06	91.45	3E-07	80.57	7E-08	86.98
CFa	2E-07	8.55	7E-08	19.43	1E-08	13.02
LERF[b]	7E-08	2.86	1E-08	3.52	2E-09	2.51
ECF	2.26E-08	0.95	9.02E-09	2.46	1.28E-09	1.67
LCF	9.25E-08	3.89	4.76E-08	13	4.81E-09	6.26
BMT	9.07E-09	0.38	5.31E-09	1.45	1.05E-09	1.36
NOISOL	3.36E-08	1.41	5.34E-09	1.46	2.22E-09	2.89
BYPASS	4.54E-08	1.91	3.89E-09	1.06	6.49E-10	0.84
Total	2.38E-06	100	3.66E-07	100	7.69E-08	100

Adapted from Proceedings of ICAPP '03, May 2003 [8] and used with permission from American Nuclear Society

[a]CF (containment failure) = ECF + LCF + BMT + NOISOL + BYPASS

[b]LERF (large early release frequency) = ECF + BYPASS

3.6 Resolution of Severe Accident Issues

3.6.1 *Severe Accident Progression*

The accident progression analysis, including in-vessel and ex-vessel melt progressions, is performed typically using MAAP4 [12] in the APR1400 to determine the physical and thermal-hydraulic behavior of accident sequences. In case any specific effects cannot be modeled properly by the MAAP code, appropriate separate effect codes are employed to evaluate the progression of a specific accident. Key events evaluated in terms of the in-vessel melt progression are core uncovery, core damage, and molten core relocation to the lower plenum. Potential consequences from core uncovery and core damage that may result in a challenge to the containment integrity include hydrogen generation and release and a temperature-induced steam generator tube rupture (SGTR). Potential consequences from core relocation include in-vessel steam explosion. In-vessel corium retention by external reactor vessel cooling (ERVC) is considered as an effective mechanism to mitigate the potential for a severe accident. Key events evaluated for the ex-vessel melt progression are melt relocation from a VB to the reactor cavity, fuel-coolant interaction (FCI), molten core concrete interaction (MCCI), and debris cooling. These events may result in challenges to the integrity of the containment area.

For the purpose of a PRA phenomenological discussion and the resolution of severe accident issues, containment failures can be classified into early and late failures. An ECF is defined as a containment failure prior to or within 1 h after core debris penetrates the reactor vessel. The above definition is relative as it is driven by severe accident phenomenological processes. For the source term and risk assessment, an ECF is driven by the severity of the potential radiological release and population evacuation concerns. In this instance, ECF implies a containment failure within 12 h of the initiation of a severe accident.

An ECF is important as these events will result in reduced warning times for initiating off-site protective measures and reduced time available for decay and deposition of radioactive materials within the containment area. The mechanisms that result in an ECF cover a range of phenomenological processes. Potential ECF modes include containment over-pressurization due to DCH, hydrogen combustion, rapid steam generation, containment structural failure due to missile generation, cavity over-pressurization, and corium debris impact on the containment wall. For the vast majority of early containment threats, a containment challenge occurs within a few minutes of the reactor VB. Similarly, the vast majority of late containment threats occur a day or more after a VB. The only exception to this rule appears to be a post-VB hydrogen burn. The 1-h post-VB time interval was chosen to denote that the ECF mode is primarily driven by a desire to separate the characteristic of a late containment hydrogen burn. A late hydrogen burn occurs after considerable core concrete interaction. This late burn can occur as quickly as 2 h after VB.

Late containment failure refers to those severe accident scenarios where containment failure occurs more than 1 h after VB and more than 24 h after event initiation. The 24-h definition of late containment failure is consistent with the deterministic containment performance goal identified in SECY-93-087. The containment performance goal is directed at ensuring that containment will maintain its role as a reliable, leak-tight barrier for approximately 24 h following the onset of core damage. Furthermore, following this period, the containment should continue to provide a barrier against the uncontrolled release of fission products. Four potential mechanisms for late containment failure are identified for the APR1400. These are:

Gradual containment over-pressurization
BMT
Temperature-induced penetration seal failure
Delayed combustion

In designing the APR1400, containment/cavity enhancements were made to the existing PWR design to minimize the risk of ECF, but also of late containment failure. The following sections provide an overview of the associated phenomenological issues and a quantitative assessment of the impact of these challenges pertaining to the APR1400 based on the severe accident issues in SECY-93-087.

3.6.2 Identification of Severe Accident Issues

USNRC policy statement on severe accident and advanced reactors states that new reactor designs must demonstrate improved severe accident characteristics compared to operating reactors. SECY-93-087 identifies the safety issues related to severe accident mitigation that are expected to be addressed for evolutionary and advanced reactor designs. Furthermore, NRC has outlined the following criteria to benchmark plant safety for advanced designs in SECY-93-087.

- Hydrogen mitigation

  Accommodation of hydrogen generation equivalent to a 100% metal–water reaction of the fuel cladding
  Limit containment hydrogen concentration to no greater than 10%
  Provide containment–wide hydrogen control for severe accidents

- Core debris coolability

  Provide reactor cavity floor space to enhance debris spreading
  Provide a means to flood the reactor cavity to assist in the cooling process
  Protect the containment liner and other structural members with concrete, if necessary
  Ensure that the best-estimate environmental conditions resulting from core-concrete interactions do not exceed Service Level C for steel containment or the factored load category for concrete containment for approximately 24 h. Also ensure that the containment capability has a margin to accommodate uncertainties in the environmental conditions from core-concrete interactions

- High-pressure melt ejection

  Provide a reliable depressurization system
  Provide cavity design features to decrease the amount of ejected core debris that reaches the upper containment area

- Containment performance

  The containment should maintain its role as a reliable, leak-tight barrier for approximately 24 h following the onset of core damage under the more likely severe accident challenges. Following this period, the containment should continue to provide a barrier against the uncontrolled release of fission products

SECY-93-087 also recommends that the equipments responsible for the mitigation of a severe accident maintain functional reliability during relevant events. In evaluating equipment survivability, the phenomena associated with the above four severe accident safety issues should be considered. Through the APR1400 design development process, the above issues were implemented in the design and assessed in a deterministic manner. The following sections give a detailed description of each issue in the APR1400 SSAR [1, 13].

3.6.3 Hydrogen Control

In 10 CFR 52.47 (a)(1)(ii), the NRC requires applicants for a standard design certification to demonstrate compliance with any technically relevant portions of the TMI requirements in 10 CFR 50.34(f). In 10 CFR 50.34(f)(2)(ix), the NRC requires a system for hydrogen control that can show with reasonable assurance that uniformly distributed hydrogen concentrations in the containment area do not exceed 10% during and following an accident that releases an amount of hydrogen equivalent to the amount that would be generated from a 100% fuel-clad metal–water reaction, or in which the post-accident atmosphere will not support hydrogen combustion. In SECY-90-016, the NRC staff recommended that the Commission approve the staff's position that the requirements of 10 CFR 50.34(f)(2)(ix) remain unchanged for evolutionary LWRs.

The generation and combustion of large quantities of hydrogen is a severe accident phenomenon that can threaten containment integrity. The major source of the hydrogen generated is from the oxidation of zirconium metal with steam when the zirconium reaches temperatures well above normal operating levels. Experiments on core degradation indicate that in-vessel hydrogen generation associated with core-damage can vary over a wide range. The specific amount of oxidation is dependent on a variety of parameters related to the sequence progression. These include the RCS pressure, the timing and flow rate of reflooding if it occurs, and the temperature profile of the reactor core during the course of the accident sequence. In addition, ex-vessel hydrogen generation must be considered. Hydrogen is produced as a result of ex-vessel core debris reacting with steam or concrete, or both.

3.6.3.1 AICC Pressure Calculation

To satisfy the requirements specified in IOCFR50.34(f)(3), the pressure rise with a hydrogen burn is reviewed based on the complete combustion of hydrogen generated by the oxidation of 100% zircaloy cladding in the active core. The maximum pressure rise is reviewed with the adiabatic isochoric complete combustion (AICC) model for various flammable gas mixtures of H_2-air-H_2O. For a mixture of H_2-air-H_2O, the steam content must be below its inerting concentration which is, from the experimental findings, nearly 56%. Furthermore, the mixture should exceed the lower flammability limit of this ternary mixture. The lower flammability limit curve could be generated with the MAAP4 model. With the MAAP4 methodology, this steam concentration was determined to be approximately 47%, as shown in Fig. 3.6. Based on the bounding results of the deterministic evaluation of the containment hydrogen threat, complete combustion of the hydrogen produced due to 100% oxidation of the zircaloy cladding in the active fuel region will result in a peak containment pressure of 0.63 MPa for a dry condition and 0.81 MPa for a wet condition. Those values do not exceed the Factored Load Category for the APRI400 containment.

3.6.3.2 Hydrogen Distribution

During a degraded core accident, hydrogen will be produced at a greater rate than that of the design basis LOCA. The basic approach for the selection of accident sequences for APR1400 hydrogen control analyses involves an assessment of how probable the occurrence of an accident sequence results in core damage. The following accident sequences were selected based on analytical results from probabilistic safety analyses.

- Loss of feedwater (LOFW)
- Small break LOCA of RCS cold leg failure (SLOCA)
- SBO

The MAAP 4.0.3 code is used to predict the quantities of invessel and ex-vessel hydrogen generation. The calculation continues over the time when an amount of the hydrogen equivalent to a 100% metal–water reaction is met. During this process, a best-estimate prediction is applied for accident scenarios. Considering the uncertainties associated with severe accident phenomenological modeling and accident progression for the hydrogen generation, conservative calculations were performed. The analysis result demonstrates that with the exception of the IRWST and reactor cavity, the hydrogen is well mixed and does not accumulate at a high concentration. Figure 3.7 shows the typical results of the hydrogen concentration of major compartments in a small-break LOCA. The natural circulation flow paths established inside the containment area facilitate the mixing of the hydrogen gas considerably. Depending on the main release location, the peak hydrogen concentration appeared in the compartments of hydrogen release, are as follows: the reactor cavity compartment and IRWST for all accident scenarios. Additionally, depending on the accident scenario, the hydrogen concentrations exceeded 10% for some time intervals. A high concentration of hydrogen is expected in the vapor space above the water in the IRWST because only hydrogen in the primary coolant loop is directly released into that compartment. To prevent unintended detonation or combustion of hydrogen in this area and to maintain a hydrogen concentration of less than 10%, passive autocatalytic recombiners (PARs) and igniters are located in the vapor space of the IRWST and throughout the containment area.

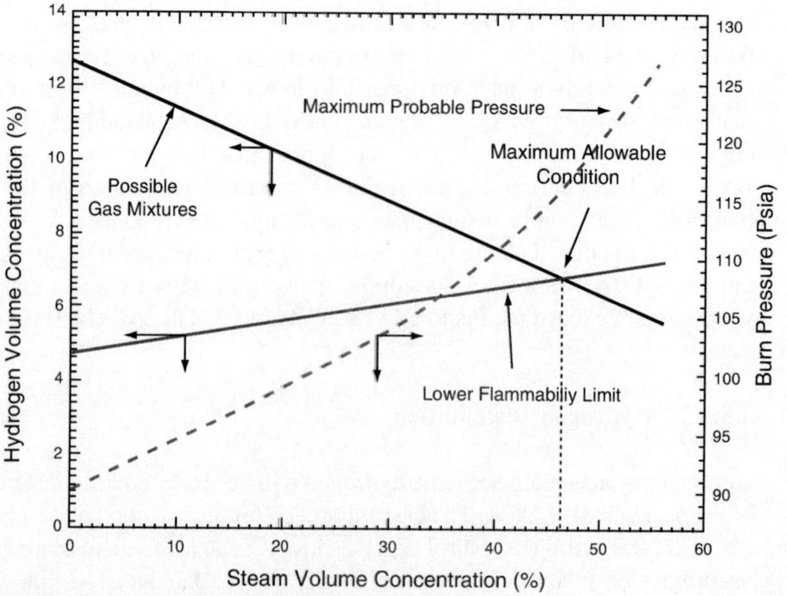

Fig. 3.6 Hydrogen combustion potential in APR1400 containment

3.6.3.3 Hydrogen Mitigation System

As a first-step for the APR1400 HMS design, a feasibility study of hydrogen mitigation devices is performed. A system of PARs complemented by glow plug igniters is selected for the APR1400 HMS. Considering the performance characteristics of the PARs and igniters, the location and number of APR1400 HMSs is determined. In this process, hydrogen control analyses are performed with the MAAP4 code to determine whether the requirements can be met. In addition, upon the feedback, the results are reflected on the HMS location and required capacity to justify. This evaluation sets these regions of containment in which the APRI400 HMS is to be located. Considering the technical criteria of the flow path, enclosed spaces, and anticipated hydrogen source release locations, final required number and locations of PARs and igniters are determined to maintain the hydrogen concentration below a controllable level within the containment area under severe accident conditions. If PARs and igniters share the same volume, the most effective placement of the igniter would be to place it below the elevation of the PARs, as igniters are ineffective at low hydrogen concentrations and do not burn completely. To avoid deterioration in the PARs performance level due to a radiating diffusion flame and to confirm the PARs performance level even when the hydrogen gas entering the PARs is at a high temperature, the igniter and PARs are properly separated from each other. In the APRI400, 26 PARs and 10 igniters were finally distributed throughout the containment area, as presented in Fig. 3.3.

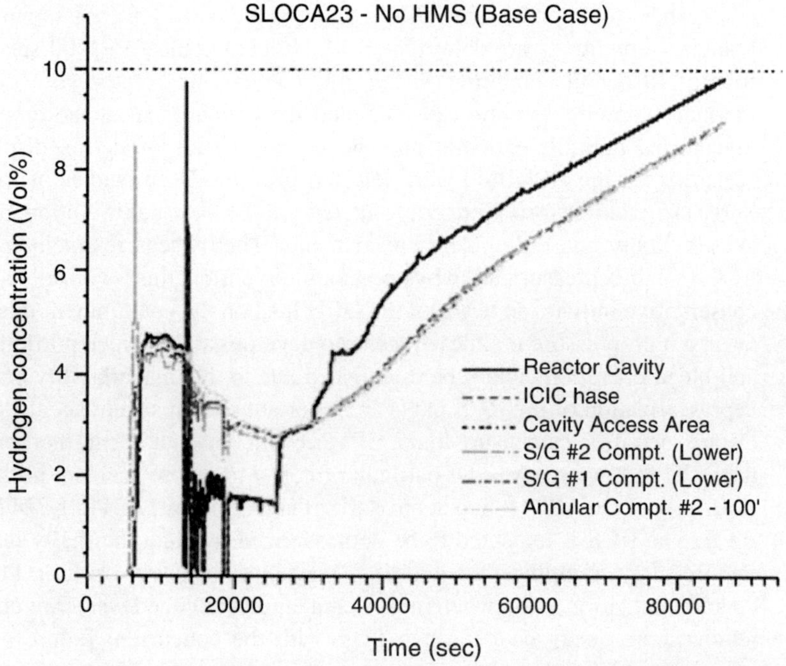

Fig. 3.7 Hydrogen concentration of major compartments in SLOCA 23 sequence

The location and exact number of PARs and igniters will be finalized during the construction of Shin-Kori 3 and 4 units using GOTHIC code [14] calculation and expert judgment.

3.6.4 Direct Containment Heating

During certain types of severe accidents, such as those initiated by a SBO or a small-break LOCA, degradation of the reactor core can take place while the reactor coolant system remains pressurized. If unmitigated, core materials will melt and relocate to the lower regions of the RPV and ultimately melt through the RPV lower head. Once the RPV is breached, fragmented core debris will be ejected from the RPV and transported directly to the containment atmosphere. During the ejection process, metallic constituents of the ejected material, principally zirconium and steel, exothermically react with oxygen and steam to generate chemical energy and (in the case of reactions with steam) hydrogen. Concomitant with the HPME process, there is the potential for hydrogen combustion and vaporization of available water. The sensible heat loss to the containment atmosphere and its associated features are typically referred to as "DCH." By directly transferring large quantities of sensible energy from the corium and corium-steam reactions into the containment atmosphere, the containment may pressurize to a point where failure is possible.

For the evaluation of HPME/DCH loads on the APRI400 containment, initial and boundary conditions are determined, which include the APRI400-specific geometries of RCS and containment, the initial inventories of the core materials, the thermal-hydraulic condition of RCS and the containment at the time of VB, the inventories of molten debris and the characteristics of debris dispersal. Three scenarios for the APR1400 were selected from the DCH studies of Zion [15] and Surry [16]; their efforts were consolidated into DCH issue resolution guidelines for Westinghouse and CE nuclear power plants. These scenarios include small-break LOCAs with repressurization by operator intervention; this is considered as a type of conservative initiator in terms of the DCH load on the containment integrity. In the event a high-pressure meltdown scenario develops and the feed portion of the feed-and-bleed operation cannot be established due to the unavailability of the SIS, the depressurization of the RCS plays an important role in severe accident mitigation. Depressurization can ensure that a HPME event does not occur, thereby preventing the DCH following a VB. Of particular interest to severe accident mitigation is the capability of the APRI400 to depressurize the RCS from 17.24 to 1.724 MPa prior to a VB. The RCS is expected to be depressurized by (1) a thermally induced creep rupture of RCS piping (hot leg/surge line) or (2) manual opening POSRV(s) by operator actuation, even when normal and emergency feed water are unavailable to remove core decay heat through SGs, with the concurrent failure of the safety injection. RCS depressurization analyses were conducted using the MAAP4 computer code. The MAAP4 calculation results show that the RCS can be sufficiently depressurized prior to a VB to the point that HPME/DCH is prevented or mitigated in all cases.

DCH analyses were also done using the CONTAIN 2.0 computer code [17] for the same base cases of three scenarios used to evaluate DCH loads induced by a HPME upon the failure of the RPV lower head. Table 3.4 shows the CONTAIN results of the peak containment pressure, considering that the fraction of debris entering the upper compartment of the containment area are between 0.1 and 0.2 from the CONTAIN calculation. Scenarios V and Va are full-pressure cases (17 MPa), while the third scenario (Scenario VI) is a partially depressurized (9 MPa) case in which the second case is the only case with the containment sprays functional. The results indicate that the maximum containment pressure rise associated with a large creep failure in the RPV lower head is 0.54 MPa, which verifies that APRI400 containment is designed not to exceed the factored load category as required by SECY-93-087.

3.6.5 Steam Explosion

A steam or vapor explosion refers to a boiling process in which steam or vapor production occurs at a rate larger than the ability of the surrounding media to relieve the resulting pressure increase acoustically, leading to the formation of a shock wave and the production of strong impulsive loadings on adjacent structures. Steam explosions within the primary system are considered to be a potential failure mechanism in both the primary system and the containment area, thereby generating a direct release path for fission products. An in-vessel steam explosion, known as an initiation event causing an alpha-mode containment failure, has been studied for many decades; it was included in the conclusion of NUREG-1524 [18] by the NRC-sponsored Steam Explosion Review Group. In that report, it was concluded that the potential for an alpha-mode containment failure is negligible; hence, the issue of this failure mode has been resolved from a risk point of view. The APR1400 design is very similar to existing PWR plants. Therefore, no new phenomena or configurations are introduced.

Ex-vessel steam explosions can also occur during the progression of a severe accident. Debris should be discharged from the reactor vessel into a pool of water. Within the containment area, the occurrence of a steam explosion would impose shock waves on submerged surfaces and subcompartment walls. These must be evaluated to determine if the resulting loads challenge the integrity of interior walls

Table 3.4 Results of CONTAIN 2.0 for the APR1400 DCH analysis

Case	Peak pressure (MPa)	Peak temperature (K)	Total H_2 burn (kg)[a]	Debris carryover (%)
Scenario V	0.484	789	214/260	18.8
Scenario Va	0.543	1,190	511/956	17.6
Scenario VI	0.385	693	169/303	11.5

and the containment boundary in terms of the application to the reactor/containment design for the APR1400. To verify the integrity of the reactor cavity as to whether it is adequately designed to withstand the effects of dynamic overpressure loads by explosions after a VB, the RV lower head failure area, the mass of the molten corium, and initial/boundary condition of the coolant are calculated. The most likely mode of a lower head failure is caused by the failure/ejection of an ICI tube. This failure mode initiates as a small hole (the size of the ICI tube outer diameter) and grows in size due to the thermal ablation of the lower head in the vicinity of initial hole. A mathematical description of the dynamics of this process was initially developed for the MAAP code. In addition, necessary transient reactor vessel failure area was computed for a conservative calculation. Once the area is calculated, the instantaneous mass flow rate can be calculated. The corium involved in an excore vessel steam explosion is estimated to be the corium mass contained in a submerged cylinder with dimensions equal to the ablated hole area at the time of discharge of the molten corium and the height of the water pool.

3.6.5.1 Integrity of Reactor Cavity Structure

To verify that the reactor cavity should be adequately designed to withstand the effects of dynamic overpressure loads by a steam explosion after a VB, the impulse load arising from an ex-vessel steam explosion is calculated with the trinitrotoluene (TNT) equivalent method. This method assumes that the stored thermal energy within a superheated mass of corium can be converted to the TNT charge. The shock wave characteristic from the TNT explosion is known to have a steeper leading edge compared to those from a steam explosion; hence, the TNT explosion will have more of an impact on the surrounding of concern. The TNT equivalent analogy, therefore, would provide a more conservative assessment when applied to steam explosion phenomena. A corium mass of 2,113 kg was used for the purpose of the bounding calculation. This value is based on the estimate of the RV failure hole area and the depth of the cavity pool. Next, from the results of the reactor cavity structure analysis and design, the dynamic pressure capacity of the reactor cavity is derived. Finally, these two quantities are compared to determine if the integrity of the APR1400 reactor cavity is maintained. The ultimate dynamic pressure capacities at various locations of the reactor cavity wall are compared with the ex-vessel steam explosion dynamic pressure loads. Table 3.5 indicates the impulse pressure loads and the dynamic pressure capacity on various locations of the cavity wall, as shown in Fig. 3.8. Based on these results, it was verified that the integrity of the APR1400 reactor cavity can be adequately maintained.

3.6.5.2 Steam Spike Analysis

When the corium from the reactor vessel failure falls to the reactor cavity floor, nonenergetic or energetic FCI (Fuel Coolant Interaction) resulting in rapid quenching, freezing, and fragmentation of the core debris may occur under certain circumstances. The rapid steam generation from this interaction or mixing between the corium and the water is known as a steam spike. A steam spike result from both the discharge of high-pressure water/steam from the reactor vessel and the generation of steam associated with the quenching of superheated core debris, as these processes can provide a very rapid steam addition to the containment area followed by a modest pressure spike. Even when the reaction time scale is not smaller and the resultant pressure rise is not as large as with a steam explosion, a moderate steam spike following a reactor breach can lead to a relatively rapid pressure increase, which challenges the integrity of the reactor cavity and containment sub-compartments. The accident sequences considered are low-pressure scenarios such as a large-break LOCA, small-break LOCA, a SBO with a hot leg creep rupture and a v-sequence LOCA. The high-pressure scenarios should be modeled as a DCH event; they are not considered as a traditional steam spike event. To estimate the pressure loads to the reactor cavity and containment lower compartment shells, CONTAIN 2.0 [17] is used with two-cell modeling. Figure 3.9 shows the pressure peak in the reactor cavity due to a steam spike for a large-break LOCA sequence. Although there are some uncertainties in the calculation modeling and physical conditions, the pressure rises due to steam spike do not exceed the factored load category for the APR1400 containment.

3.6.6 *Molten Core Concrete Interaction*

The potential hazard of MCCI is the integrity of the containment building due to the possibility of BMT, containment over-pressurization by noncondensable gases, and hydrogen burn of combustible gases. If the safety features of the reactor system fail to arrest an accident within the reactor vessel, the corium will fall into the reactor cavity and attack the concrete walls and floor. BMT refers to the process of concrete decomposition and destruction associated with a corium melt interacting with the reactor cavity basemat. The accident progression is slow (taking from a few days onward to penetrate the reactor cavity basemat and foundation). Even if the corium ablates the basemat concrete and reaches the containment subsoil, the corium

Table 3.5 Evaluation of the APR1400 reactor cavity integrity by ex-vessel steam explosion (ERVC)

Nearest wall	Radius at midpoint (m)	TNT impulse press. Load (MPa-s)	Design ultimate dynamic press. Capacity (MPa-s)	Margin (%)
AB	3.768	0.0103	0.0121	17.3
BC	2.752	0.01362	0.01653	21.4
CD	2.159	0.0169	0.0211	24.7
DE	2.159	0.0169	0.0211	24.7
EF	2.766	0.01356	0.01645	21.3
FG	3.66	0.0106	0.0124	17.6
GH	5.756	0.0071	0.0078	12.2

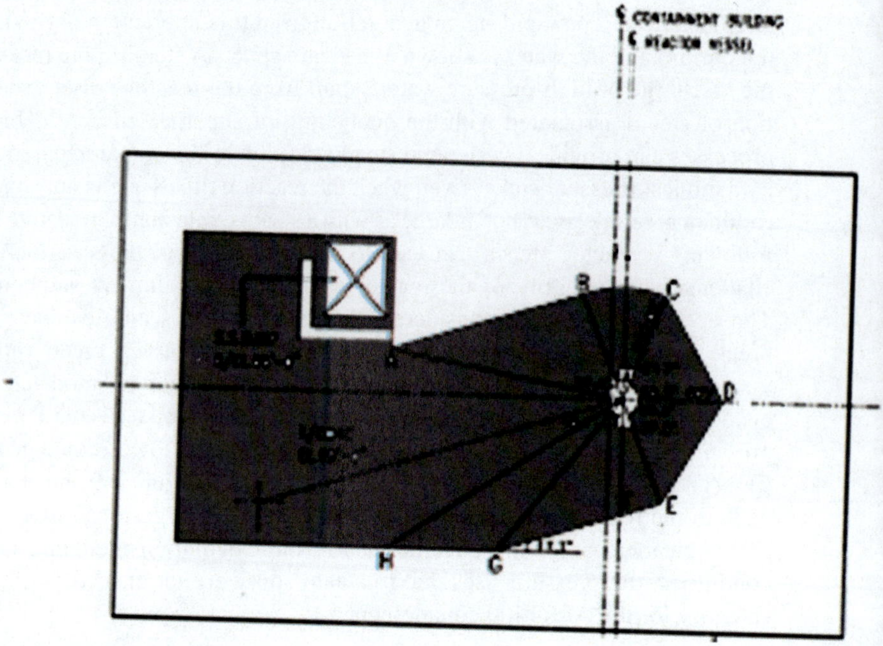

Fig. 3.8 APR1400 reactor cavity bottom floor

release to the environment is negligible. Once in contact with the subsoil, most of the corium is expected to vitrify into a relatively impermeable substance.

Corium-driven concrete erosion has been studied parametrically using MAAP 4 [12]. MAAP models MCCI phenomenology using the DECOMP module. The studies performed for the APR1400 have been directed toward qualifying the concrete erosion progression following a MCCI scenario with various imposed upper crust-water heat flux limits. MAAP4 was exercised so as to simulate a controlled concrete erosion and heat flux condition. Heat flux from the upper crust to the overlying water pool is changed by varying the MAAP FCHE parameter to control the pool boiling heat flux. By properly controlling these parameters, maximum nucleate boiling heat flux limits can be specified at the corium-water interface. The base transient analysis for this evaluation was a LOCA with a 0.15 m diameter pipe break. Following a pipe break, the reactor trip and high-pressure SIS are not available to deliver water from the IRWST to the cold legs. The only water available to make up the primary side is the inventory of four safety injection tanks. For in-vessel high-pressure sequences such as a SBO or a total loss of feed-water transient, some of the melted corium will be ejected out of the cavity at the time of reactor vessel failure via a process known as the HPME mechanism. This reduces the amount of corium remaining in the reactor cavity. Hence, a low-pressure accident sequence was selected as a base case. The reactor vessel failure time of the selected base case is approximately 5.2 h, which appears to be acceptable as a typical vessel failure time.

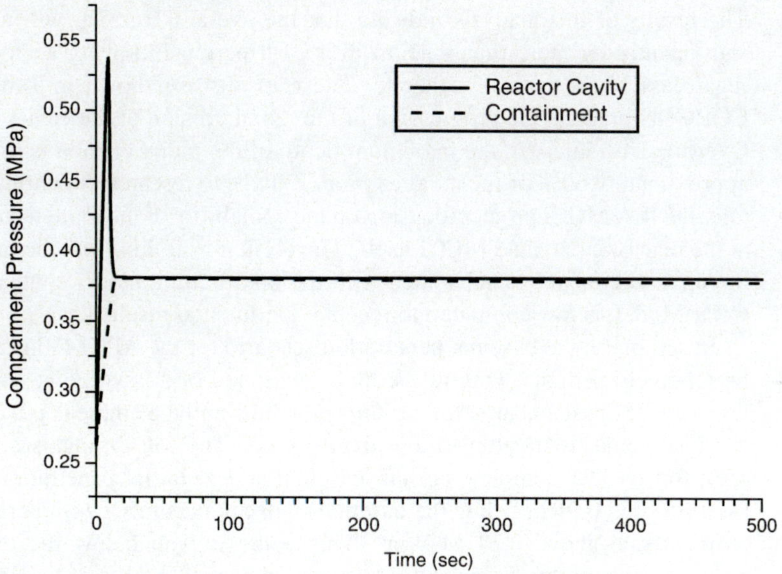

Fig. 3.9 Compartment pressure due to steam spike for large break LOCA

The APR1400 reactor cavity was designed with a large basemat area (0.02 m^2/MWt), as shown in Fig. 3.8, as well as a cavity flooding system to ensure the presence of water in the reactor cavity following severe accident scenarios. The calculations conservatively assume that 100% of the corium is involved in the MCCI process. The MAAP calculation indicates that even when 100% of the corium interacts with the cavity basemat, the concrete erosion can be permanently arrested within 24 h and the depth of erosion can be held below 0.91 m (which is the basemat concrete thickness above the embedded containment steel liner), provided the heat flux from the melt to the overlying water pool exceeds approximately 240 kW/m^2 (FCHF:0.015). As FCHF is reduced to 0.01 (approximately 230 kW/m^2), MCCI is predicted to proceed with no obstructions for 24 h. At this rate of concrete erosion, full basemat penetration to the containment subsoil is estimated to take at least 216 h (9 days). Depending on the parametric value of FCHF (heat flux from the upper crust), the maximum erosion depths of the basemat concrete are tabulated in Table 3.6.

An analysis of core/concrete interaction inside the flooded cavity region was carried out using the CORCON-Mod3 model, which is embedded in the MELCOR1.8.4 code [15]. The physical system considered by CORCON-Mod 3 consists of an axi-symmetric concrete cavity containing debris with one or multiple composite layers. The model allows for several possible configurations in each layer. The layer may be completely molten, it may have a solid crust, or it could be completely solid. For the APR1400 analysis, it is tacitly assumed that the corium crust is formed around the melt; thus, the melt is impermeable to water ingression. Furthermore, the corium melt is assumed to be in the form of continuous layered slag.

The results of this analysis indicate that the average basemat will not be eroded significantly by more than 1.16 m in a 24-h period following accident initiation regardless of the type of concrete. These results are shown in Table 3.7. Here, CORCON predicts the progression of the radial erosion of the cavity wall as well. Over the 24-h interval, the maximum depth of the radial erosion is predicted to be approximately 60% of the axial erosion. Extensive research is continuing with the OECD/NEA MCCI project focusing on the coolability of the molten core spread out at the reactor cavity and MCCI itself. The corium heat flux from the molten core to the upward coolant, as calculated with the pool boiling model contained in CORCON-Mod 3, is low compared to recent experimental results [20].

In summary, the basemat penetration scenario for the APR1400 is considered to be relatively benign owing to the high likelihood of an overlying water pool, the large surface basemat area for corium spreading, and the ample depth of the reactor cavity basemat foundation (more than 6.4 m). The MCCI analyses indicate that even for a 100% complete corium-basemat attack, initial penetration of the steel liner (located 0.91 m below the basemat) will not occur in a 24-h period for a heat removal rate above 0.24 MW/m^2. This value is well below the expected heat removal capacity of the overlying water pool as well as that of typically observed in experiments involving crust formation.

3.6.7 Equipment Survivability

According to SECY-90-016 and SECY-93-087, plant design features provided only for severe accident mitigation are not subject to the environmental qualification requirements of 10CFR50.49, the quality assurance requirements of 10CFR50, Appendix B and the redundancy/diversity requirements of 10CFR50, Appendix A. However, SECY-90-016 and SECY-93-087 state that mitigation features must be designed to provide reasonable assurance that they will operate in the severe

Table 3.6 Effect of the heat flux parameter, PCHF, in the MAAP MCCI calculation

FCHF	0.10	0.02	0.015	0.01
Max. downwards erosion (m)	0.019	0.238	0.244	1.46
Final corium temperature (K)	457	459	850	1660
Time concrete attack end (h)	5.4	10.8	16.9	>24

Table 3.7 Summary of MELCOR MCCI calculation depending on the concrete type

Concrete type	Peak axial erosion at 24 h ft. (m)	Peak radial erosion at 24 h ft. (m)
Limestone	1.75 (0.53)	1.03 (0.31)
Limestone/common sand	2.96 (0.9)	1.62 (0.49)
Basaltic	3.78 (1.15)	2.15 (0.66)
Typical Korean local concrete	3.62 (1.1)	2.13 (0.65)

accident environment for which they are intended and over the time span for which they are needed. In instances where safety-related equipment provided for design basis accidents is relied upon to cope with severe accident situations, there should be a high level of confidence that this equipment will survive in severe accident conditions for the period that it is needed to perform its intended function. In the midst of a core damage sequence, operators are confronted with multiple failures of essential safety equipment and/or operator errors which often result in a damaged core condition. For the operator to cope effectively with this plant condition and protect the general welfare of the public, he must be provided with an equipment subset in which he should be trained to use and interpret, with the ultimate goal of achieving an "in-vessel" or "in-containment" safe stable state.

The following is a summary of the safety and mitigation function for the survivability of instrument and equipment during a severe accident:

- RCS (reactor coolant system) inventory control
- RCS heat removal
- Reactivity control
- Containment integrity

The goal of the RCS inventory control safety function is to assure that a continuous and inexhaustible supply of water can be delivered to the RCS so that the core region will be covered. Inventory control is primarily provided via the APR 1400 SIS. Should the SIS not be available and the RCS become depressurized below about 1.38 MPa, inventory control can also be provided via a realignment of the containment spray or SCS pumps to operate in injection mode.

Successful RCS heat removal requires that a pathway be developed to reject heat from the RCS. Typical RCS heat removal pathways following a severe accident scenario will most likely be through the SGs via the establishment of auxiliary feed-water (AFW) to at least one SG or once through core cooling (OTCC), in which the operator feeds liquid inventory in the RCS via SI and bleeds off steam and/or water (also known as a feed-and-bleed operation). Once a sufficiently low pressure has been established in the RCS, long-term heat removal can also be accomplished via the SCS using either a SCP or a containment spray pump and associated heat exchanger. As core uncovery may occur with nearly intact or damaged core geometry, it is important that the core be prevented from achieving criticality. A return to criticality under these circumstances will likely strain the meager plant inventory and heat removal capabilities and compromise the establishment of a safe stable state. Reactivity control is provided by insertion of control rods and by assuring the delivery of sufficiently borated water into the RCS. Reactivity control is typically assured early in a transient via the insertion of control rods.

During the "in-vessel" recoverable sequence, containment integrity is necessary to prevent significant radioactivity releases to the environment. Given that the highly reliable containment isolation systems function, containment integrity for the APRI400 containment requires that pressure and temperature challenges within the containment have a low probability of causing a containment failure. If the partial operation of one train of containment sprays can be guaranteed, most containment threats can be averted. If sprays are nonfunctional for an "in-vessel" recoverable sequence and the RCS continues to reject heat to the containment, a containment failure cannot be averted unless the containment heat removal function is restored. For "in-vessel" recoverable sequences with sprays available, the only containment threat is that associated with hydrogen combustion. Based on this, instrumentation and equipment which are necessary to function during a severe accident are reviewed and identified.

The equipment/instrument required to achieve and maintain safe shutdown condition and containment integrity should survive under severe accident environmental conditions. The pre-VB environment is applied to all equipment/instrument necessary to achieve and maintain a safe shutdown condition. It consists of two components; one in the in-vessel and one in the ex-vessel (associated with the containment during the time frame in which the RV lower head is intact). The post-VB environment is typically associated with a more restricted instrument list and a containment environment which may be harsher than that earlier in the sequence. In contrast, the role of the equipment and the environment prior to a VB is safe shutdown, and the equipment required post-VB is intended to mitigate and/or prevent containment failure. A summary of the minimum instrumentation and equipment necessary to function during a severe accident, consistent with 10CFR 50.34(f), is identified in Table 3.8.

Survivability of the instrumentation and equipment is evaluated based on design basis event qualification testing, severe accident testing, and the survival time required, following the initiation of the severe accident. With minor exceptions, existing design basis class IE equipment qualification pressures are sufficient to provide a reasonable level of assurance that this equipment will function during a severe accident. The temperature and pressure survivability requirements before and after a VB are summarized in Tables 3.9 and 3.10, respectively. These tables provide the maximum containment thermal conditions along with supplemental severe accident equipment procurement and location requirements. In summary, the APR1400 equipment specifications for the prevention and/or mitigation of severe accidents can provide reasonable assurance that this equipment will survive in severe accident conditions for the period in which it is required to perform its intended function.

3.7 Conclusions

The development of the APR1400 standard design was launched at the end of 1992 and organized in three phases related to the development status. The third phase ended in December of 2001 and the design certification was issued in May of 2002 by the Korean regulatory agency. Currently, two units of APR1400, Shin-Kori 3 and 4, are under construction. They are scheduled for operation starting in September of 2013 and 2014, respectively.

Table 3.8 A summary of the instruments and equipment necessary to function during a severe accident

Instrument	Required pre-VB	Required post-VB
UHJTC on UGS	O	X
Pressure of RCS and PZR	O	X
Safety injection flow	O	X
Auxiliary feedwater flow	O	X
S/G water level	O	X
IRWST water level	O	O
Hydrogen monitors	O	O
Radiation monitor	O	O
Containment pressure	O	O
Containment temperature	O	O
Containment spray flow	O	O
Safety injection system	O	X
Auxiliary feedwater system	O	X
Containment isolation system	O	X
SDVS	O	X
Cavity flooding system	O	O
Hydrogen mitigation system	O	O
Containment penetrations	O	O
Containment spray system	O	O
Shutdown cooling system	O	O
Post-accident sampling system	O	O

Table 3.9 Maximum containment pressure/temperature conditions prior to a vessel breach (VB)

Transients	Temperature (K)	Pressure (MPa)
Local burning with 100% hydrogen	<450	<0.21
Global burning with 100% hydrogen	<605	<0.52
Without 100% hydrogen (LB LOCA)	<450	<0.41

Table 3.10 Maximum containment pressure/temperature conditions after a VB (vessel breach)

Transients	Temperature (K)	Pressure (MPa)
LBLOCA with containment sprays functional	<400	<0.3
SBO with containment sprays disabled and cavity flooding system activated prior to VB	<440	<0.8
SBO with containment sprays disabled and cavity flooding system disabled	<460	<0.4

The APR1400 was designed to withstand design basis events beyond its original specifications due to features such as a large containment, a large robust reactor cavity with thick concrete walls and floors, an in-containment refueling storage tank for cavity flooding, and a rapid Depressurization system for the reactor coolant system. The severe accident issues and mitigation features of the APR1400 are reviewed here along with their impact on the phenomenological response of the plant to beyond design basis accidents along with SECY-93-087. Bounding deterministic calculations indicate that early containment challenges associated with VB phenomena and hydrogen combustion results in peak loadings below the ASME factored load category containment limit and hence provide a high degree of confidence that containment integrity can be maintained. Steam explosion loadings were also quantified deterministically. These assessments suggest that the APR1400 cavity design can withstand impulsive loadings associated with a steam explosion involving 5–10% of the ejected corium mass without serious damage to the reactor cavity. Deterministic BMT scenarios were also analyzed. These analyses assumed that 100% of the corium debris was cooled with the overlying water within the reactor cavity as a layered impermeable media. Under these circumstances, a local below-ground penetration of the containment shell in the area of the basemat will be delayed for more than 24 h after the onset of a core melt event. Through plant-specific deterministic analyses of severe accident issues, it was concluded that the APR1400 design is robust and is capable of mitigating the consequences of a wide spectrum of severe accident scenarios while maintaining containment integrity and minimizing radiation release to the general public.

In conjunction with developing the APR1400 severe accident mitigation design, a varied R&D program was instigated to support the design and the results of the analysis in accident conditions. Through national and international cooperating research programs, numerous experiments are continuing in an effort to improve the system design and validate the analytical results in a manner that can be reflected in the further development of the APR1400. These R&D results provide the phenomenological background and data necessary for understanding the pertinent processes and examining uncertainties in the development of a severe accident management strategy for the APR1400 design.

References

1. Korea Hydro and Nuclear Power Company (2002) APR1400 standard safety analysis report (Revision 1), May 2002
2. International Atomic Energy Agency (2004) Status of advanced light water reactor designs 2004, IAEA-TECDOC-1391, May 2004
3. U.S. Nuclear Regulatory Commission (2007) Regulatory guide 1.206: combined license applications for nuclear power plants (LWR Edition), June 2007
4. Oh SJ, Choi YS (2002) APR1400 design: its safety features and associated test program, presented at workshop on advanced nuclear reactor safety issues and research needs, Paris, France, 18–20 Feb 2002
5. U.S. Nuclear Regulatory Commission (1996) Status of integration plan for closure of severe accident issues and status of severe accident research, SECY96-088, April 1996
6. Nuclear Energy Institute (1994) Severe accident closure guidelines NEI91-04 Rev.1, Dec 1994
7. Fauske and Associate Inc. (1992) Severe accident management guidance technical basis report, EPRI TR-101869, Dec 1992
8. Kang SK et al (2003) Results of insights from probabilistic safety assessment for the APR1400 standard design. In: Proceedings of International Congress on Advances in Nuclear Power Plants(ICAPP '03), Cordova, Spain, 4–7 May 2003

9. U.S. Nuclear Regulatory Commission (1982) PRA procedure guide, NUREG/CR-2300
10. U.S. Nuclear Regulatory Commission (1984) Probabilistic safety analysis procedure guide, NUREG/CR-2815
11. U.S. Nuclear Regulatory Commission (1989) Severe accident risks: an assessment for five U.S. nuclear power plants, NUREG-1150
12. Fauske & Associates Inc. (1994) User's manual for maap4: modular accident analysis program for LWR power plants, May 1994
13. Lim JY, Byun JY (2007) APR1400 severe accident mitigation design. In: Proceedings of international congress on advances in nuclear power plants (ICAPP '07), Nice, France, 13–18 May 2007
14. George et al TL (2001) GOTHIC containment analysis package technical manual, NAI 8907-06, Numerical Applications, Richland, WA, April 2001
15. Pilch MM, Yan H, Theofaneous TG (1994) The probability of containment failure by direct containment heating in Zion, NUREG/CR-6075. Sandia National Laboratories, Albuquerque, NM
16. Pilch MM et al (1995) The probability of containment failure by direct containment heating in Surry, NUREG/CR-6109. Sandia National Laboratories, Albuquerque, NM
17. Murata KK et al (1997) Code manual for CONTAIN 2.0: a computer code for nuclear reactor containment analysis, NUREG/CR-6533. Sandia National Laboratories, Albuquerque, NM
18. U.S. Nuclear Regulatory Commission (1996) A reassessment of the potential for an alpha-mode containment failure and a review of the current understanding of broader fuel-coolant interaction issue, NUREG-1524
19. Gauntt RO et al (2000) MELCOR computer code manual, NUREG/CR-6119, Rev.2. Sandia National Laboratories, Albuquerque, NM
20. Farmer MT et al (2000) Results of MACE core coolability experiments M0 and M1b. In: Proceedings of 9th international conference on nuclear engineering (ICONE-8), Baltimore, MD, 2–6 April 2000

Chapter 4
Development and Design of the EPR™ Core Catcher

Dietmar Bittermann and Manfred Fischer

4.1 Introduction

The EPR™ is an evolutionary pressurized water reactor in the thermal range of 4,500 MWth, designed and marketed by AREVA NP. Currently, there are four EPR™ plants under design and construction: Olkiluoto-3 (OL3) in Finland, Flamanville-3 (FA3) in France, and Taishan 1&2 (TSN) in the People's Republic of China.

The EPR™ strategy to avoid severe accident conditions rests on the proven and improved defense-in-depth approaches of its predecessor plants, the German Konvoi and the French N4. Beyond this, the EPR™ takes measures to drastically decrease the potential consequences of a postulated Severe Accident (SA) with core melting. The target is to eliminate the need for an evacuation of the surrounding population and for long-term restrictions with respect to the consumption of locally grown food. This requires significant reductions in the magnitude of the activity release into the environment, and in particular in the frequency of large releases, under SA conditions.

EPR™ achieves these targets by design provisions that prevent early as well as late containment failure. These provisions address, in a comprehensive and balanced approach, all relevant containment challenges, including: containment overpressure failure, steam explosion, hydrogen detonation, and basemat melt-through.

This contribution describes the challenges for the development and the approach followed for the design of the core melt stabilization system (CMSS) [1,2] and the state after detailed design, as well as the industrial realization of the key components, including their implementation in the plant design.

D. Bittermann (✉) and M. Fischer
AREVA Nuclear Power GmbH, Erlangen, Germany
e-mail:Dietmar.Bittermann@areva.com

T. Saito et al. (eds.), *Advances in Light Water Reactor Technologies*,
DOI 10.1007/978-1-4419-7101-2_4, © Springer Science+Business Media, LLC 2011

4.2 Overview of the EPR™ Severe Accident Mitigation Features

Complementing the proven three-level safety concept for the prevention of severe accidents, the EPR™ implements a new, fourth safety level, aimed at preserving short- and long-term containment integrity even in case of a severe accident with core melting. Under the related extreme conditions, the integrity of the containment is challenged by the following events that need to be addressed and safely avoided.

4.2.1 High-Pressure Core Melting

High-pressure core melting is prevented by dedicated pressure relief valves. Safe depressurization of the primary circuit eliminates the risks related with missile generation and Direct Containment Heating (DCH) after failure of the Reactor Pressure Vessel (RPV) at high internal pressure. In addition, early depressurization significantly reduces the likelihood of steam generator tube rupture.

The dedicated valves provide a total discharge capacity of 900 t/h at the design pressure of the Reactor Coolant System (RCS), which ensures a fast pressure reduction to below 0.5 MPa at the time of vessel failure. The dedicated valves will be activated manually by the operator, at the latest when the core outlet temperature exceeds 650°C.

4.2.2 Hydrogen Deflagration/Detonation

To mitigate the risk related with the formation of combustible gases, the EPR™ uses a dedicated hydrogen control and mixing system. It is based on catalytic recombiners and effective H_2-dilution. The latter is achieved by: (1) the large free containment volume of about 80,000 m^3, (2) flow connections between inner and outer containment rooms that open passively in case of SA, and (3) steam discharge into the lower containment following the release of the RCS inventory during primary depressurization.

The chosen arrangement of 46 large and small recombiners enhances atmospheric convection and mixing in the containment already early in the accident and leads to a homogeneous distribution of combustible gases. This way the maximum average H_2-concentration is kept safely below 10 vol%, even for highly conservative assumptions regarding the rate and location of H_2-release out of the RCS.

Though fast deflagrations cannot be completely excluded in regions of high local H_2-concentrations close to the release location, such fast deflagrations are shown to slow down in surrounding regions of lower concentration. As a consequence, detonations and fast deflagrations that could impose critical pressure loads on the containment structure are avoided.

4.2.3 Energetic Steam Explosions

No specific design measures are taken to prevent or mitigate in-vessel steam explosions. This is because, for all relevant scenarios and boundary conditions, the mechanical loads are predicted to be too low to challenge the stability of the RPV vessel and in particular to detach its upper head. This eliminates the risk of missile generation leading to induced early containment failure.

Different from the strategy with respect to the in-vessel steam explosion risk, dedicated design measures are taken in the EPR™ to avoid ex-vessel steam explosions. These include: (1) provisions to ensure a dry reactor pit and a dry core catcher, (2) the addition of silica-rich, viscous slag to the core melt, and (3) the strategy to flood the melt from the top, at low flow rate, and only after melt spreading into the core catcher is complete.

4.2.4 Basemat Penetration

The EPR™ involves design measures that prevent the attack of the molten core on the basemat, as such an attack could result in:

- The penetration of the embedded containment liner
- The heat-up and mechanical deformation of the containment civil structure
- A sustained release of noncondensable gas into the containment

No attempts are made to prevent RPV melt-through by outside vessel cooling because – at the high power rating of the EPR™ – the margins for In-Vessel Retention (IVR) are considered too low.

Instead, the EPR™ relies on an ex-vessel strategy to stabilize the molten core. After its release from the RPV, the molten core debris is first accumulated and conditioned in the reactor pit. Then, it is spread into a large core catcher, to increase the melt's surface-to-volume ratio and to take benefit from the related increase in the efficiency of quenching and cooling after flooding.

In the following, the function of the EPR™ Core Melt Stabilization System (CMSS) and the state of design of its main components are presented, complemented by information regarding the industrial realization in current EPR™ projects.

4.3 Challenges for the Development of a Core Catcher System and Approach Followed for the EPR™

4.3.1 Challenges for the Design of Measures for Melt Stabilization

Mitigation measures for design basis accidents can take benefit of the fact that the components are more or less intact and that the conditions are not far from normal operating conditions. In severe accident scenarios, the situation is completely different. The corresponding conditions are characterized by the fact that the integrity of the fission product barriers – fuel rod and RPV – are lost and that pressure and temperature levels in the containment strongly deviate from operating and design basis accident conditions. In addition, a large number of different scenarios and boundary conditions must be dealt with.

These facts are extremely challenging for the designer who develops concepts for the mitigation and control of the consequences of severe accidents. Therefore, before starting the design process it is necessary to think about the consequences of different design methods.

One possible approach is to envelop the range of scenarios and potential loads and to define this envelop as the design basis for the mitigation measures. For many cases, this approach may be appropriate. However, for the severe accident issue, it has two important disadvantages: first, the uncertainty to really include all potential scenarios and loads despite the numerous combinations of phenomena and conditions to deal with and despite the fact that the selected scenarios may be more or less arbitrary. In addition, such an "enveloping load approach" may lead to extreme requirements on the technology to be provided and to extremely expensive design solutions.

The other approach – which is to be nominated as "plant-state controlled approach," and which is described in more detail in this chapter – intends to influence, from the beginning and by dedicated design-based features, the type of the severe accident scenario and the corresponding course of events. The idea is to generate states, which are characterized by rather well-defined conditions and which can be mitigated by applying known technology. It is required that for these states, either assured knowledge already exists or is to be acquired with limited specific R&D. As a consequence, it is possible to evaluate the effectiveness of the developed mitigation measures with high confidence.

4.3.2 Characteristic Features of the Approach Proposed to be Followed

The task to elaborate a concept for the control of each severe accident phenomenon can be subdivided into the following steps:

1. Identification of the governing issues
2. Identification of the dependence between relevant parameters
3. Definition of design conditions and requirements

The characteristic of the proposed approach is to implement design provisions for the first two steps to influence the third one, namely, the conditions for which design requirements have to be defined.

In the first step, the provisions shall either reduce the number of problems or prevent conditions which may lead to extensive design efforts, or at least to a minimization of thermal and mechanical loads. In the second step, design provisions shall reduce the importance of critical parameters or influence the required extent of analysis of individual parameters and consecutively the required R&D work.

The result of the proposed design measures is a reduction of the range of the conditions that have to be mitigated. Also, the design conditions and requirements for the measures to be introduced can now be realized on the basis of proven technology and appropriate cost.

In order to be able to identify governing issues and parameters that can be influenced, a close cooperation between R&D teams, analytical experts, and the designers is necessary. It is evident that this approach must be iterative as the situation may change by the ongoing evolution of R&D results.

4.3.3 Design Principles for the Core Melt Stabilization Measures

On the basis of the general strategy described, detailed design measures were developed. The design principles and important details of requirements and associated measures are described below.

The measures to be implemented to generate states with well-defined conditions are selected according to the following priorities:

1. Prevention of inadmissible events and conditions
2. Minimization of effects and loads

The provided design measures that lead to the intended specific conditions and that are implemented as mitigation measures are selected under consideration of the following principles:

- Separation of function
- Use of passive means to appropriately consider the plant state in case of severe accidents
- Simple and robust design

4.4 General Strategy for Core Melt Stabilization

The EPR™ core catcher, within which the melt will ultimately be contained and cooled, is located in a compartment lateral to the reactor pit. During plant operation, the connection between pit and this compartment is closed. In a SA, it will be opened passively by the core melt via thermal destruction of a separating plug.

Thanks to the spatial separation between pit and spreading compartment, the core catcher is protected from potentially critical loads related to the failure of the RPV.

The relocation of the melt into the core catcher is preceded by a phase of temporary melt retention in the reactor pit. Its introduction responds to the prediction that the release of molten material from the RPV will, most likely, not take place in one pour, but in stages. Temporary retention is based on the provision of a sacrificial concrete layer that must be penetrated by the melt prior to its escape from the pit.

The related time delay ensures the accumulation of the core inventory in a single molten pool. As shown in [3], the admixture of sacrificial concrete further makes the characteristics of the molten pool predictable and independent of the preceding accident scenario and the uncertainties associated with in-vessel melt pool formation and RPV failure.

After penetrating a melt plug at the bottom of the reactor pit, the molten corium–concrete mixture is finally released into the lateral core catcher, where it

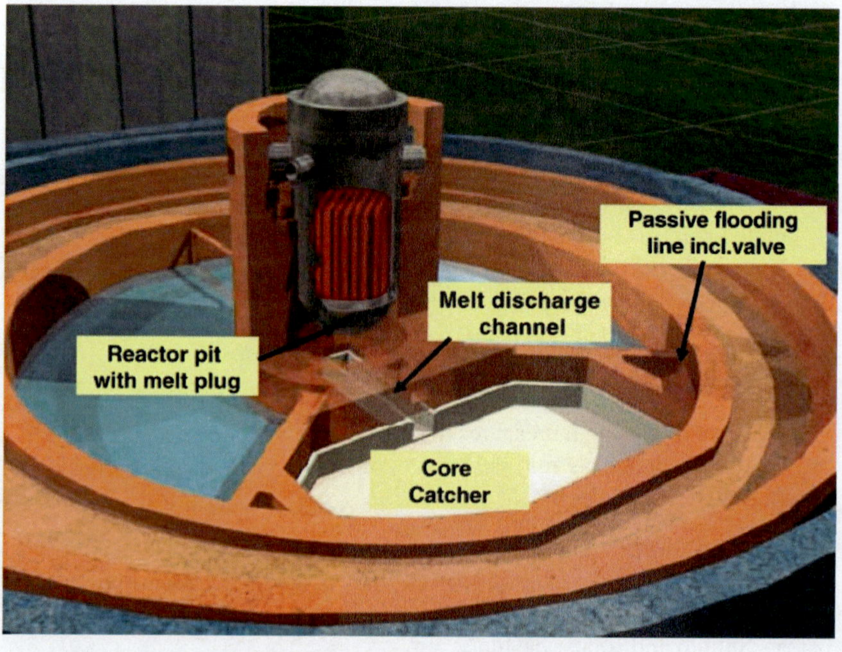

Fig. 4.1 Main components of the CMSS system. (Taken from Proceedings of ICAPP'09 paper 9061, May 2009 [6] and used with permission from Atomic Energy Society of Japan)

spreads on the provided large surface and initiates the opening of flooding valves that start the passive, gravity-driven overflow of water from the In-containment Refueling Water Storage Tank (IRWST). The water cools the metallic structure of the core catcher and eventually pours onto the melt's free surface from above. Decay power is extracted by evaporation and steam release into the containment. The steam is later recondensed by sprays supplied by the dedicated SA containment heat removal system (CHRS). The melt stabilization process thus involves the following stages:

- Temporary melt retention & accumulation in the pit
- Opening of the gate and melt spreading
- Flooding and quenching of the spread melt
- Long-term cooling and heat removal to the water

The arrangement of the components of the CMSS in the lower containment of the EPR™ is sketched in Fig. 4.1.

Along the described sequence, the transformation of the molten core into a coolable and cooled configuration is achieved passively on the basis of simple physics and without requiring operator action or the use of internal or external active systems.

4.5 Description of Components

In the following sections, relevant details of the individual components and the reasoning of selection are described, together with their effect on the functioning of the melt stabilization system.

4.5.1 Components used for Temporary Melt Retention in the Reactor Pit

4.5.1.1 Sacrificial Concrete

The sacrificial material is provided in the form of concrete because of its easy fabrication, transport, and installation. Concrete combines high mechanical stability with high decomposition enthalpy. The latter is favorable, as it reduces the amount of sacrificial material needed to perform the retention function in the pit, which reduces the melt volume (after concrete incorporation), the layer depth in the core catcher and, as an ultimate consequence, the time to completely solidify the spread melt.

Among the aggregates investigated for the sacrificial concrete in the reactor pit, a mixture of siliceous pebble and iron oxide has been found most suitable, because of:

- The low resulting melt temperature
- The low viscosity for spreading
- The fast oxidation of the core melt

- Good fission product retention
- The low content of gases

The finally chosen aggregate is characterized by a low maximum grain size (to promote homogeneous erosion), a high porosity (to reduce the risk of spalling under thermal load), and a standard cement fraction (to provide sufficient mechanical stability).

The thickness of the sacrificial concrete layer is 50 cm, which according to [3] is sufficient to achieve complete melt accumulation for all considered scenarios and conditions of initial melt release from the RPV. Locally, this thickness is reduced by embedded structures, namely, the pit ventilation ducts and the lower ends of the core neutron detector tubes.

At the top, the sacrificial layer extends into four concrete "bumpers" arranged around the melt plug in the center of the pit bottom, see Fig. 4.2.

These concrete structures are aimed at limiting the drop height of the lower head and protect the melt plug in case of creep-induced circumferential rupture. The reinforcement of the bumpers is connected with that of the sacrificial concrete to increase mechanical stability and to better transmit the impact loads.

The reinforcement of the sacrificial concrete is also connected with the structural concrete along the offset at the top of the layer. This connection mechanically fixes the sacrificial layer against the civil structure and limits its displacement in case of earthquake.

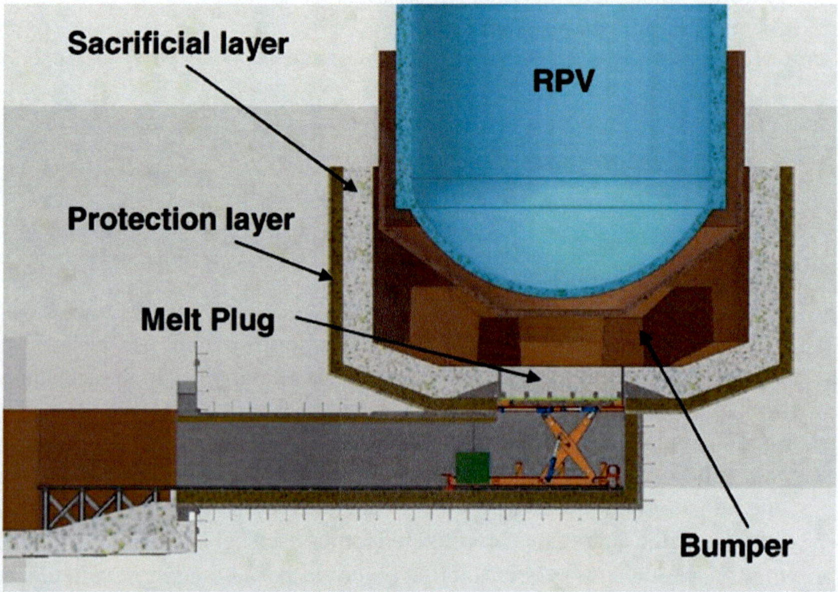

Fig. 4.2 Provisions for temporary melt retention in the reactor pit. (Taken from Proceedings of ICAPP'09 paper 9061, May 2009 [6] and used with permission from Atomic Energy Society of Japan)

The sacrificial concrete applied has undergone extensive optimization and testing, including the pouring of large-scale samples. The measured stability is equal or better than that of standard-grade silicate concrete.

4.5.1.2 Protection Layer

All along its outer surface, the sacrificial concrete layer is backed by a melt-resistant protection layer (see Fig. 4.2). Its purpose is to ultimately enclose the melt during interaction with the sacrificial concrete and to avoid direct melt contact with the civil structure. The protection layer homogenizes the erosion process by stopping any locally faster melt progression. It, thus, makes the prediction of the concrete erosion process independent of uncertainties in the heat flux distribution in the molten pool.

In consequence, most of the sacrificial concrete will become incorporated in the molten pool which makes the state and properties of the melt at the end of the temporary retention phase predictable.

As material for the protection layer, a zirconia-based ceramic is used, which showed high thermochemical stability against both the metallic and oxidic corium melt under relevant conditions [2]. Sufficient resistance against thermal upshock and mechanical deformation were achieved by optimizing the size of the bricks, the stabilizer, and the characteristics of the sintered powder.

All bricks are connected with each other by tongue-and-groove joints and ceramic mortar. After being assembled, they form a free-standing, vault-shaped structure, see Fig. 4.3. The gap between the assembled layer and the civil structure, which allows adaptation to civil tolerances, is later filled with zirconia-based

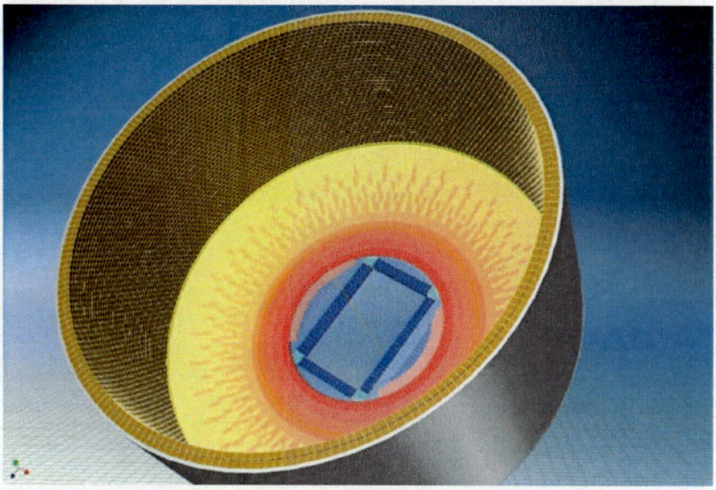

Fig. 4.3 Protection layer inside reactor pit. (Taken from Proceedings of ICAPP'09 paper 9061, May 2009 [6] and used with permission from Atomic Energy Society of Japan)

refractory castable. After pouring of the sacrificial concrete, the protection layer is no longer visible.

4.5.1.3 Melt Plug

Under plant operational conditions, the melt plug acts as a part of the sacrificial concrete layer. During plant outages, it can be removed to open a pathway for inspecting the pit and the outside of the RPV.

In addition, the melt plug serves as the predefined weak point in the enclosure of the melt during temporary retention. This is achieved by locally replacing the protection layer with an aluminum plate, the so-called "gate."

The gate mechanically supports the concrete and transfers pressure loads to a steel grid below, see Fig. 4.4.

When contacted by the hot melt, the aluminum gate will be quickly destroyed. Therefore, the rate of melt release is exclusively determined by the size of the initial opening in the concrete and by the speed at which this opening will grow by erosion

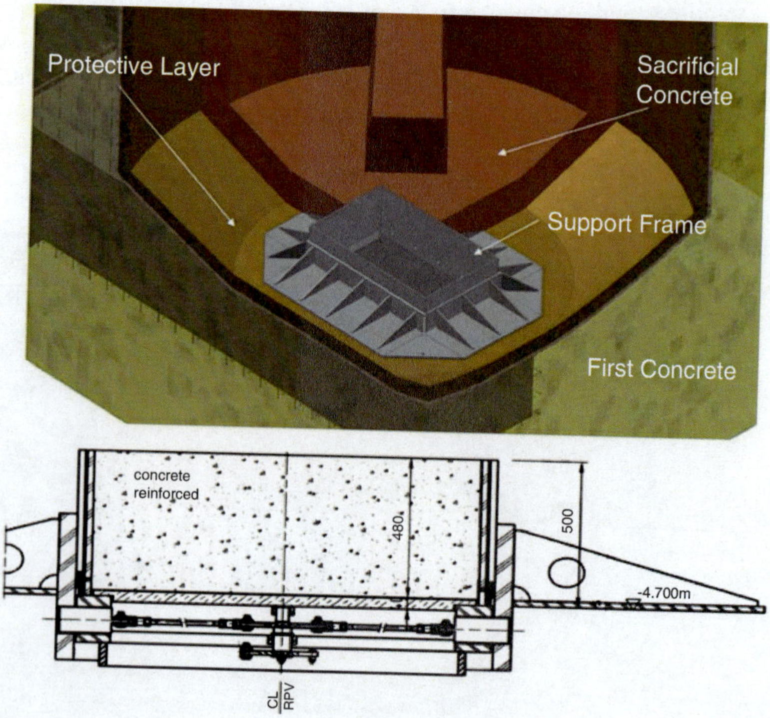

Fig. 4.4 Melt plug and support frame. (Taken from Proceedings of ICAPP'09 paper 9061, May 2009 [6] and used with permission from Atomic Energy Society of Japan)

Fig. 4.5 Protection layer in the MDC, near the position of the melt plug. (Taken from Proceedings of ICAPP'09 paper 9061, May 2009 [6] and used with permission from Atomic Energy Society of Japan)

caused by the out-flowing hot material. The steel grid does not act as an obstacle in this process due to the large open cross-section between the individual bars.

The steel grid carries the passive components of the mechanism needed to mechanically fix the melt plug in its outer support frame, see Fig. 4.4.

The fixation is realized by ten simultaneously moved steel cylinders (locking bolts). The support frame rests on the surface of the protection layer and is fully embedded in the sacrificial concrete layer.

Steel grid, support frame, and locking bolts are made of stainless steel and designed to withstand a maximum pit pressure of 2 MPa. For the melt plug's cross-section of ~2 m^2, this translates into a static force of ~4 MN.

4.5.2 *Components used for Melt Spreading*

4.5.2.1 Melt Discharge Channel

After leaving the reactor pit, the melt is guided into the core catcher through the Melt Discharge Channel (MDC), a steel channel embedded in the civil structure below the reactor pit. The MDC is connected with the reinforcement of the surrounding concrete via welded studs on its outside.

On the side of the pit, the entrance into the MDC is closed by the melt plug. All loads acting on the melt plug and on the top of the channel must be borne by the MDCs steel structure. Due to this, the thickness of the channel roof below the pit is higher than in other regions.

On the side of the core catcher, the level of the outlet is above the maximum melt depth. This makes the spreading process independent of the development of the melt depth in the core catcher.

To insulate and protect the inner surface of the MDC from the high thermal and erosive loads during the outflow of the core melt, the protection layer in the pit is extended into the MDC, see Figs. 4.5 and 4.6.

The vertical wall, exemplarily shown in Fig. 4.5, is located below the melt plug support frame, see Fig. 4.3. A part of the loads imposed on this frame has, therefore, to be borne by the brickwork inside the channel. The related requirement on the mechanical stability of the bricks has been considered when selecting the ceramic material.

The protection layer covers the inner surface of the MDC and all melt-facing surfaces of the cooling elements at the entrance into the core catcher, see Fig. 4.6.

All vertical brick walls inside the channel are supported by anchors welded against the MDCs steel structure, as shown in Fig. 4.5. Special types of anchor are used at the bottom and top facing surfaces.

After completion of the work on the brick assembly, the protection layer in the channel will be completely covered by a stainless steel liner to avoid damage during the later installation and operation of the melt plug transport system.

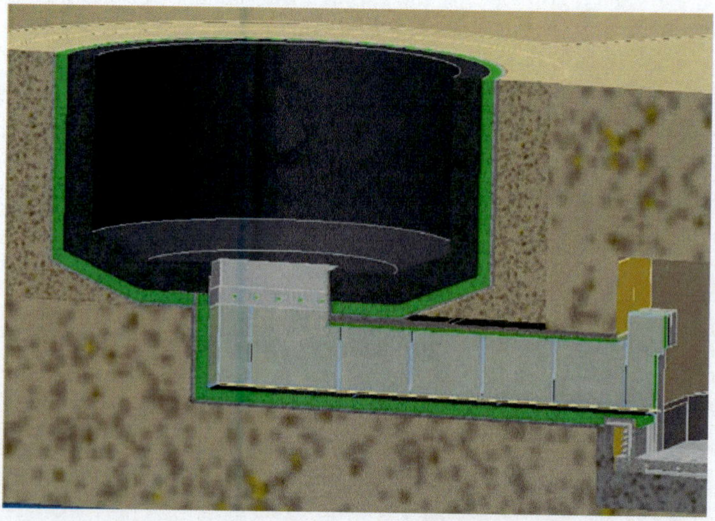

Fig. 4.6 Areas covered by the protection layer. (Taken from Proceedings of ICAPP'09 paper 9061, May 2009 [6] and used with permission from Atomic Energy Society of Japan)

Fig. 4.7 Melt plug on top of transport cart during testing. (Taken from Proceedings of ICAPP'09 paper 9061, May 2009 [6] and used with permission from Atomic Energy Society of Japan)

4.5.2.2 Melt Plug Transport System

To facilitate the transport of the melt plug in and out of its operational position, a dedicated transport system is provided. It consists of a transport cart (Fig. 4.7), rails, and a shunting station and electric cabinet located in front of the MDC outlet inside the core catcher (Figs. 4.8 and 4.13). The system operates as a "black box" and is not connected with the plant's I&C.

The rails are mounted on a steel plate that is fixed against the protection layer at the bottom of the MDC. This fixation is necessary to achieve the required accuracy in positioning the cart during remote-controlled lift-up of the melt plug. The cart is equipped with position sensors and limit switches for the hydraulic system. The latter help to prevent damage in case of incorrect positioning of the cart.

During plant operation, the (empty) cart is parked on the shunting station inside the core catcher, while the exit of the MDC is closed by a removable neutron shield. This shield allows using standard equipment for the electric and hydraulic components of the transport system.

4.5.2.3 Core Catcher Assembly

The function of the core catcher (CC) is to ultimately contain and stabilize the molten core debris after spreading using water drained from the IRWST.

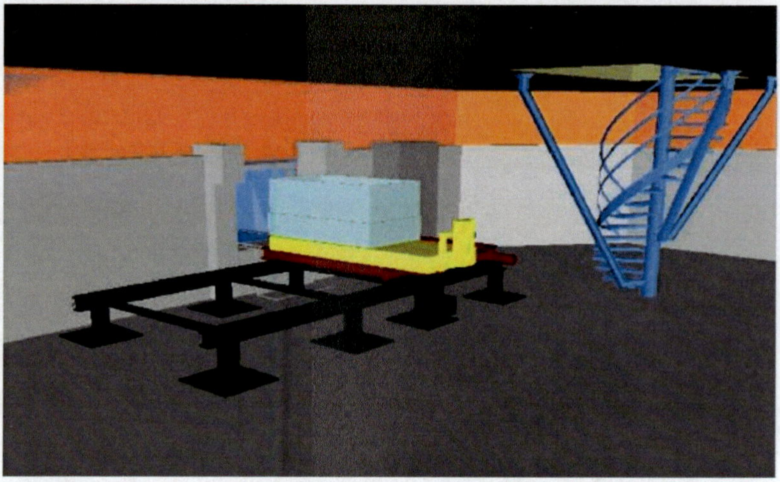

Fig. 4.8 Shunting station inside the core catcher. (Taken from Proceedings of ICAPP'09 paper 9061, May 2009 [6] and used with permission from Atomic Energy Society of Japan)

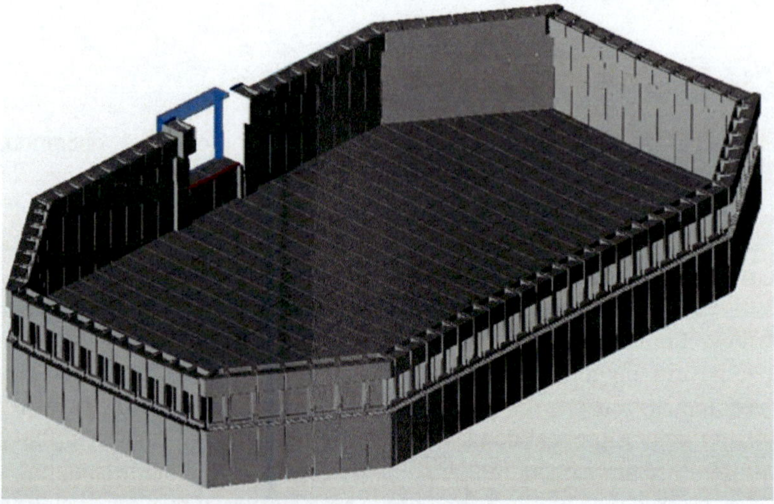

Fig. 4.9 3D view of the core catcher. (Taken from Proceedings of ICAPP'09 paper 9061, May 2009 [6] and used with permission from Atomic Energy Society of Japan)

Fulfilling this function requires that the size of the core catcher be sufficiently large to reduce the heat fluxes to the water below relevant CHF limits. In addition, the cooling structure must be capable of absorbing all related thermal and mechanical loads, both during initial melt contact and long-term heat removal. For this, the

EPR™ core catcher has been designed as a robust metallic crucible, consisting of a large number of individual elements, see Fig. 4.9. The latter avoids excessive thermal stresses and mechanical deformation during heat-up.

All elements are flexibly connected with each other by tongue-and-groove joints. The entire structure is made of ductile cast iron, which combines a high mechanical robustness with a high thermal conductivity.

The chosen thickness of the structure of 25 cm provides sufficient thermal inertia to withstand the transient contact loads even under temporarily noncooled conditions.

The chosen width of the gaps between the cooling elements allow thermal expansion and deformation without impacting neighboring elements up to temperatures at the melt- and water-facing surfaces of 1,000°C and 100°C, respectively. These values bound the temperatures obtained from the analyses of the thermal response of the structure after melt contact.

The inner side of the core catcher is covered by a layer of sacrificial concrete. It protects the cooling structure during melt spreading and becomes incorporated into the melt later on. The added concrete changes the melt's properties in a similar way as the sacrificial concrete in the reactor pit.

Contrary to the pit, a concrete with high silica content is chosen for the core catcher. Admixture of this concrete improves the long-term retention of fission products and reduces the density of the oxidic phase, which further enhances the stability of the layered melt configuration.

As a consequence, water poured on the melts surface will interact with the lighter oxidic phase, while the concrete at the bottom will react with the denser metallic phase during Molten Core Concrete Interaction (MCCI). The high heat transfer from the molten metal to the concrete is predicted to result in partial freezing of the metallic layer before it contacts the cooling structure. This reduces the risk of liquid melt ingress into the space between the cooling elements.

The risk is further diminished by the choice of overlapping tongue-and-groove connections between the elements, and by seals of compressed ceramic felt inserted into the gaps above the steel tongues, see Fig. 4.10.

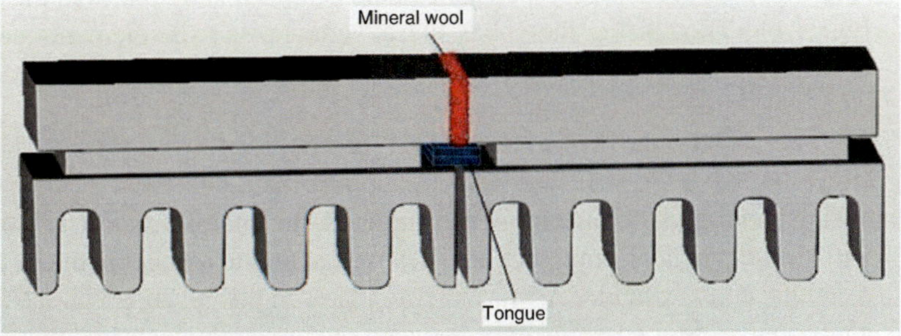

Fig. 4.10 Provisions to avoid melt ingress into the gaps between cooling elements. (Taken from Proceedings of ICAPP'09 paper 9061, May 2009 [6] and used with permission from Atomic Energy Society of Japan)

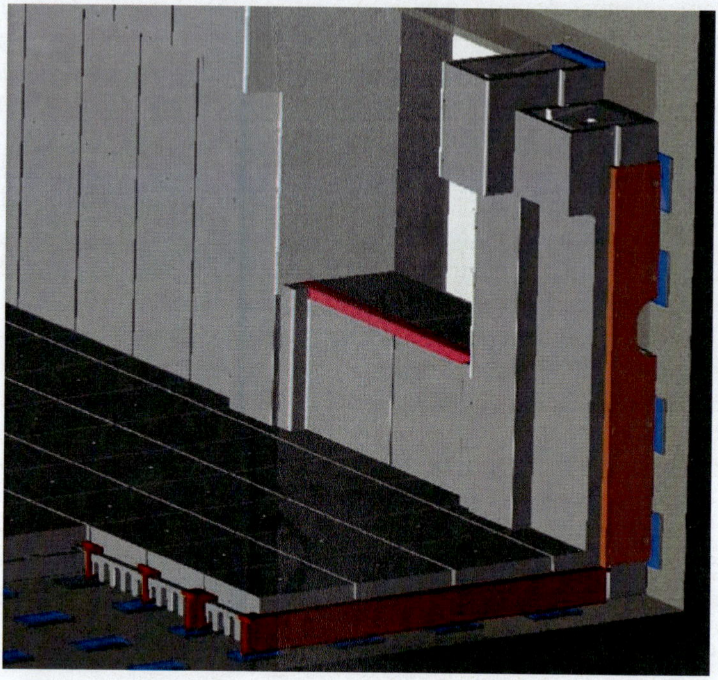

Fig. 4.11 CC fixed bottom cooling elements. (Taken from Proceedings of ICAPP'09 paper 9061, May 2009 [6] and used with permission from Atomic Energy Society of Japan)

No effort is made to make the gaps water-tight. This is because of the positive results of the experiments performed in the frame of the COMET concept [4] on the effect of water injection into the melt from below. These experiments demonstrated that water entering the melt at a low rate can significantly enhance melt coolability without the risk of energetic fuel coolant interactions (FCI).

Two different designs for the bottom cooling elements have been developed. One type stands freely on the concrete floor, see Figs. 4.10 and 4.12. The other is suspended on horizontal T-beams, welded against anchor plates in the concrete, see Fig. 4.11. Though the latter has the advantage that the elements can be locked against the civil structure, the design is more complex, and specific "closing elements" are needed in each row.

Other than for the bottom elements, T-beams are always used at the lateral cooling elements, see Figs. 4.12 and 4.13 to support them and prevent them from falling during erection. After assembly is completed, the lateral structure is ultimately fixed with the help of horizontal sectional beams that fit into notches in the rear of the upper elements and are welded against anchor plates in the concrete behind. These beams additionally close potential gaps between the upper elements and the concrete wall.

The arrival of the melt in the core catcher thermally destroys metallic receptors which relieves prestressed steel cables that lead to spring-loaded valves, located in the neighboring valve compartments, see Figs.4.1 and 4.13.

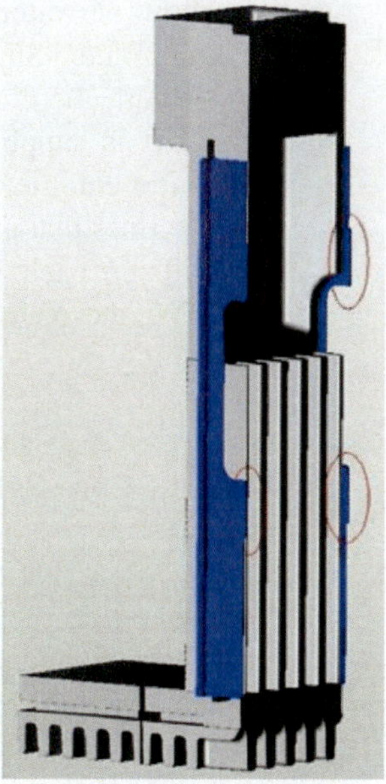

Fig. 4.12 Lateral element with vertical support beams (the marks indicate the welding positions). (Taken from Proceedings of ICAPP'09 paper 9061, May 2009 [6] and used with permission from Atomic Energy Society of Japan)

Fig. 4.13 Location of the valve compartments at the two sides of the core catcher. (Taken from Proceedings of ICAPP'09 paper 9061, May 2009 [6] and used with permission from Atomic Energy Society of Japan)

These valves, which open and stay open after triggering, start the gravity-driven overflow of water from the IRWST into the core catcher compartment via the Central Water Supply Duct (CWSD), a steel channel embedded in the concrete underneath the core catcher. The CWSD continues into the valve compartments where it is supplied by one flooding valve on each side. Inside the spreading compartment, the channel is U-shaped and open on top.

The lines that provide water to the valves are dimensioned to supply a flow rate of >50 kg/s each. Considering that the maximum decay power in the melt is about 30 MW, the opening of one valve alone would therefore be sufficient for melt quenching and cooling.

4.5.2.4 Core Catcher Cooling Structure

The cooling elements have integrated fins on their water-facing side that form parallel, rectangular cooling channels. All elements are aligned in a way that their channels combine into continuous flow paths for the coolant. The fins enhance the heat transfer to the water. For the horizontal elements, they additionally avoid the negative consequences of the formation of an insulating steam layer at high heat fluxes, because they allow the heat to enter the water through their sidewalls. The resulting high critical heat flux has been confirmed in dedicated experiments [2].

The bottom cooling structure, like the concrete below, has an inclination of ~1°. The resulting V-shape establishes a preferred flow direction for the steam–water mixture in the channels, from the CWSD to the adjacent lateral walls.

After triggering the flooding valves and filling the CWSD, the water successively submerges the cooling channels and all space below and around the core catcher.

The 1° inclination of the bottom and the nonrectangular shape of the room complicate the design of the EPR™ cooling structure and result in more than 30 different element types. A further increase of this number is avoided by the two axial symmetries of the room.

All lateral cooling elements (except the transition elements next to the MDC outlet, see Fig. 4.11) consist of two parts, an upper and lower one, see Fig. 4.13. Two different designs for the lower cooling elements have been developed: one with open and the other with closed channels (see Figs. 4.12 and 4.13). While the latter stand on the concrete floor, those with open channels stand on top of corresponding bottom elements.

Independent of these differences the lower parts are always arranged in a way that the vertical channels extend the adjacent horizontal channels. Only at the two small sides of the core catcher, the vertical channels are supplied separately, because not horizontal channels end here.

The height of the lower elements has been chosen to exceed the predicted maximum melt level in the core catcher. This makes it possible to omit the cooling fins in the upper elements and thus to create sufficient open space for delaying the start of water overflow onto the melt.

As the consequence, the speed, at which the water level behind the cooling structure rises during flooding, is slowest at the end, and highest in the beginning. The latter is advantageous as it quickly ensures cooled conditions in all potentially melt-contacted lower regions. The sum of the free volumes is adjusted so that the water overflow onto the melt will only start after the predicted end of melt spreading.

Water overflow takes place through the vent holes in the upper lateral cooling elements, see Fig. 4.13. The levels of these vent holes are the same for all lateral elements, except above and near the CWSD, where the levels are lower, due to the 1° inclination of the concrete bottom. Therefore, the water will start pour onto the melt at these two sides, which are furthest away from the MDC outlet.

At the beginning, most of this water will evaporate after contact with the hot surface. Therefore, it will take a certain period of time before a continuous water layer will develop atop the melt and before this layer reaches the MDC outlet position. Up to this time, potential late melt releases out of the MDC remain undisturbed by the onset of flooding.

The water overflow will take place in parallel with the interaction of the melt with the sacrificial concrete. Under corresponding conditions, prototypic MCCI experiments [5, 6] showed enhanced coolability and the absence of energetic FCIs.

Irrespective of the latter, the robust design of the core catcher is capable to deal with the consequences of FCIs during melt spreading and flooding. The cast iron walls can absorb significant pressure loads, and the vent openings at the top are protected against the ingression of dispersed melt by "melt splash guards," see Fig. 4.13. In addition, the design of the top cooling elements avoids the intrusion into the vertical cooling channels of any material potentially splashed against the concrete walls above by an FCI.

4.5.2.5 Interface with the CHRS

With the flooding valves being open, overflow of water from the IRWST will continue until the hydrostatic pressure levels within the spreading room and the IRWST are balanced. Under these equilibrium conditions, also the MDC and the lower pit are submerged, see Fig. 4.14, while the decay heat produced inside the spread melt will be carried to the water either via the cooling structure or across the melt's free surface.

Under the related saturated conditions, all decay heat is used to generate steam. The steam enters the containment through the vertical steam exhaust chimney in the roof of the spreading compartment, see Fig. 4.14. The evaporated water is resupplied by overflow from the IRWST.

In this passive mode of operation, the CMSS function can be fulfilled during an unlimited period of time. However, because the created steam will have to be recondensed, active systems are needed in the long-term, to avoid containment overpressure failure.

For this purpose the EPR™ is equipped with a dedicated SA containment heat removal system (CHRS). Its active components and heat exchangers are located in

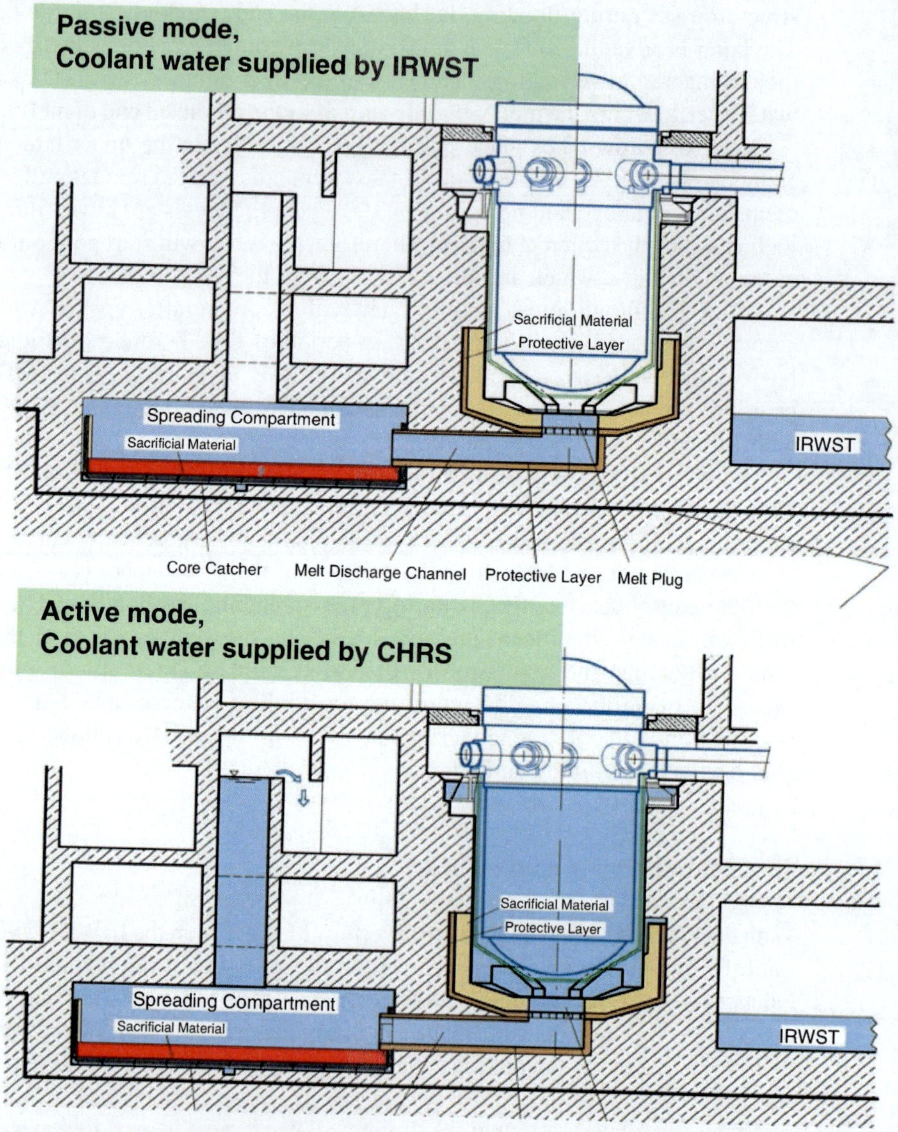

Fig. 4.14 Passive and active mode of CC cooling. (Taken from Proceedings of ICAPP'09 paper 9061, May 2009 [6] and used with permission from Atomic Energy Society of Japan)

protected areas outside the containment. Water is taken from the IRWST and reinjected via spray rings in the upper containment.

Thanks to the large open volume of the containment and the high thermal capacity of the passive heat sinks, the activation of the CHRS and the start of containment spraying is only required 12 h after scram, at the earliest.

As an alternative to the combination of passive CMSS operation with evaporation and containment spraying, the CHRS can also be used to directly feed water into the core catcher (Fig. 4.15).

In this active CMSS mode, the water level in the spreading compartment and in the connected reactor pit will rise up to the overflow level of the steam exhaust chimney. From there, the water will circulate back into the IRWST where the CHRS takes suction.

Because of the high capacity of the EPR™ CHRS and the steadily decreasing decay power, the water in the cooling channels and in the pool atop the spread melt will soon become subcooled under these active conditions.

This allows transporting the decay heat out of the containment by single-phase flow, instead by evaporation and recondensation. Thanks to the related decrease in steam pressure, the active CMSS mode provides a way to achieve ambient pressure conditions inside the containment in the long-term without the need for venting.

To avoid a shortcut flow into the IRWST during active injection via the (open) passive flooding line, a unidirectional flow device, the Passive Outflow Reducer, denoted as "flow limiter" (FL) in Fig. 4.15, is added to the line, between CHRS injection point and IRWST.

Because of the favorable absence of movable parts and the large open cross-section, a vortex diode is used to fulfill this function. As illustrated in Fig. 4.16,

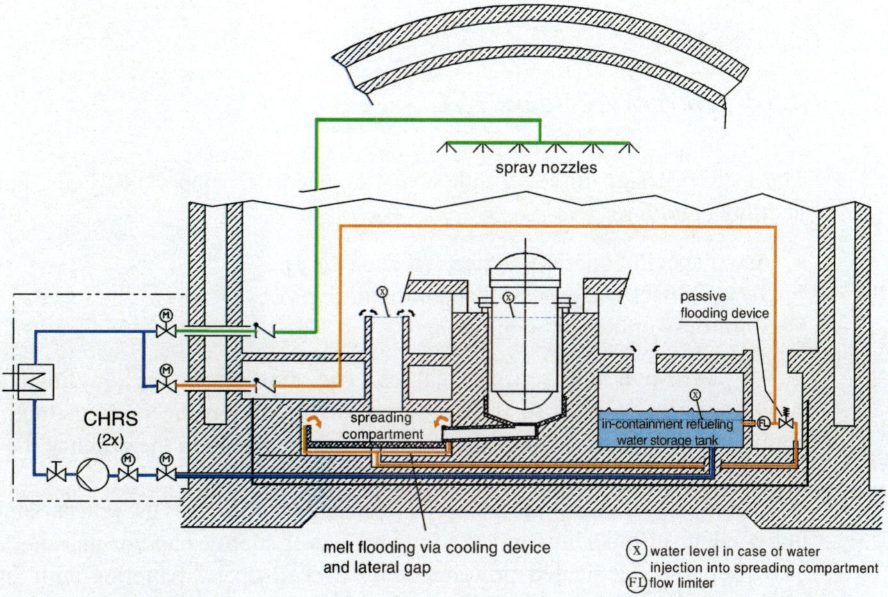

Fig. 4.15 CHRS flow diagram. (Taken from Proceedings of ICAPP'09 paper 9061, May 2009 [6] and used with permission from Atomic Energy Society of Japan)

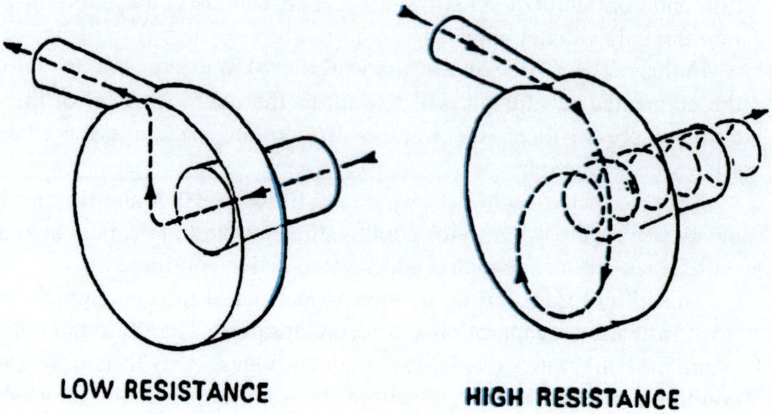

Fig. 4.16 Vortex diode, functional principle. (Taken from Proceedings of ICAPP'09 paper 9061, May 2009 [6] and used with permission from Atomic Energy Society of Japan)

it achieves the required high resistance under reverse flow conditions by centrifugal forces acting on the fluid.

The overflow rate in forward direction can be adjusted by modifying the cross-section.

4.5.3 Severe Accident I&C

The EPR™ design to cope with severe accidents is supported by an appropriate instrumentation to:

- Assist specific operator actions (if necessary)
- Survey the effectiveness of the mitigation process
- Monitor overall plant condition

All corresponding sensors, cables, and connectors are qualified for SA conditions. In addition to this, other requirements on the instrumentation may apply, which depend on the specific licensing situation in the country the EPR™ is built.

These may involve that the SA instrumentation and its power supply are independent of other instrumentation and power supplies and/or that the SA I&C has a separate, dedicated power supply, backed up by batteries with sufficient capacity, and/or that decoupling devices must be provided to ensure the independence from operational I&C.

All SA-relevant information is displayed on a dedicated control panel. It includes signals that allow monitoring:

- The depressurization of the primary circuit
- Core degradation and relocation
- Hydrogen control
- Core melt stabilization
- Containment heat removal system functions
- The activity distribution within the plant and potential releases to the environment

The CMSS is a completely passive system, so its function cannot be impacted by the operator. Consequently, all related information is strictly informative in the sense that it allows to follow the course of events and to detect deviations from the mitigation path, up to a potential failure of the CMSS function.

The CMSS-related part of the SA instrumentation consists of thermocouples located:

- Close to the outer side of the RPV lower head, to detect whether RPV failure is imminent and/or has occurred
- In the chimney above the spreading area to detect melt arrival in the core catcher
- In the core catcher's central water supply duct to detect core catcher melt-through and threat of basemat penetration

4.6 Conclusions

The design of the EPR™ involves a complete and balanced set of systems and components for severe accident mitigation and control, including the stabilization of the molten core. The function of the core melt stabilization system is based on physical principles that are simple and sufficiently well understood. The potential impact of remaining uncertainties is eliminated by a robust design of the components. The applied materials are commonly known and also used in other industrial applications.

For those EPR™ plants for which construction is underway, the design of the components of the CMSS has either been already approved by the costumer and licensing authorities, or these reviews are in progress.

References

1. Fischer M. (2003), Severe accident mitigation and core melt retention in the European pressurized reactor (EPR) Proceedings of ICONE-11 paper 36196, Tokyo, Japan, April 2003
2. Fischer M. (2006), The core melt stabilization concept of the EPR and its experimental validation. Proceedings of ICONE-14 paper 89088, Miami, USA, July 2006
3. Nie M. Fischer M. (2006), Use of molten core concrete interactions in the melt stabilization strategy of the EPR. Proceedings of ICAPP-2006 paper 6330, Reno NV, USA, June 2006

4. Alsmeyer H. et. al (1998), The COMET concept for cooling of ex-vessel corium melts. Proceedings of ICONE-6, San Diego CA, USA, May 1998
5. Farmer M.T. et al. (1998), Status of large scale MACE core coolability experiments. OECD workshop on ex-vessel debris coolability, Karlsruhe, Germany, 15–18 November 1999
6. Fischer M. Henning A. (2009), EPRth engineered features for core melt mitigation in severe accidents. Proceedings of ICAPP'09 paper 9061, May 2009

Chapter 5
Nuclear Power Development and Severe Accident Research in China

Xu Cheng

5.1 Introduction

From a technological viewpoint, the development of nuclear power technology worldwide has undergone four generations. Most of the nuclear power plants (NPPs) operating nowadays belong to the second generation. After the accidents of TMI and Chernobyl, intensive efforts were made to improve the safety features of the second-generation NPPs, and the third generation of nuclear power technology was developed. The main motivation driving the further development of light water reactor (LWR) technology consists of three aspects, i.e., safety, sustainability, and economics, as indicated in Fig. 5.1.

Sustainability is a key issue for long-term nuclear power development and consists of several aspects, such as fuel utilization and waste management. Nearly all NPPs in operation today as well as those coming online in the near future use reactors with a thermal neutron spectrum. The conversion ratio is low, e.g., in a conventional PWR, the conversion ratio is about 0.6. This low conversion ratio restricts the fuel utilization, which is less than 1%. For countries with a shortage of uranium resources such as China, fuel utilization is a key factor affecting long-term nuclear power development. In addition, low fuel utilization leads to a high production of nuclear waste. This results in a challenging task for nuclear waste management. In China, efforts are being made to explore advanced LWRs beyond generation III with the purpose to improve the fuel utilization and to transmute high-level nuclear waste [1].

Economics is one of two key criteria for the selection of NPPs by utilities. The continuous improvement in safety features makes their system more complicated and expensive which strongly affects economic competitiveness. Various measures have been taken to improve the economics, such as increasing power density. Also, there are many research activities in China on new reactor concepts with supercritical water [2].

X. Cheng (✉)
Shanghai Jiao Tong University, Shanghai, China
e-mail: chengxu@sjtu.edu.cn

T. Saito et al. (eds.), *Advances in Light Water Reactor Technologies*,
DOI 10.1007/978-1-4419-7101-2_5,

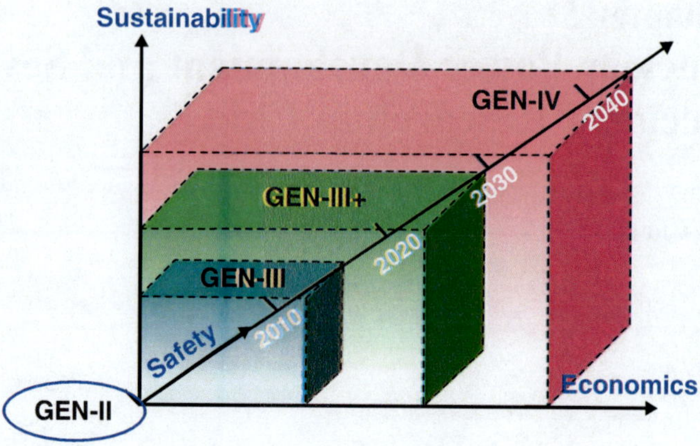

Fig. 5.1 Factors driving the LWR technology development

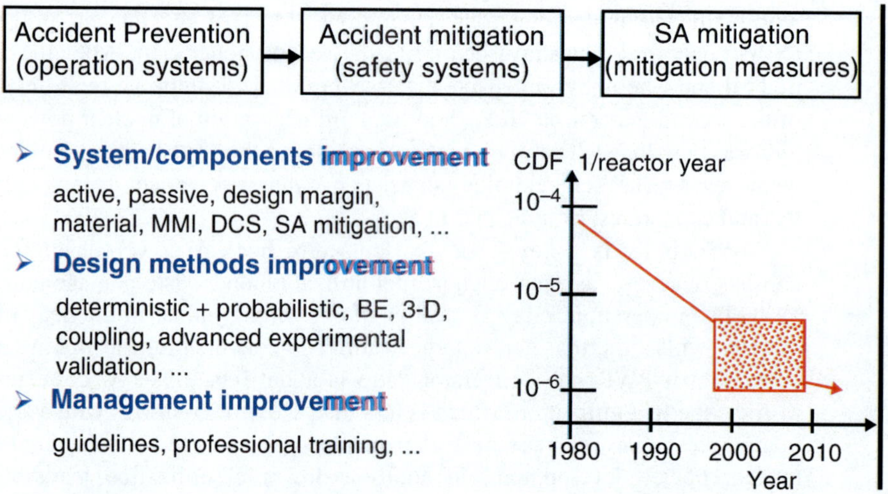

Fig. 5.2 Measures to improve NPP safety

This chapter concentrates on the other driving motivation for the development of future LWRs, i.e., safety. After the accidents of TMI and Chernobyl, intensive efforts were made to improve the safety features of LWRs. Compared to the second generation, the third generation owns a much higher safety level. The core damage frequency (CDF) is lower than 10^{-5} per reactor year. Safety improvement was made in various stages, from accident prevention to severe accident mitigation, as illustrated in Fig. 5.2.

– Improvement of individual components as well as systems, including active/passive systems, advanced materials, and man–machine interactions.

– Improvement of design methodology, including advanced design tools and experimental verification and validation.
– Improvement in management, such as advanced guidelines for operation and maintenance, and enhancement of professional training.

5.2 LWR Development in China

Since the start of economic reforms in the 1970s, the Chinese economy has been undergoing rapid development. One of the bottleneck issues in its economic development is a sustainable and environment-friendly energy supply. By the middle of this century, the primary energy demand in China will be four times that of today. For the time being, more than 70% of the primary energy comes from fossil fuel. The highest portion (about 80%) is dedicated to electricity production. Development of environment-friendly energy supplies is thus becoming a crucial issue in the future Chinese economy. Due to the well-known limitations in renewable energy and hydro power sources, nuclear power is considered as a safe, clean, sustainable, and economic energy source.

In November 2007, China issued an ambitious program for midterm nuclear power development [3]. The report predicted total nuclear power installations will reach 40 GWe or higher by 2020. According to the estimation of the Chinese nuclear experts, nuclear power installations will be around 250 GWe by the middle of this century. That will be about 15% of the total electricity production at that time. Figure 5.3 schematically shows the expected nuclear power development in China. In the last 2 years, the predicted capacities have changed continuously. More recently, it has been reported that the nuclear power capacity will be more than 80 GWe by 2020. In spite of various numbers being reported, Fig. 5.3 clearly indicates that the future nuclear power development will only go much faster. Water-cooled reactors of

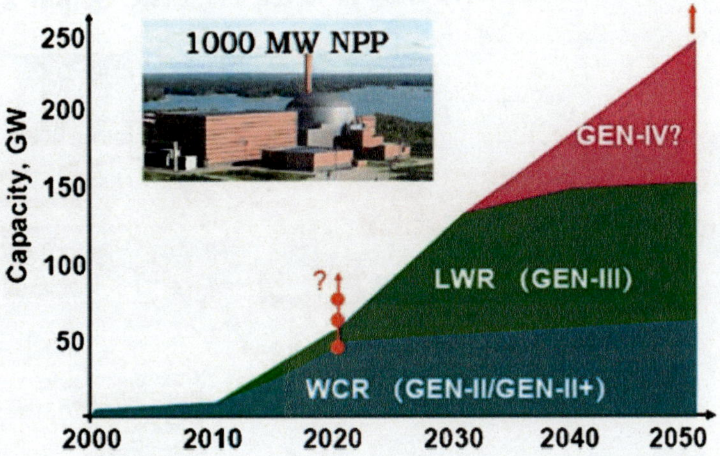

Fig. 5.3 Expected nuclear power development in China

GEN-II or GEN-II extension will make the major contribution to the nuclear power generation until 2020. After that, LWRs of GEN-III will start to be built on a large scale. Clearly, LWRs will still remain as the most important reactor type for nuclear power generation in the next few decades.

Based on the experiences gathered worldwide in the nuclear power development of the past five decades, attention has to be paid to the following issues, to ensure a safe, economic, and fast development of nuclear power.

- Selection of technology lines.
- Realization of self-reliant technology.
- Nationwide coordination.

5.2.1 *Selection of Technology Lines*

It is well agreed that realization of a ambitious nuclear power program urgently requires decision of the technology lines for the future NPPs. As indicated in Fig. 5.4, at present 11 units are under operation with a total installed capacity of 9 GWe, and 12 units are now under construction with an installed capacity of 12 GWe. There are 18 additional units, for which construction will start in the next 3 years. All these NPP units use water-cooled reactors; therefore, water-cooled reactors have clearly been selected as the main reactor type for the next few decades.

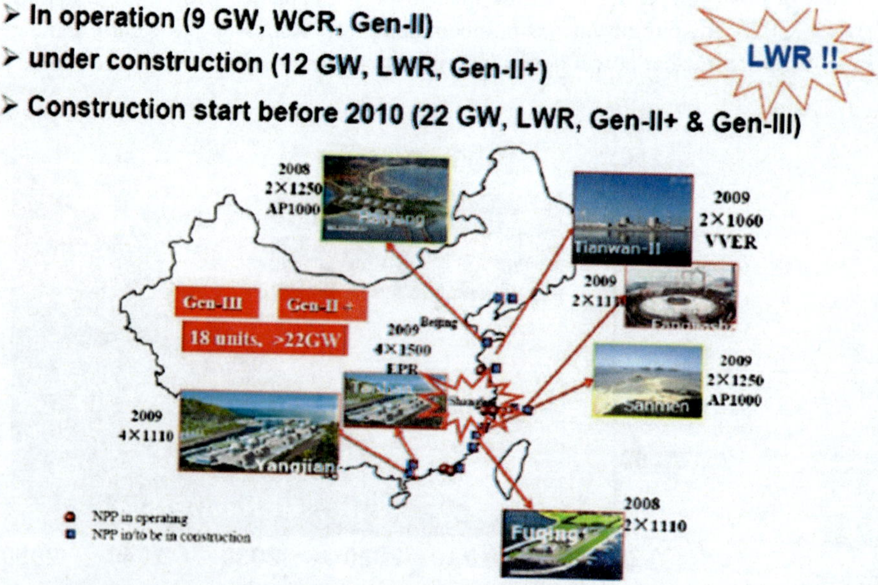

Fig. 5.4 Status of NPPs in China (Current in September 2008)

The operating NPP units are from four different technology lines, i.e., the Chinese PWR of 300 MW/600 MW class, the Canadian CANDU of 700 MW, the French PWR of 900 MW, and the Russian WWER of 1,000 MW. Existing experience emphasizes the necessity to reduce the number of technology lines for future NPPs, and it is highly desired to define a single major technology line for future Chinese nuclear power generation. Considering China's specific situation and the experience gathered in the national and international nuclear community, it has been decided that water-cooled reactors of GEN-III will be the main reactor type for the future Chinese nuclear power generation, at least for the midterm. Passive safety systems should be key features of the Chinese GEN-III PWR. In addition, it should fulfill the following requirements.

– System simplicity.
– Economical competitiveness.
– Operating reliability and easy maintainability.
– Advanced passive engineering safety features.
– Compliant with the latest safety codes for severe accident prevention and mitigation measures as issued by the China National Nuclear Safety Administration (NNSA) and IAEA.
– Digital instrumentation and control systems.
– Advanced human factor engineering techniques and an advanced main control room.

The above technology requirements justify the choice of the AP1000 technology of Westinghouse as the reference technology for the Chinese GEN-III PWR.

5.2.2 Self-Reliant Technology

As soon as the future technology lines are defined, extensive efforts should be made to develop self-reliant technology, so as to reduce the strong technology dependence on other countries, as is the present case in China. To achieve the midterm target, China is using a twofold strategy. On one side, construction of NPPs based on existing GEN-II PWR technology will be continued. Modification of the GEN-II PWR power plants will be undertaken with respect to reactor fuel management and safety performance. The improved GEN-II PWR power plants will make the main contribution to the newly installed NPPs in the next 10 years. Most of the NPPs nowadays under construction or receiving their construction license do belong to this category, e.g., Qinshan Phase-II extension, which is based on Chinese PWR technology of 600 MW class and CPR1000 (improved reactor type based on the French M310).

On the other side, large efforts are being made to accelerate the process for self-reliance of the GEN-III PWR technology. The Chinese government has instigated a large national program to develop technology of advanced large-scale PWRs [4, 20] and to accelerate the self-reliance of Chinese nuclear technology.

Table 5.1 Advanced LWRs in China

Company	Name	Size	Type	Status
NPP in operation: varioustypes of GEN-II				
CGNPC	CPR1000	1000	PWR	Construction
SNPTC	CAP1400	1400	PWR	Design
SNPTC	CAP1700	1700	PWR	Preliminary design
National consortium	SCWR-M	1500	SCWR	Pre-conceptual design

The Nuclear Power Self-reliance Program has been launched with the Sanmen Project in Zhejiang Province and the Haiyang Project in Shandong Province as supporting projects [4]. Three steps will be taken for the development of the "Chinese large-scale advanced PWR NPP."

1. Transfer of AP1000 technology: In this stage, design and construction of 4 units AP1000 will take place under the guidance of Westinghouse. Chinese engineers and scientists will actively participate in this procedure.
2. Design of the modified AP1000 NPP: Based on the experience gathered in the first stage, the existing AP1000 will be modified. This work will be carried out by Chinese engineers and scientists in collaboration with Westinghouse.
3. Design and construction of a self-reliant large-scale PWR: The Chinese AP1000 will be extended with respect to enlarging its reactor power (larger than 1,400 MW) and improving its economics. At the end of this stage (2020) a prototype reactor of the Chinese self-reliant GEN-III PWR will be constructed and put into operation.

In accordance with the self-reliant technology of LWRs, research activities on LWRs of GEN-IV, i.e., supercritical water-cooled reactors, are being promoted, to ensure the sustainability of LWR technology development. Table 5.1 gives an overview on advanced LWRs in China.

5.2.3 *Nationwide Coordination*

Realization of self-reliance in nuclear technology requires high-quality coordination, across various institutions for design, research, manufacturing, and education/training. For this purpose, a new organization, the State Nuclear Power Technology Corp. Ltd. (SNPTC), was founded in 2007. SNPTC is responsible for the self-reliance of the Chinese GEN-III PWR technology and has established separate subcompanies for research, design, and manufacture. In addition, SNPTC is also the direct partner with Westinghouse for the AP1000 technology transfer. Contracts between SNPTC and Westinghouse were signed in July 2007 and came into force in September 2007. Four AP1000 units will be put into commercial operation from 2013 to 2015.

Severe accident research is well recognized as a main task in the large-scale national project related to self-reliance of LWR technology. Research programs related to severe accidents of the CAP1400 have been issued and research work will start in the near future. Furthermore, some basic studies on severe accident related phenomena have been identified as important cross-cutting issues for future investigations.

5.3 Severe Accident Research in China

Various phenomena can be identified according to the progression of severe accidents (SAs) (Fig. 5.5). After a SA is triggered, the core melt process starts. During this process, fission products are released. The chemical reaction between water and zirconium is the main source of hydrogen production. At the early stage, the reactor pressure vessel (RPV) is still intact. Main events happen inside the RPV. Core melt collects in the lower head of the RPV and forms a melt pool. Melt cooling and RPV cooling become the key tasks to keep the integrity of the RPV and to restrict the core melt inside the RPV (in-vessel retention, IVR). IVR is the key methodology of SA mitigation in several advanced LWR designs, such as the AP1000. Cooling of the RPV from outside (ex-vessel cooling of IVR, ERVC-IVR) is often an important measure to realize the RPV integrity.

In case the RPV fails, core melt spreads in the containment. In this case, a core catcher has to be designed to accommodate and cool core melt. Design of the core catcher has to consider the core spreading behavior, interaction of core melt with other materials, and cooling capability. There are a number of phenomena occurring during the entire SA procedure, such as fuel-coolant interaction (also referred

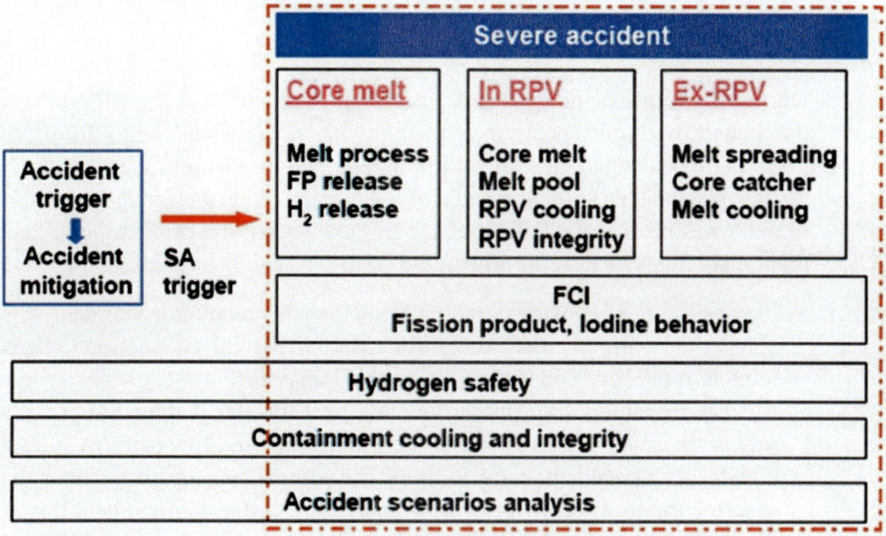

Fig. 5.5 Procedures and phenomena involved in SA

Table 5.2 Ranking of SA phenomena

Phenomena	R&D performed	Existing knowledge	Priority
Core melt property	Low	Low	Medium
Core melt process	Medium	Medium	medium
Melt pool behavior	High	Medium	Medium
Ex-vessel cooling of IVR	Medium	Low	High
Melt spreading	Medium	Medium	low
Core catcher	Medium	Medium	Medium
FCI	High	Low	Medium
Hydrogen	High	Medium	High
Fission products, iodine	High	Medium	Medium
Containment	Medium	Medium	High

to as steam explosion), behavior of fission products and aerosols, hydrogen safety, containment cooling, and integrity that must be considered.

Since the 1980s, a large number of SA studies have been carried out worldwide. Table 5.2 summarizes the state of the art related to various phenomena and their priority for the Chinese nuclear research community. The present author identifies three phenomena as high priority, i.e., ERVC-IVR, hydrogen safety, and containment integrity. In the following sections, research activities on these three phenomena and FCI are briefly presented.

It has to be pointed out that regarding SA research, China is still at the beginning stage. In addition to well coordinated nationwide activities, in which many institutions are involved, the Chinese nuclear community is looking for enhanced international exchange and collaboration.

5.3.1 IVR

During the transient phase of SA progression, integrity of the RPV lower head is threatened by a wide spectrum of phenomena, e.g., various melt relocation scenarios, potential steam explosion, jet impingement, etc. A limiting case and strategy in the late phase of SA is maintenance of lower head integrity through external cooling of the lower head of the RPV to reach in-vessel retention of the molten pool. Some main challenges of IVR are illustrated in Fig. 5.6 and summarized below:

- *Properties and behavior of melt pool*: This is a calculation problem with multi-components, multi-phase, three-dimensional natural convection. There are no reliable models or codes to describe this behavior.
- *Ex-vessel cooling*: Two-phase flow and heat transfer is the main process occurring in the gap between the RPV and the insulation. Flow patterns and local heat transfer are dependent on the individual design. There are no methods to reliably describe the local behavior of flow and heat transfer. Critical heat flux (CHF) is an important criterion in the design of ERVC-IVR. However, CHF depends on local parameters and shows a complex dependence on individual designs.

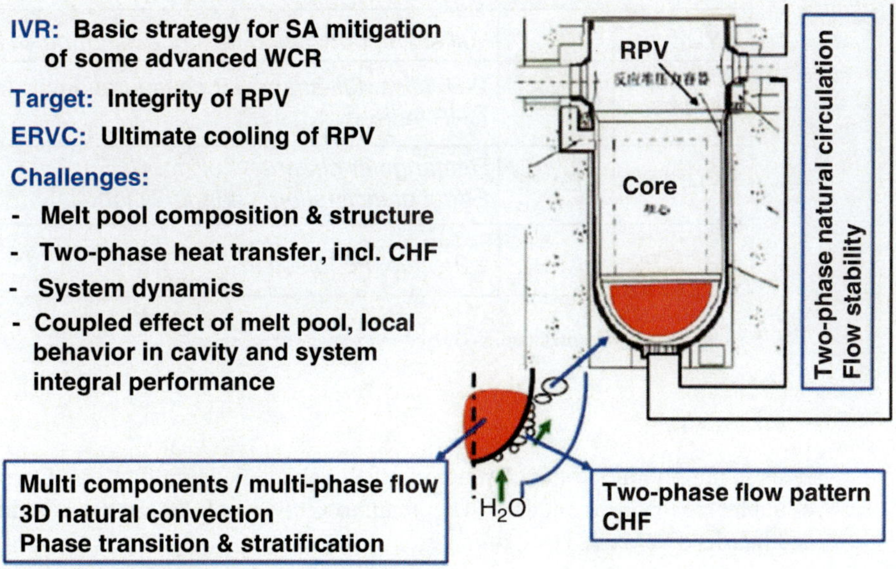

Fig. 5.6 Main challenges involved in IVR

– *The entire heat removal system based on natural circulation*: Stability is the key issue of a two-phase natural circulation system. Ensuring flow stability and at the same time providing sufficiently large natural circulation capability is a challenging research task.

The IVR–ERVC concept was first investigated and explored for the Loviisa PWR in Finland. It was accepted as the major accident management measure by the Finnish regulatory agency. In the USA, the design of the AP1000 employs reactor ex-vessel flooding as an accident management scheme [5]. The safety strategy of the AP1000 is to keep the RPV intact under any conditions, including SA core melt conditions. There is no core catcher outside the RPV. Later on, the IVR-ERVC was also proposed for other PWRs and BWRs such as the Korean APR-1400 [6] and the German SWR1000 [7].

In China, the Shanghai Nuclear Engineering Research and Design Institute (SNERDI) has adopted the IVR-ERVC concept in the design of the Chashima-2 300 MW NPP. An engineering investigation has been conducted during the design phase. Furthermore, the China Guangdong Nuclear Power Corporation (CGNPC) is also considering application of the IVR-ERVC strategy in the CPR1000 design. To extend the reactor power of the AP1000 to a higher level, e.g., 1,400 MW, the feasibility of the passive IVR-ERVC concept becomes one of the bottleneck factors and it has attracted very strong attention from the Chinese nuclear community.

Although several studies were carried out at various organizations, as illustrated in Fig. 5.7, it was concluded that the existing results cannot be easily extrapolated to new designs. Experimental studies are highly required for each specific design.

CYBL	SNL	*full scale, downward facing, boiling, no CHF*
ULPU	UCSB	*2-D, slice, full-scale, AP configurations, CHF tests*
SYLTAN	CEA	*Rectangular channel, boiling, CHF, Effect of inclination, channel height*
HERMES	KAERI	*1:2 scale, air injection*

Fig. 5.7 Some experimental studies carried out worldwide

This is mainly due to the complexity of the phenomena involved and the lack of reliable mechanistic models. The main objectives of the ongoing studies are summarized below [21]:

- To investigate the feasibility of ERVC-IVR and to optimize engineering designs for Chinese LWR designs such as the CPR1000 and CAP1400.
- To reveal more details of the mechanistic processes and to develop more reliable mechanistic models for prediction.

The present ongoing research work is mainly being carried out by Shanghai Jiao Tong University (SJTU), in cooperation with SNERDI and CGNPC. Both experimental and theoretical studies were initiated at SJTU 2 years ago, in collaboration with SNERDI and CGNPC. Figure 5.8 schematically shows the REPEC test facility built at SJTU.

The experimental study consists of three phases.

- *Phase I*: Cold tests: In this phase, air is used to simulate steam. The main purpose is to study two-phase flow characteristics in the test section and the natural circulation capability of the passive cooling system.
- *Phase II*: Hob tests: The test section is electrically heated to produce steam. The main purpose of this test phase is to study two-phase flow and heat transfer behavior, including critical heat flux, in the gap and on the surface of the RPV. Stability of the natural circulation is also one of the main phenomena under consideration.
- *Phase III*: Small scale three-dimensional ERVC assessment test and scaling law.

The first object under investigation is ERVC-IVR of CPR1000. The CPR1000 vessel/insulation configuration is shown in Fig. 5.9. The RPV vessel outer diameter is about 4 m; the elevation from the insulation bottom plate to the cold leg is about 7 m, and the gap between the RPV and insulation is about 250 mm. Unlike the AP1000, there is a cylindrical flange in the lower head of the CPR1000. The flange makes the local flow change abruptly and might decrease the local critical heat flux in this region. The heat flux along the flange is as high as 0.8 MW/m^2, which may

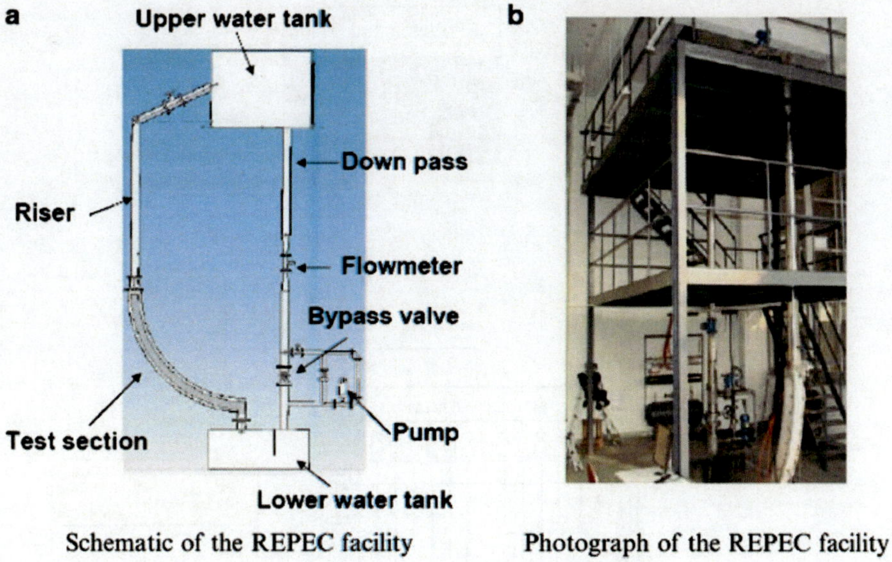

Fig. 5.8 REPEC test facility at SJTU

Fig. 5.9 Schematic diagram of vessel/insulation cooling system in CPR1000

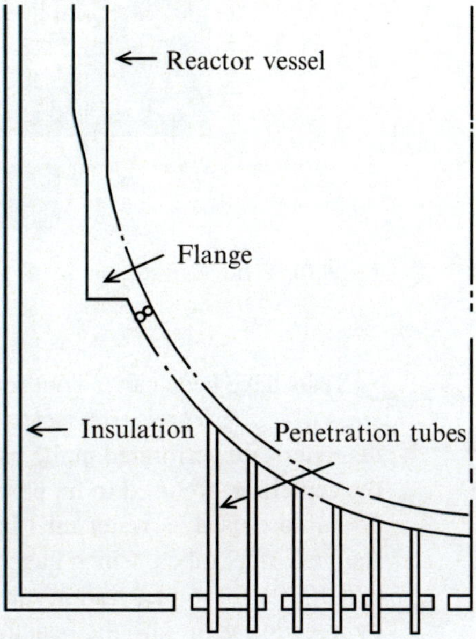

exceed CHF and result in the RPV wall failure. Fifty penetration tubes, which distort the flow field, are arranged at the bottom of the RPV vessel. An insulation baffle is going to be introduced to improve the flow and heat transfer behavior. The main outcomes from the study are the geometrical optimization of the flange, the penetration plate, and the insulation baffle.

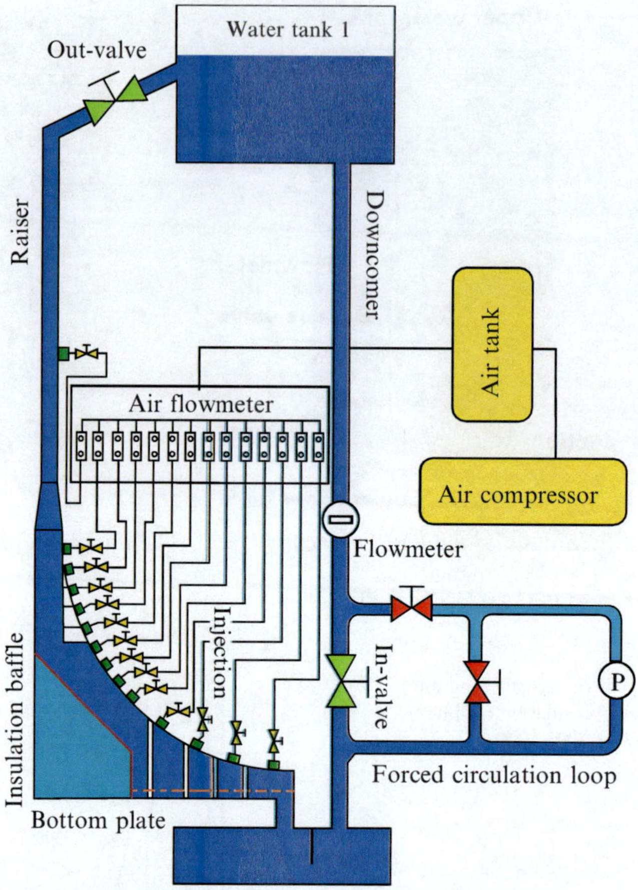

Fig. 5.10 Perforated structure of the inlet plate

Tests have been carried out for four insulation baffles of different minimum gap sizes (d_A, d_m). Three arrays (16 columns per array, Fig. 5.10) of holes of different diameters are perforated in the baffle. The small holes with a diameter of 35 mm in the centerline are used to fix penetration tubes and the big ones with a diameter of 50 mm are used as water inlet holes. The penetration tubes are simulated by five stainless steel tubes, which have the same diameter as the penetration tubes in the CPR1000. These penetration tubes are located according to their original position by inserting them into the penetration plate.

In the cold tests, bubbles in the test section are produced by air injectors. Fourteen individually controlled air injectors are used in the test section (Fig. 5.11). Since heat flux along the lower head in the top region is much higher than that in the bottom region, more injectors are installed in the top region. The air injector is made of porous plate sealed in a stainless steel box, which can generate fine air bubbles to simulate the stream bubbles.

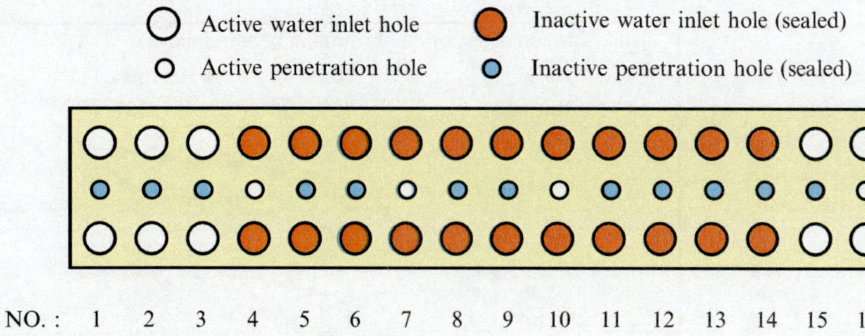

Fig. 5.11 Schematic diagram of non-heating facility

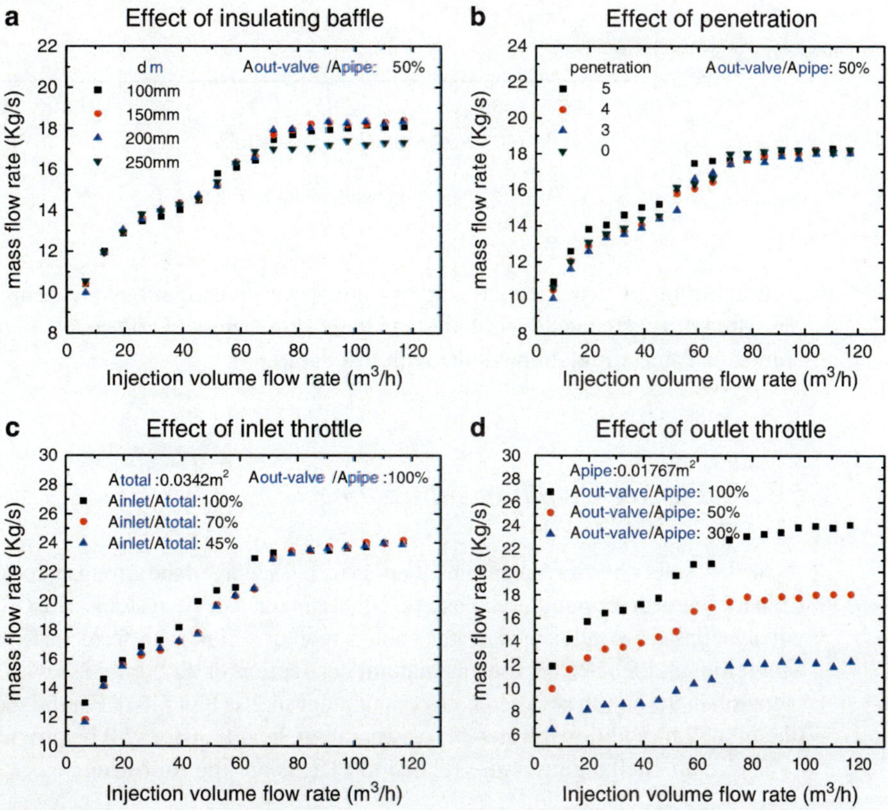

Fig. 5.12 Induced water mass flow rate at various test conditions

Figure 5.12 show example test results presenting the induced water mass flow rate veZrsus the injected air flow rate at various test conditions. Obviously, the opening of the valve at the test section exit has the strongest effect on the induced water flow rate.

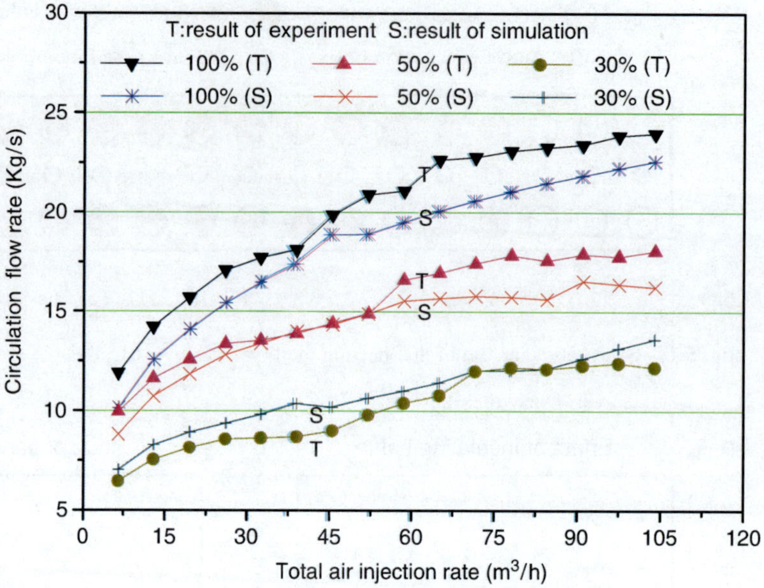

Fig. 5.13 Comparison of RELAP simulation results with test data

In addition to experimental studies, numerical investigations are being carried out with both system analysis codes and three-dimensional CFD codes. Figure 5.13 compares some simulation results with test data.

5.3.2 Passive Containment Cooling

As the last safety barrier, containment integrity has received the strong attention of the Chinese nuclear community. Passive containment safety systems (PCCSs) were widely applied to advanced water-cooled reactors. For long-term passive decay heat removal, the AP1000 uses the natural convection of air combined with thermal radiation in the annuli between both containment shells (Fig. 5.14). For the short-term (the first 72 h) additional water-film evaporation, heat transfer will be provided [5].

The main challenging issues related to PCCSs are the following.

- Coupling of multimechanisms: Many phenomena are involved in PCCSs, such as mixed convection, thermal radiation, water film distribution and evaporation, buoyancy driven stratification and condensation. Coupling is a process involving multicomponents, multiphases, and multiphenomena. Figure 5.15 schematically shows heat transfer at mixed convection conditions. There exists a region with impaired heat transfer. Design of PCCSs should minimize operating conditions in this impaired region. Previous experiments at the PASCO test facility in

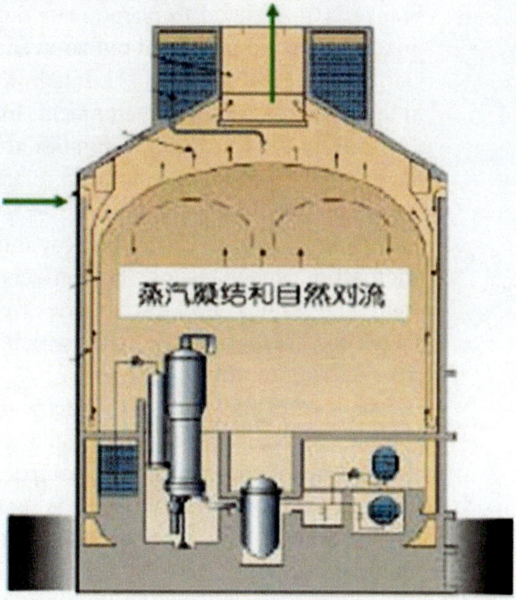

Fig. 5.14 PCCSs of AP1000

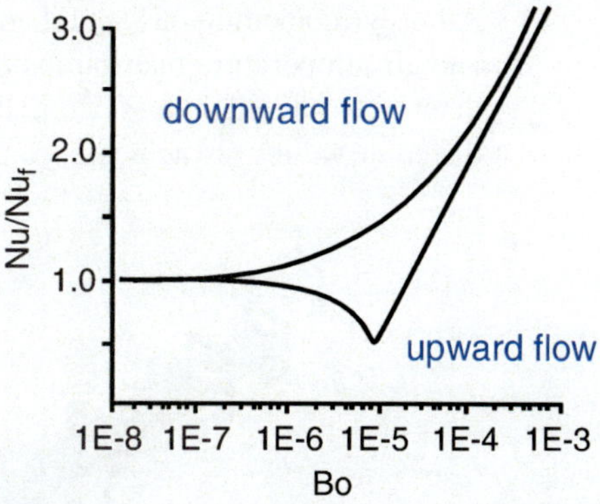

Fig. 5.15 Heat transfer at mixed convection conditions

Germany [8] indicated that water film distribution remains a complex phenomenon and needs further investigations.

– Various geometric scales. Containment consists of compartments of various sizes. Resolution of various scales in thus necessary to understand and predict the complex processes. Strong three-dimensional effects occur related to all important phenomena, such as stratification at natural convection conditions.
– Deficiency in reliability of prediction models.

In the past, investigations were performed by other international partners. Related to phenomena inside the containment, separate effects on condensation, distribution of gases were studied. Furthermore, integral tests using model containment were carried out in several countries, such as by BMC in Germany [9] and MISTRA in

France [10]. Related to phenomena outside containment, separate effects tests on mixed convection, thermal radiation and water film evaporation were carried out in Germany [8] and in Italy [11]. Integral tests were also performed by Westinghouse in the frame of AP600 development. In Europe, a specific project DABASCO was financed by the European Commission in the Fourth Framework Program to study the thermal-hydraulic phenomena related to PCCSs [12].

Fundamental investigations on PCCS-related phenomena are being initiated at SJTU, to investigate the cooling capability of the passive systems and the involved microscopic mechanisms. The university is carrying out both experimental and numerical studies. Two test facilities were built with different purposes. Figure 5.16 shows the test section MICARE, which is a square flow channel with the maximum cross-section of 400 × 250 mm.

One side of the channel is electrically heated. This side consists of 16 heating plates, which are separately heated and controlled to achieve a good uniform distribution of the heated wall temperature. The orientation of the flow channel can be changed arbitrarily. The test section has a total height of 8 m, of which 6 m (in the middle) can be heated. The test section can be connected to auxiliary equipment to realize a forced flow of air into the test channel using a compressor. The wall temperature can be varied up to 200°C. The test facility is equipped with a large number of thermocouples to measure the distribution of wall temperatures. A hot-wire anemometer and thermocouples are applied to measure the air velocity and air temperature distribution in the flow channel. Calibrations are performed to determine the heat loss from the heated wall to the ambient surroundings at different values of the heated wall temperature.

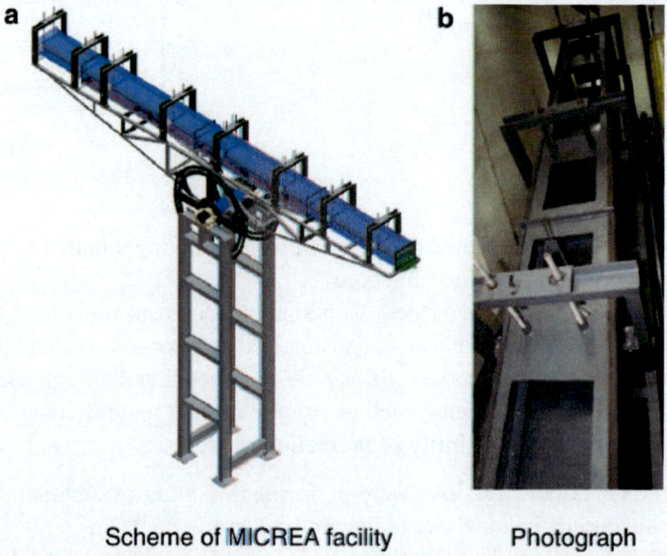

Fig. 5.16 MICARE test facility at STJU

The main study objectives of the MICARE test facility are the following:

- Cooling capability of PCCSs.
- Flow pattern under various orientation.
- Contribution of thermal radiation to PCCSs.
- Effects of various parameters on water film distribution.
- Test data for validation of CFD codes.

The main test parameters are:

- Heat wall temperature, 30–15°C;
- Test section orientation, 0–360°;
- Wall emissivity, 0.1–1.0; and
- Bo number, $1.0e^{-8}$–$1.0e^{-3}$.

The main measurements taken are:

- Air velocity and temperature;
- Wall temperature;
- Air mass flow rate;
- Water film thickness;
- Flow visualization.

The second test facility WAFIP is devoted to study water film behavior, as shown in Fig. 5.17. The test section has a height of 5 m and a width of 2 m. Effect of various parameters on water film distribution and dynamics will be investigated, such as water injection modes and surface properties.

Figure 5.18 shows examples of test data presenting the temperatures at both the heated wall and side wall at various test conditions, i.e., orientation and heated

Fig. 5.17 WAFIP test facility at STJU

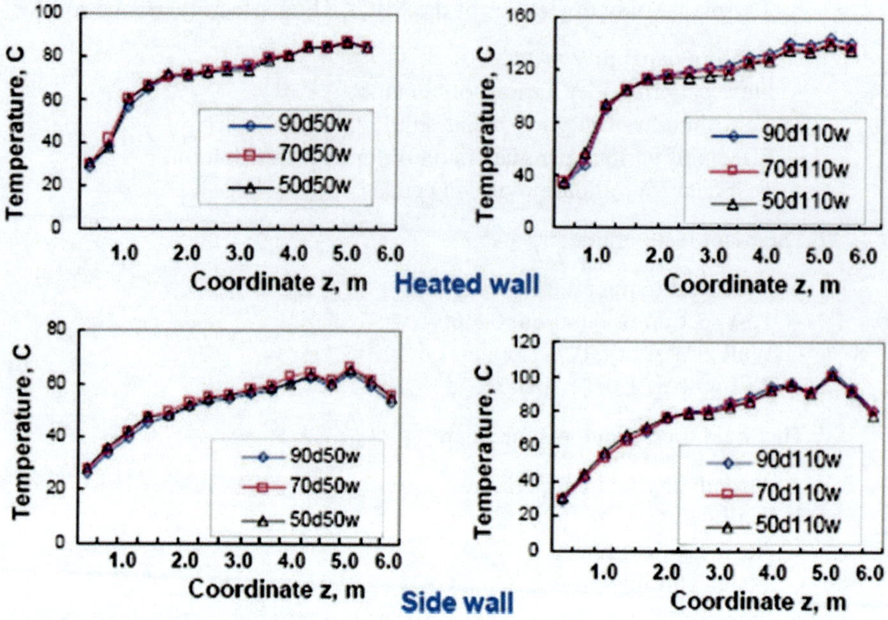

Fig. 5.18 Measured wall temperature at various test conditions (d: inclination degree; w: heating power in watts)

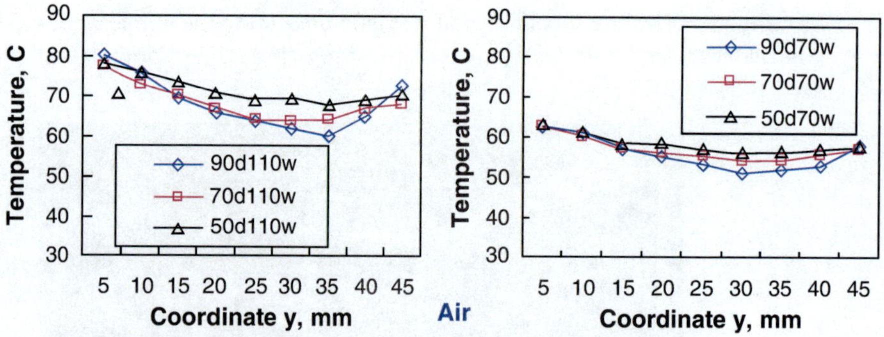

Fig. 5.19 Measured air temperature distributions at various test conditions

power. For the test conditions considered, the test channel orientation has less effect on the wall temperature distribution.

Figure 5.19 presents the air temperature distribution at various test conditions. Air temperature shows a minimum in the bulk region. It increases on approaching the walls (both the heated wall and the rear wall). Due to thermal radiation, the temperature at the rear wall is higher than the bulk air temperature. In a vertical test channel, the bulk air temperature is lower than that at other orientations.

In addition to the experimental work, numerical simulations are carried out to understand the microscopic phenomena involved in the mixed convection in a square channel with various orientations. Numerical simulation of three-dimensional turbulent natural convection coupled with thermal radiation requires high capabilities of a computer code. The present numerical simulation has two main features.

5.3.2.1 Low Reynolds $k - \varepsilon$ Turbulence Model

The RANS expressions of the momentum and energy conservation equations are:

$$\rho\frac{\partial\overline{u_i}}{\partial t}+\rho\overline{u_j}\frac{\partial\overline{u_i}}{\partial x_j}=-\frac{\partial\overline{p}}{\partial x_i}+\mu\frac{\partial}{\partial x_j}\left(\mu\frac{\partial\overline{u_i}}{\partial x_j}-\rho\overline{u_i'u_j'}\right)+\rho g_i \tag{5.1}$$

and

$$\rho\frac{\partial\overline{\varphi}}{\partial t}+\rho\overline{u_j}\frac{\partial\overline{\varphi}}{\partial x_j}=\Gamma\frac{\partial}{\partial x_j}\left(\mu\frac{\partial\overline{\varphi}}{\partial x_j}-\rho\overline{u_i'\varphi'}\right) \tag{5.2}$$

Both terms of fluctuation correlation are presented via the eddy viscosity approach.

$$\rho\overline{u_i'u_j'}=-\mu_t\frac{\partial\overline{u_i}}{\partial x_j} \tag{5.3}$$

$$-\rho\overline{u_i'\varphi}=\Gamma_t\frac{\partial\overline{\varphi}}{\partial x_j} \tag{5.4}$$

$$\Gamma_t=\frac{\mu_t}{\sigma} \tag{5.5}$$

Low Reynolds $k - \varepsilon$ turbulence models are expressed as the following.

$$\mu_t=f_\mu C_\mu\frac{k^2}{\varepsilon} \tag{5.6}$$

$$\frac{\partial(\rho\,u_j\,k)}{\partial x_j}=P_k+G_k-\rho\,\varepsilon+\frac{\partial}{\partial x_j}\left[\left(\frac{\mu+\mu_t}{\sigma_k}\right)\frac{\partial k}{\partial x_j}\right] \tag{5.7}$$

$$\frac{\partial(\rho\,u_j\,\varepsilon)}{\partial x_j}=C_1\,\frac{\varepsilon}{k}\,(P_k+G_k)\left(1-C_3\,\frac{G_k}{P_k}\right)-C_2\,\frac{\rho\,\varepsilon^2}{k}+\frac{\partial}{\partial x_j}\left[\left(\frac{\mu+\mu_t}{\sigma_\varepsilon}\right)\frac{\partial\varepsilon}{\partial x_j}\right] \tag{5.8}$$

The difference between the various models is the selection of the eight coefficients, $f_1, f_2, f_\mu, C_\mu, C_{\varepsilon 1}, C_{\varepsilon 2}, \sigma_k, \sigma_\varepsilon$.

5.3.2.2 Advanced Models for Thermal Radiation

The convective heat transfer is coupled with thermal radiation by the thermal boundary condition at unheated walls, where the net radiative heat must be transferred by natural convection of air. A thermal radiation model with high numerical efficiency has been developed to determine the radiative heat transfer. The fluid (air) is considered radiatively nonparticipating, and the walls are gray and diffuse. The net radiative heat power of a surface element $Q_{r,i}$ is computed by the net-radiation method for enclosures [13].

$$\left(\frac{Q_{r,i}}{\varepsilon_i}\right) = \left(\frac{E_i}{\varepsilon_i}\right) - \sum_j \left(\frac{E_j}{\varepsilon_j}\right)\varphi_{j,i} + \sum_j (1-\varepsilon_j)\varphi_{j,i}\left(\frac{Q_{r,j}}{\varepsilon_j}\right) \tag{5.9}$$

The above radiation equation can be solved either directly or iteratively. The direct solution is exact and usually needs larger computing expenditure. An iterative solution, e.g., the Gauß–Seidel iteration, requires only a few iterations for intermediate or high wall emissivities. Nevertheless, at low emissivities, the direct solution method is more efficient than the iterative method.

The radiative heat power can be easily computed by solving (5.9), as long as the view factors are known. Generally, view factor can only be obtained numerically. For a flow channel in the Cartesian coordinate system where boundary walls are either parallel or perpendicular to each other, the view factor between any two surface elements has been derived analytically. Figure 5.20 shows two different cases: (a) two parallel surface elements and (b) two perpendicular surface elements. To specify the dimensions of any two surface elements and their relative positions, seven geometric parameters are needed, indicated as a to g in the figure. The view factor for two parallel surface elements (Fig. 5.20a) is derived as follows [13].

$$\begin{aligned}\frac{\varphi_{12}\pi A_1}{a^2} =& \sum_{i=1}^{4} Z_i \sum_{j=1}^{4} S_j \left\{\frac{1}{2} X_j \sqrt{1+Y_i^2}\arctan\left(\frac{X_j}{\sqrt{1+Y_i^2}}\right)\right\} \\ &+ \sum_{i=1}^{4} Z_i \sum_{j=1}^{4} S_j \left\{\frac{1}{2} Y_i \sqrt{1+X_j^2}\arctan\left(\frac{Y_i}{\sqrt{1+X_j^2}}\right)\right\} \\ &- \sum_{i=1}^{4} Z_i \sum_{j=1}^{4} S_j \left\{\frac{1}{4}\ln\left(1+X_j^2+Y_i^2\right)\right\}.\end{aligned} \tag{5.10}$$

For two perpendicular surface elements (Fig. 5.20b) the view factor can be calculated by:

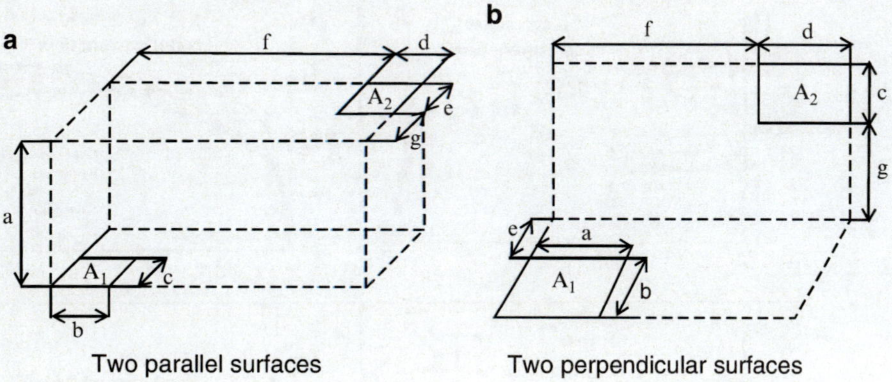

Fig. 5.20 Two cases for view factor calculation

Table 5.3 Parameters in (5.10)

	X_i	Y_i	Z_i	S_i
$i = 1$	$(f + d)/a$	$(g + e)/a$	+1	+!1
$i = 2$	$(b-f)/a$	g/a	−1	+1
$i = 3$	$(b-f-d)/a$	$(c-g-e)/a$	−1	−1
$i = 4$	f/a	$(c-g)/a$	+1	−1

Table 5.4 Parameters in (5.11)

	X_i	Y_i	Z_i	S_i
$i = 1$	$(f + d)/a$	$[g^2 + (e+b)^2/a^2]$	+1	+1
$i = 2$	$(a-f)/a$	$[e^2 + (g+c)^2/a^2]$	+1	+1
$i = 3$	$(a-f-d)/a$	$[g^2 + e^2/a^2]$	−1	−1
$i = 4$	f/a	$[(g+c)^2 + (e+b)^2/a^2]$	−1	−1

$$\frac{\varphi_{12}\pi A_i}{a^2} = \sum_{i=1}^{4} Z_i \sum_{j=1}^{4} S_j \left\{ \frac{1}{8} \left(X_i{}^2 - Y_j \right) \ln\left(X_i{}^2 + Y_j \right) \right\} + \sum_{i=1}^{4} Z_i \sum_{j=1}^{4} S_j \left\{ \frac{1}{2} \sqrt{Y_j} X_i \arctan\left(\frac{X_i}{\sqrt{Y_j}} \right) \right\}. \quad (5.11)$$

The parameters X_i, Y_i, Z_i, and S_i in (5.10) and (5.11) are summarized in Tables 5.3 and 5.4, respectively. Figure 5.21 shows an example of numerical results,

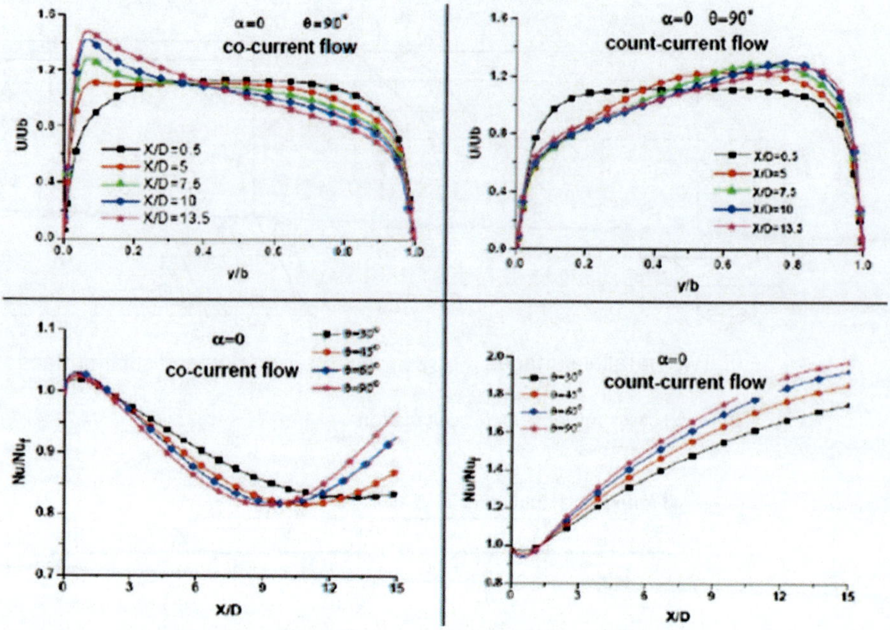

Fig. 5.21 Calculated velocity and Nusselt number distributions

presenting the distribution of air velocity at various elevations (x/D) and orientations (θ) for both cocurrent and countcurrent flows.

Figure 5.22 illustrates the contribution of thermal radiation to total heat removal versus the wall emissivity (ε) for both cocurrent and countcurrent flows. The contribution of thermal radiation becomes stronger with increasing wall emissivity and at cocurrent flow conditions. About 40% of the heat can be removed by thermal radiation at cocurrent flow conditions and at a wall emissivity of 0.9.

Figure 5.23 compares the numerical results with data for two test cases. The first test case has a small Bo number and represents conditions similar to forced convection, whereas the second test case corresponding to conditions similar to heat transfer impairment. For both test cases, reasonable agreement between the test data and the numerical results is achieved.

In addition, analysis using a lumped parameter approach was carried out at SJTU [14]. Effects of various parameters on the heat removal capability are investigated. Figure 5.24 gives an example indicating the effect of the thermal conductivity of the buffer plate on heat removal. Results are obtained with a containment temperature of 150°C and the wall emissivity of 1. It is seen that a higher thermal conductivity leads to an increase in heat removal of about 15%. A strong effect is observed in the region of low thermal conductivity (<0.5 W/m K). The maximum removable heat from the containment is about 7.5 MW.

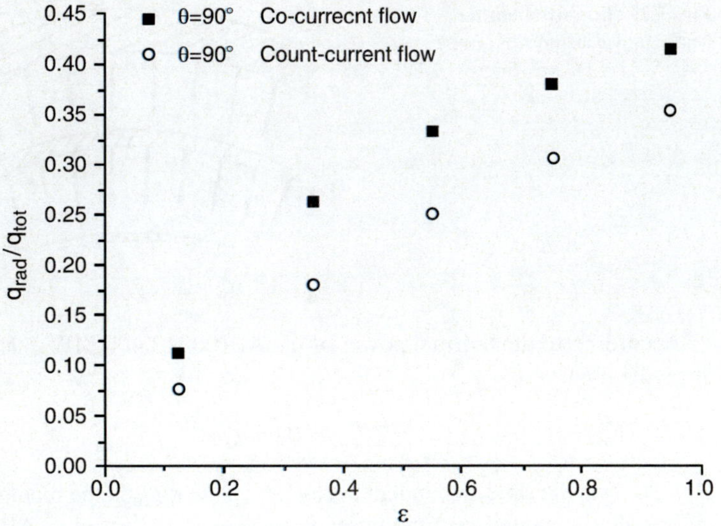

Fig. 5.22 Contribution of thermal radiation heat transfer

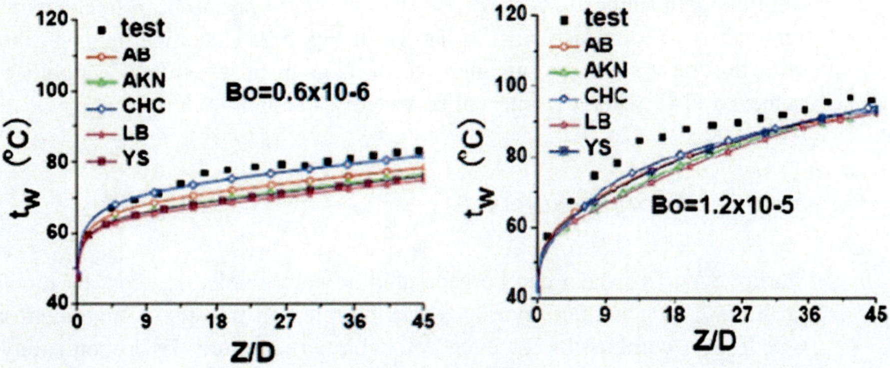

Fig. 5.23 Comparison of numerical results with test data

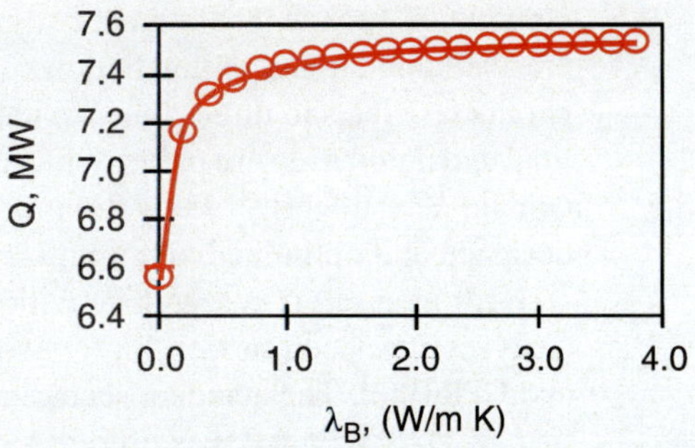

Fig. 5.24 Effect of baffle conductivity on heat removal

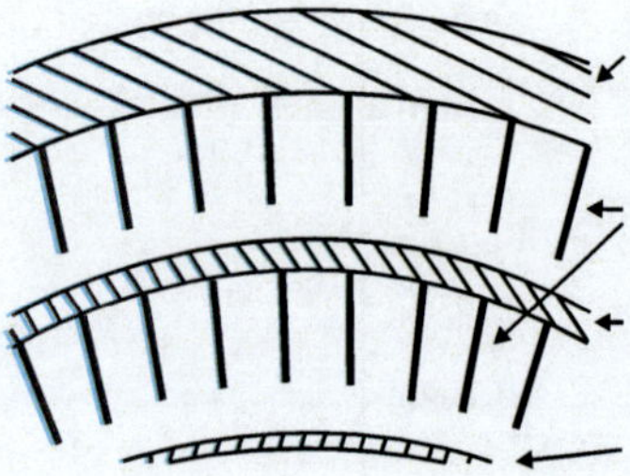

Fig. 5.25 Improvement for heat removal using ribs

According to the thermal power of the AP1000 (3,400 MW_{th}) and the simplified decay heat curve,

$$Q(t) = 0.062 Q_0 t^{-0.2}. \qquad (5.12)$$

Here, $Q(t)$ is the time-dependent decay heat power; Q_0 is the reactor thermal power before shutdown, and t is time in seconds. The decay heat in an AP1000 goes down to the level of 7.5 MW 40 days after the reactor shutdown. Obviously, this passive system is insufficient to remove decay heat and needs improvement, especially for the Chinese GEN-III PWR with a much larger thermal power. Therefore, various improvement suggestions are proposed. One of the possibilities to enhance the heat removal is to introduce ribs, as shown in Fig. 5.25. Detailed analysis shows that with this new structure, an increase of 15% in heat removal capability can be achieved [14], when the interval between ribs is about 0.5 m.

5.3.3 Hydrogen Safety

During SAs, hydrogen can be generated in water-cooled reactors by metal–steam reaction. The generated hydrogen will be released into the containment where it will form a combustible or even detonable gas mixture. Hydrogen safety covers many processes, from hydrogen production to hydrogen detonation, as shown in Fig. 5.26. Hydrogen production and release depends on accident scenarios and can be simulated using SA codes, such as SCDAP-RELAP. Released hydrogen is distributed inside the containment. This process is strongly affected by various conditions, such as containment structure and containment spray. Different codes are available for simulating hydrogen distribution inside containment, from lumped parameter codes to three-dimensional CFD codes. Hydrogen combustion, deflagration, and detonation are processes affecting the containment load and, subsequently, containment integrity. Mitigation measures, such as PAR and igniters, need to be designed and optimized, to minimize the consequence.

Hydrogen safety research activities were launched at SJTU in 2006. Up to now, analyses have been carried out for two different containments, i.e., Qinshan Phase-II and CPR1000. The accident scenarios considered were defined together with utilities. SCDAP/RELAP was used to simulate the accident scenarios and it provided

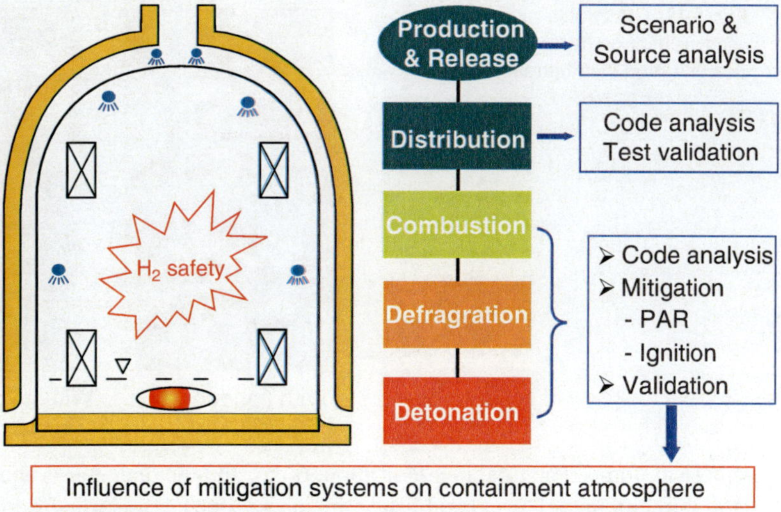

Fig. 5.26 Hydrogen safety issues under SA conditions

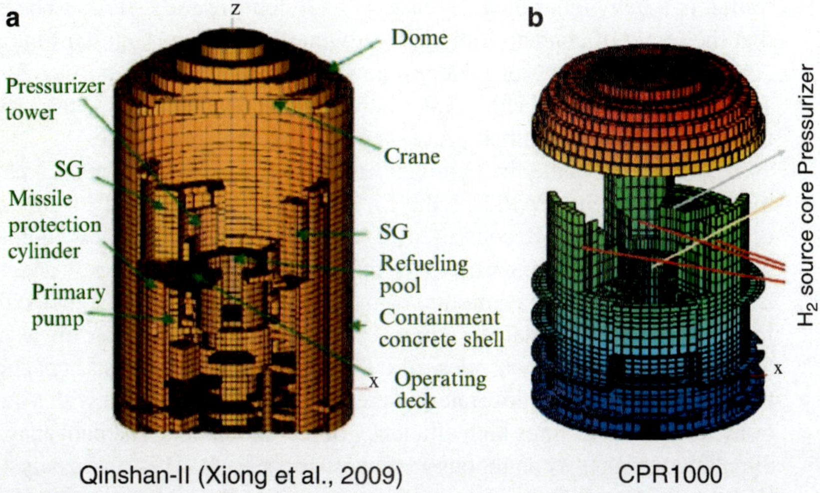

Fig. 5.27 Mesh structure for GASFLOW simulations

source term of hydrogen and steam. The 3D CFD code GASFLOW [15] was applied to simulate the hydrogen distribution inside containment. Both PAR and igniters were considered as mitigation measures. Furthermore, effects of source term distribution, steam condensation and spray system on hydrogen distribution were investigated.

Figure 5.27 illustrates the GASFLOW mesh structures of both Qinshan-II containment and CPR1000 containment. The former consists of a cylindrical part and a

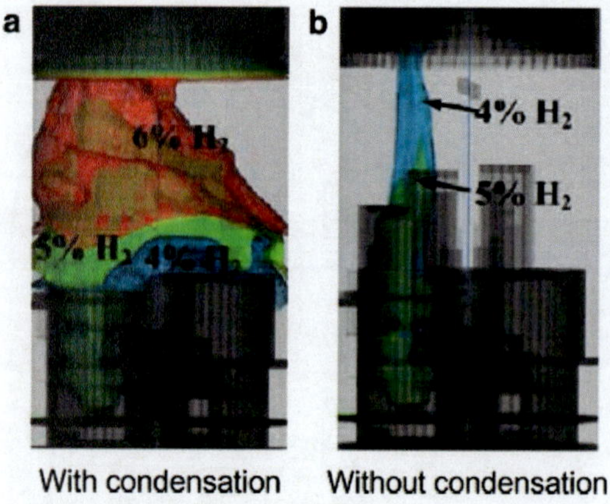

Fig. 5.28 Effect of condensation on hydrogen concentration distribution

spherical dome. The total height is about 60 m, and the diameter is about 38 m. The containment model was established in the cylindrical coordinates. More details about the containment and the corresponding mesh structure can be found in [16].

Figure 5.28 shows the hydrogen concentration inside the containment in both cases with and without condensation. With condensation, local hydrogen concentration is much higher than the case without condensation. It has to be pointed out that the effect of condensation on hydrogen distribution is similar to that of spray. An accurate modeling of condensation and spray is thus important for hydrogen safety. On the other hand, condensation and spray also reduce the total pressure inside containment and that prevents containment overpressure.

In order to mitigate the hydrogen risk during severe accidents, 22 passive automatic recombiners (PARs) of the Siemens type are installed in the containment compartments. A PAR consists of a vertical channel and stack equipped with a catalyst bed in the lower part, as presented in Fig. 5.29. In the case of SAs, the catalyst is in contact with the gas mixture of the containment. Hydrogen molecules coming into contact with the catalyst surface react with oxygen to form steam, as indicated in Fig. 5.30. The reaction heat released at the catalyst surface causes a buoyancy-induced flow accelerating the inflow rate and thereby feeding the catalyst with a large amount of hydrogen that ensures high efficiency of recombination. The buoyancy influenced circulation ensures a continuous gas supply to the PARs [17]. The catalyst sheets can be heated up to 900 K or even higher, so a considerable amount of heat is also transferred from the catalyst to the environment by heat radiation.

For small and medium recombiners of the Siemens type, both height and depth are about 15 cm. The width of the flow channel is less than 1 cm. In PARs, the gas velocity, u, is in the magnitude of 1 m/s. The gas temperature can vary from 300 to 700 K. Assuming the gas in the PAR is dry air, the Reynolds number of the flow between the catalyst sheets is Re $= uL/\upsilon = 2uD/\upsilon = 400 - -1,250$. The flow is considered as a laminar flow in the channel. Here, υ is kinetic viscosity of air.

A two-dimensional PAR model is developed to simulate the flow in the channel, the heat transfer between the catalyst sheet and gas flow, the heat conduction in the catalyst sheet and the chemical reaction on the catalyst surface, as illustrated

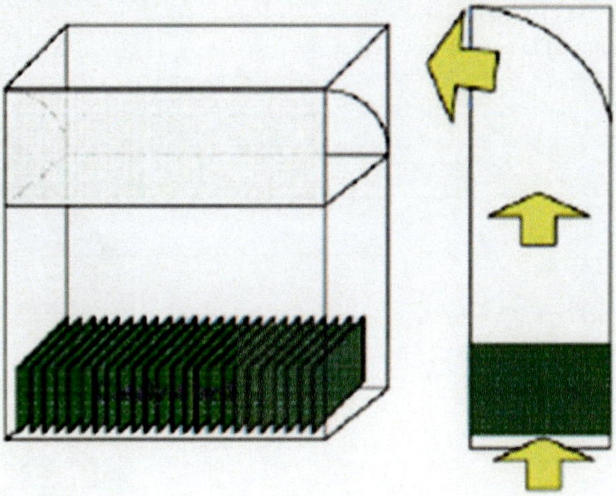

Fig. 5.29 Scheme of a PAR

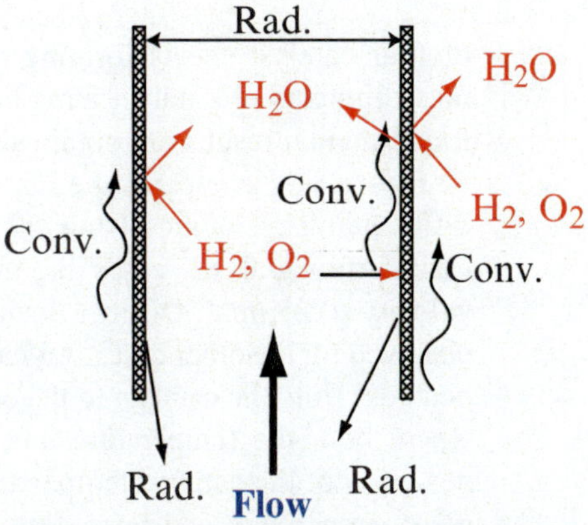

Fig. 5.30 Mechanisms involved in a PAR

in Fig. 5.31. **The variation of flow velocity, temperature, and gas concentration in** the depth direction is then neglected. The continuum equation, Navier–Stokes equation, and energy equation are coupled and solved with the SIMPLER algorithm. The Bossinesq assumption is applied to consider the buoyancy caused by the **heating up. Because the flow is laminar flow, no turbulence model is utilized in this** model. For the radiation heat transfer, the emissivity and absorption ratio of the catalyst sheet are each assumed to be 1. The view factor can be easily obtained for parallel and perpendicular plates in a two-dimensional model, as indicated in Sect. 5.3.2.2. An environment temperature is given at the inlet and outlet of the channel to calculate the radiation heat transfer between the catalyst and the environment.

The REKO-3 experiment results [18] were utilized to validate the model. REKO-3 experiments were conducted to study the process on the catalysts and in the channel between the catalysts. The test section of the REKO-3 facility consists

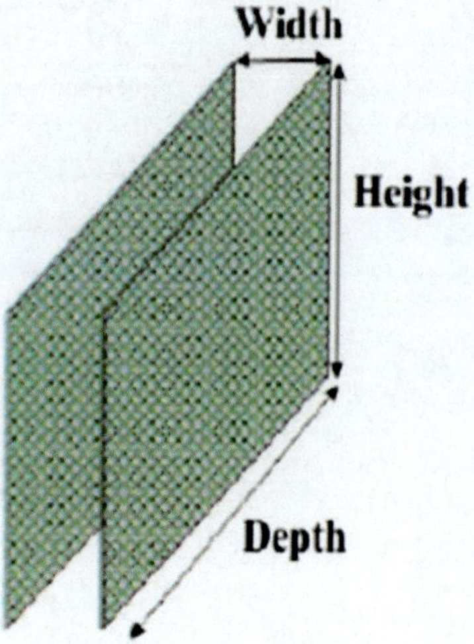

Fig. 5.31 Model for two-dimensional CFD approach

of four catalyst sheets forming three flow channels. The facility allows for the measurements of catalyst temperature and gas concentration at different heights. Experimental results are obtained at different inlet velocities.

Figure 5.32 compares the numerical results and the experimental data at three different inlet velocities, while the hydrogen volume fraction at the inlet is 4% for all cases. Among all the cases, the model gives the best prediction at the lowest inlet velocity (0.25 m/s). Distinct deviation of the catalyst temperature near the inlet is observed for the other cases. An increasing catalyst temperature leads to a significant heat loss from the catalyst to the environment, especially for the inlet neighborhood where both the temperature and view factor to the environment are high. The deviation of the catalyst temperature can be minimized by optimizing the environment temperature and by setting the exact emissivity and absorption ratio of the catalyst material. In the cases where the inlet velocities are 0.5 and 0.8 m/s, an overestimation of recombination by the model is observed. This could be caused by overestimating the chemical reaction rate on the catalyst or by overpredicting the mass transfer to the catalyst. Generally, the model gives satisfactory prediction of the experiment results.

5.3.4 Steam Explosion

Fuel–coolant interaction (FCI) is an important safety issue for both in-vessel and ex-vessel SA mitigation. In the past, many studies were carried out. However, understanding of the mechanisms leading to steam explosion and their prediction is still limited. Steam explosion consists of three phases, i.e., premixing, triggering, and explosion, as illustrated in Fig. 5.33. During the premixing phase, hot melt is divided into small particles and distributed in the coolant (water). The hot particles

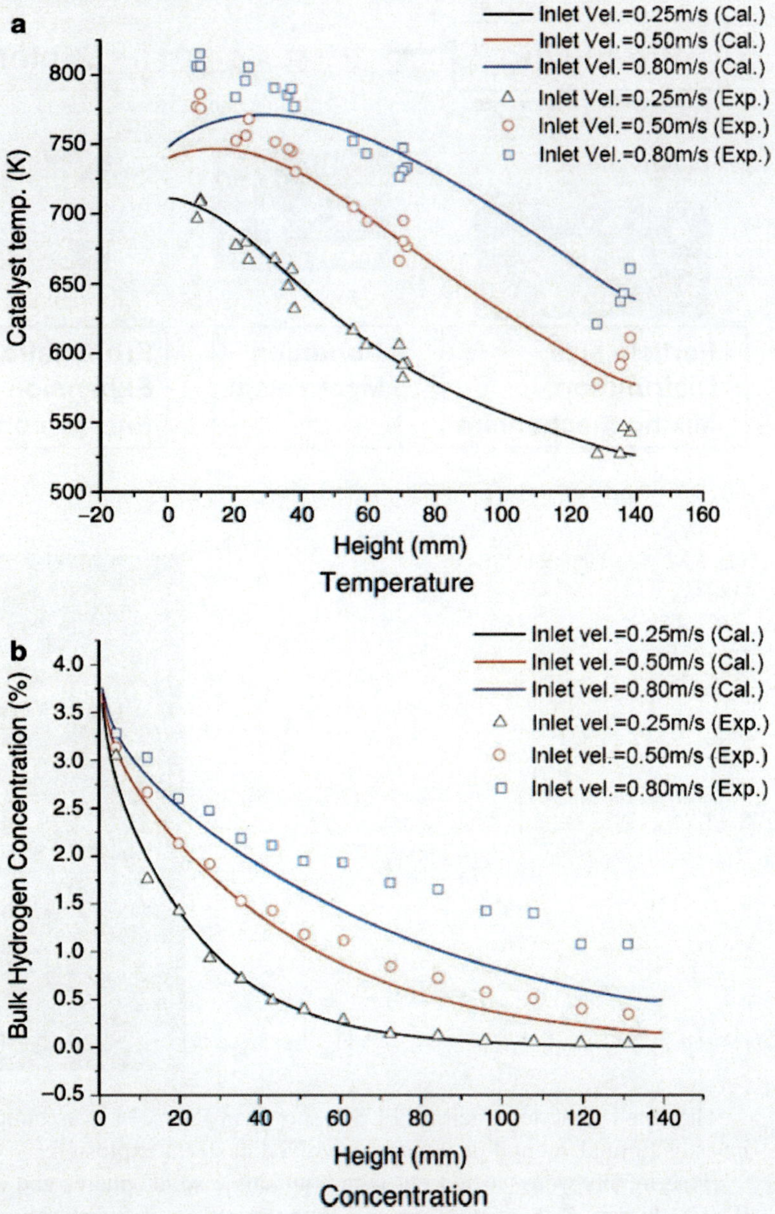

Fig. 5.32 Comparison of numerical results with test data

are surrounded by a stable vapor film. Triggering occurs, when the stable vapor film is disturbed. The direct contact of hot melt particles with liquid water fragments the hot melt into much smaller size particles and produces a pressure wave, which subsequently disturbs the stable vapor film of other melt particles.

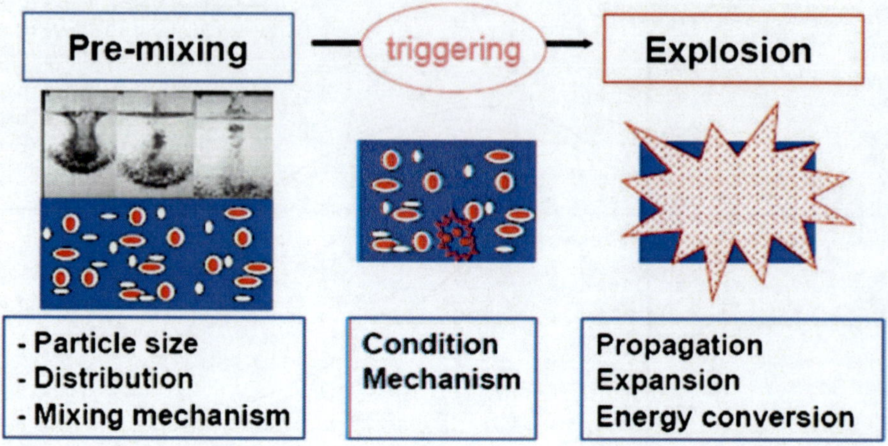

Fig. 5.33 Procedure involved in FCI

Fig. 5.34 Test facility FUSE at SJTU

The small-scale test facility, FUSE, shown in Fig. 5.34, was built at SJTU, to investigate fundamental phenomena involved in steam explosion.

Experiments were carried out with both single solid spheres and molten metal (PbTi). Figure 5.35 shows example images of single solid sphere tests. The measured distance of the hot sphere is shown in Fig. 5.36.

In addition, experiments with molten metal (PbTi) were carried out. Figure 5.37 shows an example of images obtained.

A self-produced CFD code was developed to simulate the premixing and fragmentation process [19]. The main features of the code are:

- Multiphase with heat transfer;
- Thermal nonequilibrium;

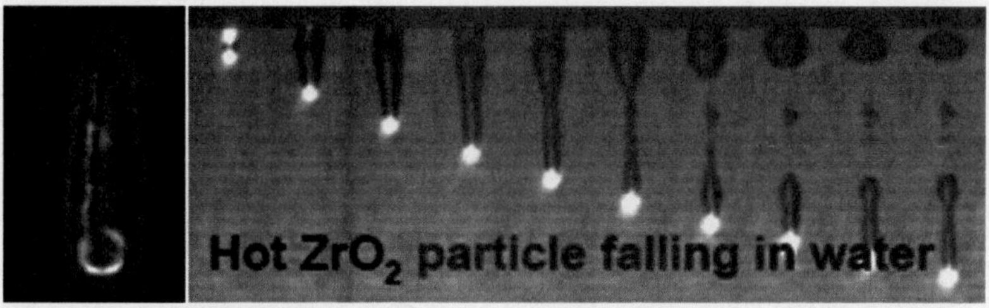

Fig. 5.35 Hot sphere test results

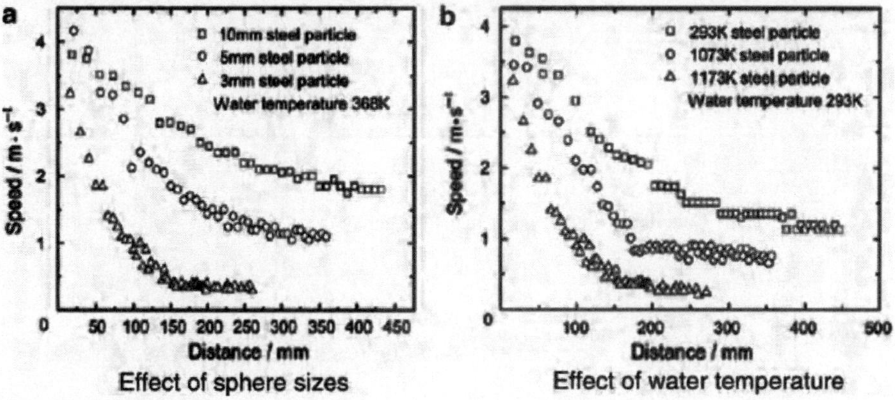

Fig. 5.36 Measured speed of hot spheres

Fig. 5.37 Tests with molten metal

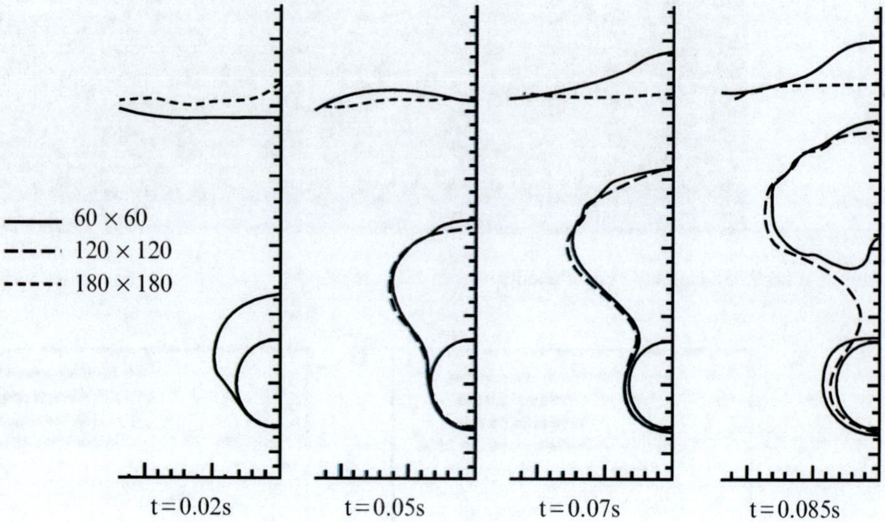

Fig. 5.38 Calculated liquid–vapor interface behavior

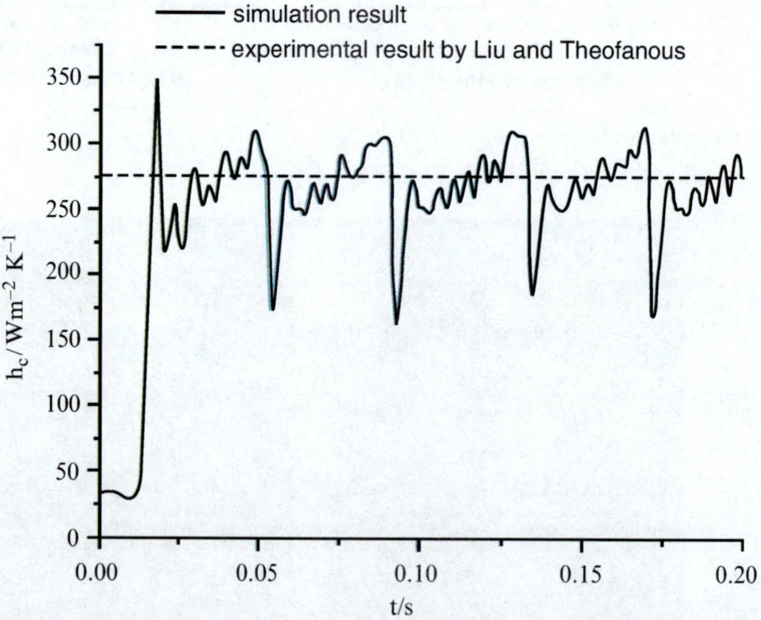

Fig. 5.39 Comparison of the calculated heat transfer coefficient with experimental data

– Nonorthogonal body-fitted coordinates;
– Extension of VOF method to tracking interface with dynamic meshes; and
– Double staggered grids with the SIMPLE method.

Figure 5.38 shows example of results presenting the liquid–vapor interface movement around a hot solid sphere.

Figure 5.39 shows that the calculated heat transfer coefficients are in good agreement with the experimental data considering the uncertainty in experiments. It should be noted that radiation from the sphere enhances the production of vapor that goes into the vapor film and affects the convective contribution of film boiling. For the sphere temperature considered in the experiment, thermal radiation only accounts for about 10–15% of the total heat flux.

5.4 Summary

Safety research is playing an important role in the Chinese LWR technology development. Although SA research is still at the beginning stage in China, its importance is well recognized in the Chinese nuclear community. Items of high priority have been identified, i.e., in-vessel retention, containment-related issues, and hydrogen safety. In the frame of a large-scale national project, SA research will become a key research subject with the goal to establish a SA research platform in China, including test facilities, simulation tools, and human resources. International collaboration is strongly encouraged.

Acknowledgment The author's colleagues of Shanghai Jiao Tong University, especially Prof. Y.H. Yang, contributed significantly to this chapter with their research results and fruitful discussions.

References

1. Cheng X (2007) Studies on advanced water-cooled reactors beyond generation III for power generation. Front Energy Power Eng China 1(2):141–149
2. Cheng X (2009) R&D activities on SCWR in China. 4th international symposium on super-critical water-cooled reactors, Heidelberg, Germany, 8–11 March 2009, Paper No. 53
3. National Development and Reform Commission (2007) Nuclear power medium and long-term program (2005–2020). China (in Chinese)
4. Ouyang Y (2008) Development strategy and process of world nuclear power states and nuclear power development in China. China Nucl Power 1(2):118–125
5. Cummins WE, Corletti MM, Schulz TL (2003) Westinghouse AP1000 advanced passive plant. Proceedings of ICAPP'03, Cordoba, Spain, 4–7 May 2003
6. Kim J, Lee U et al (2008) Spray effect on the behavior of hydrogen during severe accidents by a loss-of-coolant in the APR1400 containment. Int Commun Heat Mass Transf 33:1207–1216
7. Stosic ZV, Brettschuh W, Stoll U (2008) Boiling water reactor with innovative safety concept: the Generation III + SWR-1000. Nucl Eng Des 238(8):1863–1961
8. Tan SS, Leng GJ, Neitzel HJ, Schmidt H, Cheng X (2001) Investigations on the passive containment cooling system of an advanced Chinese PWR. Wissenschaftliche Berichte, FZKA-6622

9. Kanzleiter T (1992) Modellcontainment-Versuche zum Wasserstoffabbau bei auslegungsueberschreitenden Ereignissen, Jahrestagung Kerntechnik, Karlsruhe, Inforum GmbH, 207–210 (in German)
10. Libmann J (1997) Elements of nuclear safety. EdF Science, Les Ulis Cedex, France
11. Ambrosini W, Forgione N, Oriolo F, Vigni P, Anhorn I (1998) Surface characteristics of a water film falling down a flat plate in the laminar-wavy regime. ICMF 98, Lyon, France, 8–12 June 1998
12. Cheng X, Jackson JD, Bazin P et al (2001) Experimental data base for containment thermal-hydraulic analysis. Nucl Eng Des 204:267–284
13. Cheng X, Müller U (1998) Turbulent natural convection coupled with thermal radiation in large vertical channels with asymmetric heating. Int J Heat Mass Transf 41:1681–1692
14. Cui SW, Liu JX, Cheng X (2006) Performance analysis of passive safety containment cooling system. Annual meeting of National Key Laboratory of Bubble Physics & Natural Circulation, Chengdu, China
15. Travis JR, Royl P et al (1998) GASFLOW: a computational fluid dynamics code for gases, aerosols, and combustion, Volume 2, User's Manual, LA-13357-MS, FZKA-5994
16. Xiong JB, Yang YH, Cheng X (2009) CFD application to hydrogen analysis and PAR qualification, Sci Technol Nucl Installations, 2009: Article ID 213981
17. Bachellerie E, Arnould F, Auglaire M et al (2003) Generic approach for designing and implementing a passive autocatalytic recombiner PAR-system in nuclear power plant containments. Nucl Eng Des 221:151–165
18. Reinecke EA, Tragsdorf IM, Gierling K (2004) Studies on innovative hydrogen recombiners as safety devices in the containments of light water reactors. Nucl Eng Des 230:49–59
19. Yuan MH, Yang YH, Li TS, Hu ZH (2008) Numerical simulation of film boiling on a sphere with a volume of fluid interface tracking method. Int J Heat Mass Transf 51(2008):1646–1657
20. State Council (2006) National medium and long-term science and technology plan (2006–2020). State Council, China, in Chinese
21. Li YC, Kuang B, Yang YH et al (2009) Experimental studies on heat removal capacity of IVR-ERVC. The 13th international topical meeting on nuclear reactor thermal-hydraulics (NURETH-13), Kanazawa City, Japan, 27 September–2 October 2009, Paper N13P1030
22. Li TS, Yang YH, Li XY, Hu ZH (2007) Experimental study of high temperature particle dropping in coolant liquid. Nucl Sci Tech 18(4):252–256

Chapter 6
Full MOX Core Design of the Ohma ABWR Nuclear Power Plant

Akira Nishimura

6.1 Introduction

The first advanced boiling water reactors (ABWRs) were constructed in the early 1990s as Kashiwazaki-Kariwa Nuclear Power Plant Nos. 6 and 7 in Japan. Each ABWR generates an electric power of 1,350 MW and features the application of several advanced technologies and components, such as reactor internal pumps, fine motion control drives, and a slightly wider pitch of control rods between fuel assemblies (the N-lattice) [1]. These increase the safety margins in a loss of coolant accident or for fuel thermal stress impact and provide further flexibility in using high burn-up fuel or mixed oxide (MOX) fuel.

Ohma Nuclear Power Plant (NPP) is the world's first full MOX core ABWR plant to apply the above-mentioned features for enhancing plutonium utilization [2, 3]. MOX utilization is one of the basic nuclear energy policies in Japan for ensuring a stable energy supply and saving natural uranium resources. The full MOX application will greatly contribute to the flexible use of plutonium.

6.2 Outline of the Ohma NPP: A Full MOX Core ABWR

Construction of the Ohma NPP by the Electric Power Development Co., Ltd., (J-Power) began in May 2008. The plant is located in the town of Ohma in the northern end of Japan's Honshu Island (Fig. 6.1), and it is very close to the Higashidori NPP and the Rokkasho Nuclear Fuel Recycle Facility. Commercial operation is expected to start in November 2014.

A. Nishimura (✉)
Global Nuclear Fuel-Japan Co., Ltd, Tokyo, Kanayawa, Japan
e-mail: akira.nishimura@gnf.com

T. Saito et al. (eds.), *Advances in Light Water Reactor Technologies*,
DOI 10.1007/978-1-4419-7101-2_6, © Springer Science+Business Media, LLC 2011

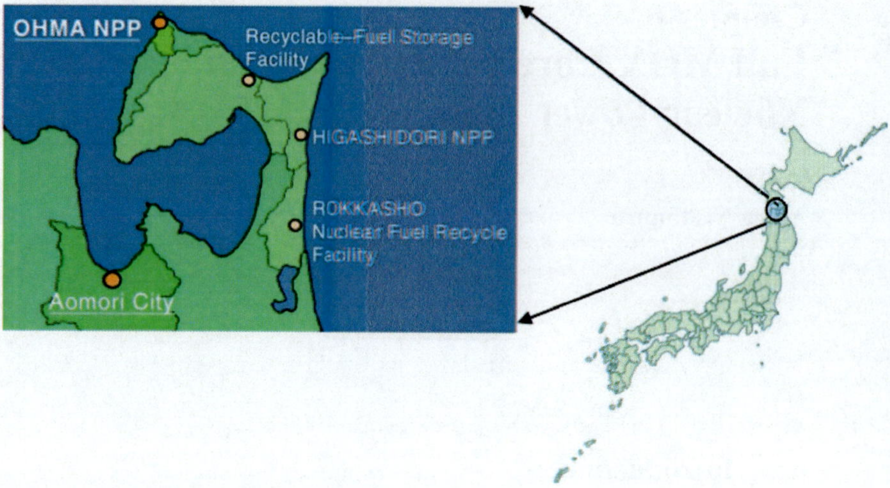

Fig. 6.1 Location of Ohma Nuclear Power Plant

Table 6.1 Basic specifications of the full MOX ABWR

Electrical output	1,383 MWe
Thermal output	3,926 MWt
Number of fuel assemblies	872
Number of control rods	205
Number of MOX fuel assemblies	
Initial core	0–264 bundles (~1/3)
Reload core	Transition to full MOX

The basic specifications of the Ohma full MOX ABWR plant are shown in Table 6.1.

Regarding the initial core, it will be allowed to change from 0 to a 1/3 MOX core for the Establishment Permit Licensing. The number of MOX assemblies will be finally decided based on various practical conditions, including the fabrication capacity, sea transportation capability, available amount of plutonium at the time of fabrication, etc.

6.3 Design of the Full MOX Core ABWR

6.3.1 Design Principles

The design principles of the full MOX ABWR are to apply the proven technology of current BWRs and not to change significantly well-established designs.

But the following improvements will be applied to enhance operation for the full MOX core [1].

- Increase capacity of recirculation pumps
- Increase standby liquid control (SLC) capability
- Partially adopt higher worth control rods by using enriched boron, etc.

6.3.2 Plutonium Characteristics Needing Consideration in the Design: Large Neutron Absorption

Plutonium has slightly different characteristics than uranium. The major differences may be summarized as follows. Due to the large neutron absorption cross-section of plutonium, the void reactivity coefficient will be more negative than for uranium fuel. This more negative value increases core pressure during an overpressure transient. Additionally, the large neutron absorption cross-section will reduce the reactivity worth of neutron absorbers, such as control rods or the SLC. This will reduce the shutdown margin reactivity. But in the MOX core, excess reactivity is small. This can cancel the reduction of reactivity worth of control rods. As is explained later, the actual shutdown margin is enough and almost comparative with that of the uranium core.

6.3.3 Plutonium Characteristics Needing Consideration in the Design: Variation in the Amounts of Pu Isotopes

Plutonium has several isotopes, Pu-239, Pu-240, Pu-241, etc. The amounts of each isotope cannot be decided at a certain fixed value as with uranium fuel. The amounts depend on the reprocessing fuel history, such as the initial enrichment, burn-up, void history and reactor type, etc. When plutonium fuel is used, a certain range of composition variation must be allowed for, especially with respect to safety characteristics. Another point is the decay of Pu-241 with a half-life of 14.4 years. This half-life is relatively short to handle it in the fuel in plants and consideration must be given to the degradation of reactivity or build-up of Am-241 effects.

6.3.4 Plant Design Modifications for the Enhancement of the Ohma NPP

The Ohma NPP design work considered the following modifications for the enhancement of operation flexibility [1].

- *Reinforcement of the shutdown system*. For this, the Ohma design increases the boric acid storage tank capacity in the SLC and provides higher reactivity worth of control rods by using enriched B-10.
- *Extension of the reactivity compensation range*. For this, the Ohma design has enhanced reactor internal pumps to get the maximum core flow rate.
- *Relaxation of higher core overpressure*. For this, the Ohma design increases the total capacity of the safety relief valves (SRVs).

6.4 Core Design

6.4.1 Design Conditions of MOX Fuel

The basic bundle structure uses high burn-up type 8 × 8 fuel that is an identical design to Step II uranium fuel, a proven technology [3, 4]. Fissile plutonium (Pu^f) enrichment in MOX pellets is less than 6% (less than 10% of total plutonium), so in the proven range. Bundle average discharge exposure is less than 40 GWd/t, so in the proven range. Plutonium isotopic composition used is reactor grade. Normally, Pu^f content is less than 77%. As shown in Fig. 6.2, current uranium fuel applies the Step III design. But MOX fuel applies the Step II lattice design and 33 GWd/t average discharged the exposure of Step I fuel, again relying on proven designs of uranium fuel.

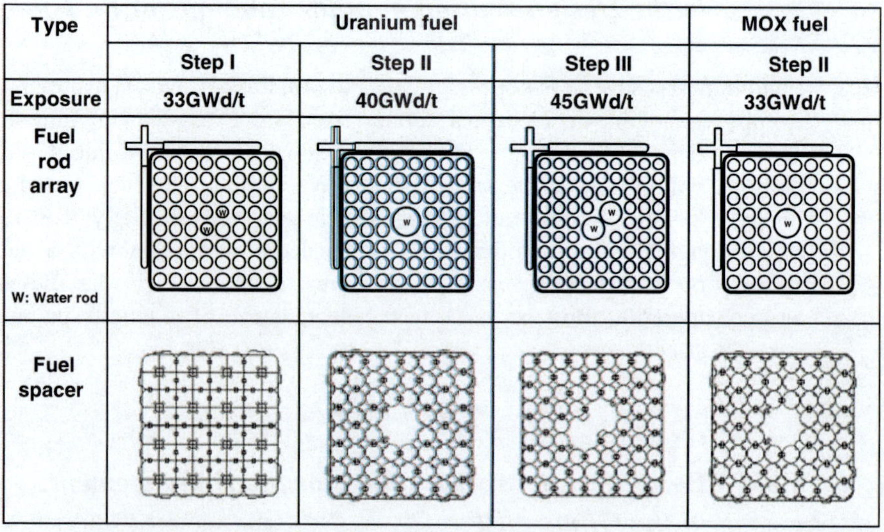

Fig. 6.2 MOX fuel applying proven designs of UO_2 fuel

Table 6.2 MOX fuel and core basic specifications

	MOX	Uranium
Array	8 × 8	9 × 9
U enrichment (%)	1.2	3.8
Pu^f enrichment (%)	2.9[a]	–
Max exposure (GWd/t)	40	55
Number of fuel rods	60	74
Pellet diameter (mm)	10.4	9.6
Pellet material		
MOX	UO_2–PuO_2	UO_2
Gd	UO_2–Gd_2O_3	UO_2–Gd_2O_3
Number of water rods	1	2

[a]Initial Pu^f fraction: 67%

6.4.2 MOX Fuel and Core Basic Specifications

MOX fuel and core basic specifications are shown in Table 6.2. The uranium content for the matrix of MOX fuel rods is 0.2% but gadolinia (Gd_2O_3) rods use enriched uranium without plutonium. Plutonium enrichment is 2.9% for the discharge burn-up of 40 GWd/t. Pellet diameter is 10.4 mm, which is standard value of 8 × 8 fuel design.

6.4.3 MOX Fuel Rod Specifications

MOX fuel rod specifications are basically identical with uranium Step II fuel except for the plenum length. In consideration of the large fission product gas release in plutonium fuel, the plenum length is extended about 15 cm and the active fuel length is shortened the same amount.

6.4.4 MOX Fuel Lattice Design

The MOX fuel lattice also has a few different plutonium enrichments of the MOX rods to flatten the local power distribution. Figure 6.3 shows that the Ohma MOX fuel has four types of MOX enrichment and one type of uranium rod as a Gadolinia-containing rod. Gadolinia is used mainly to compensate for excess reactivity at an early stage of burn-up of fresh fuel.

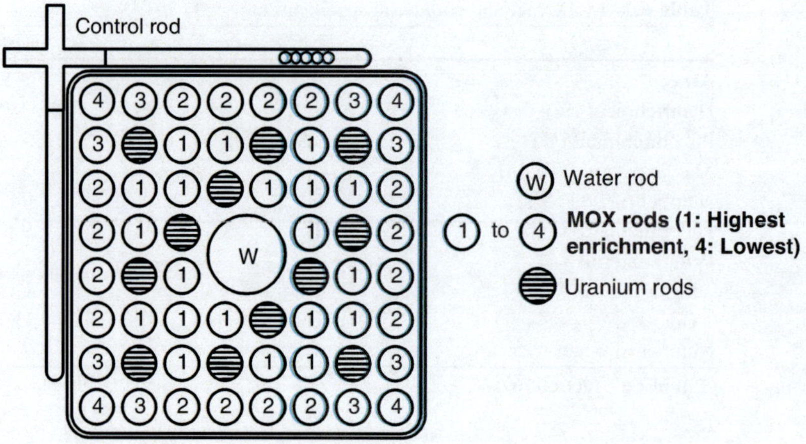

Fig. 6.3 MOX fuel lattice design

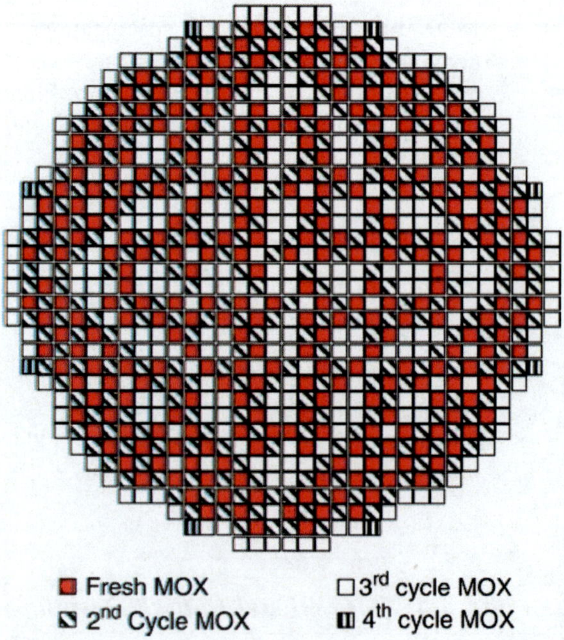

Fig. 6.4 Typical fuel loading pattern of full MOX core

6.4.5 Full MOX Fuel Loading Pattern

Figure 6.4 shows a typical fuel loading pattern of the full MOX core. The MOX fuel assemblies are loaded in a scattered pattern in the core to flatten power distribution, the same as in the uranium core.

6.5 Core Characteristics

6.5.1 *Void Coefficient and Dynamic Parameters*

Figure 6.5 shows void coefficient and dynamic parameters, such as the Doppler coefficient and delayed neutron fraction. Void coefficient is about 20% more negative in the full MOX core than in the uranium core. The Doppler coefficient is almost the same magnitude for various MOX fuel fractions in the core. The delayed neutron fraction is about 20% smaller in the full MOX core than in the uranium core. As mentioned later, when increasing the MOX fuel ratio, these differences in dynamic parameters do not have large impacts on transient or accident behaviors.

6.5.2 *Control Rod Worth in MOX Core*

Maximum control rod worth decreases with increasing MOX fuel ratio but total control rod worth decreases very slightly as shown in Fig. 6.6. These relatively mild impacts on dynamic parameters and control rod worth come from the BWR lattice configuration, which has a water gap between fuel assemblies. The water gap has a large volume of water as neutron moderator. In plutonium fuel, neutron spectrum is harder than in uranium fuel. But the large capacity for moderation by the water gap compensates for and softens the neutron spectrum hardened by plutonium. The control rods are located in the water gaps in a BWR, this means in a highly moderated neutron area.

As shown in Fig. 6.7, the ABWR has wider pitch lattice (called the N-lattice). This provides a large potential for compensation in control rod worth and void coefficient, etc.

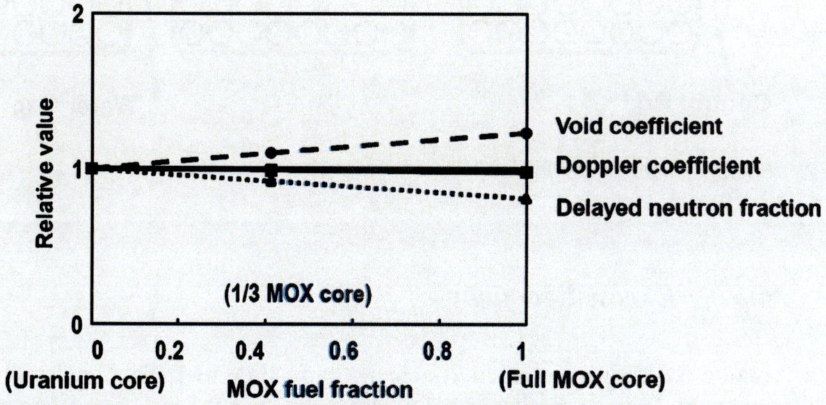

Fig. 6.5 Void coefficient and dynamic parameters

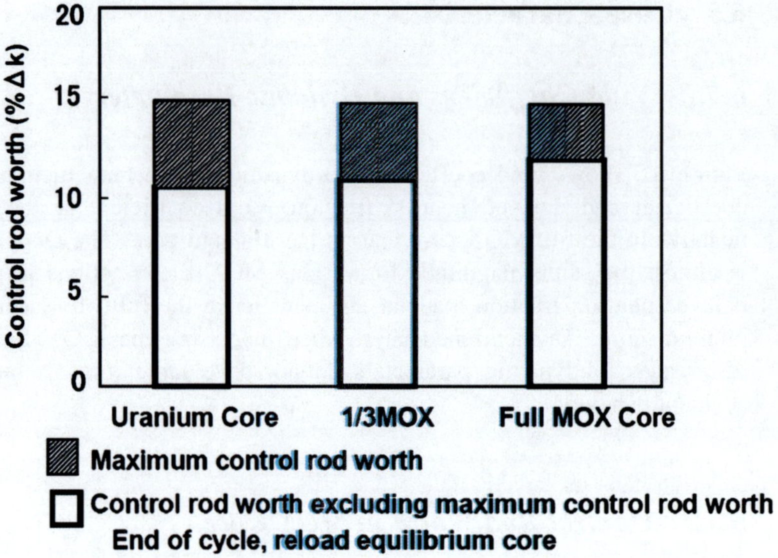

Fig. 6.6 Control rod worth in MOX core

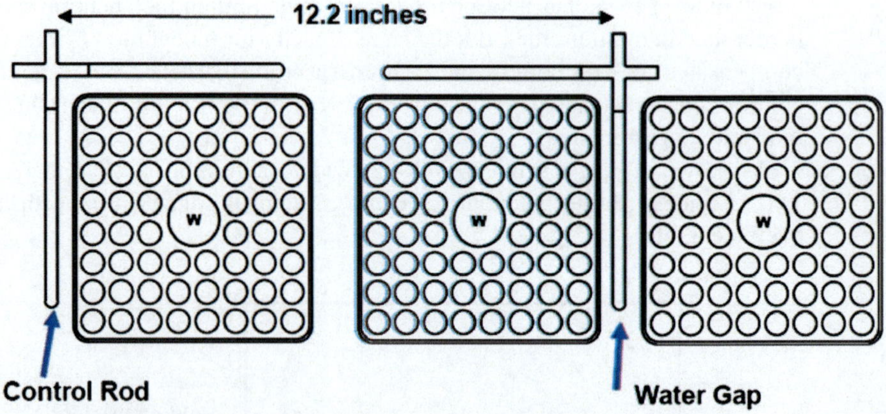

Fig. 6.7 ABWR wider pitch lattice (N-Lattice)

6.5.3 Excess Reactivity

Figure 6.8 shows the reactivity changes in MOX fuel. This moderate change of reactivity leads to smaller radial power peaking in the core and a larger margin to shutdown with compensation for control rod worth reduction in the MOX core.

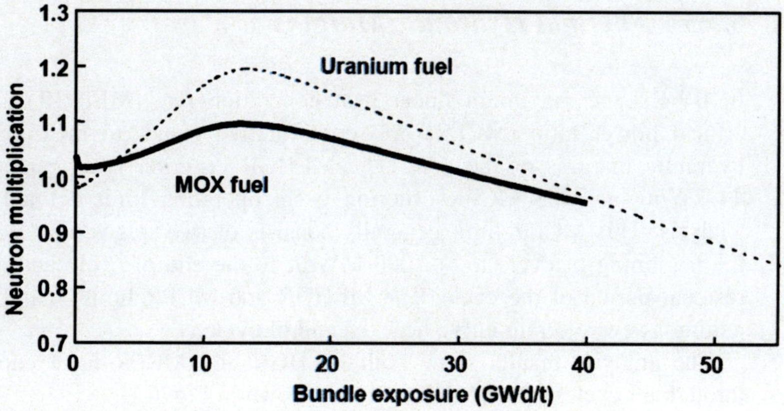

Fig. 6.8 Reactivity changes more moderately in MOX fuel

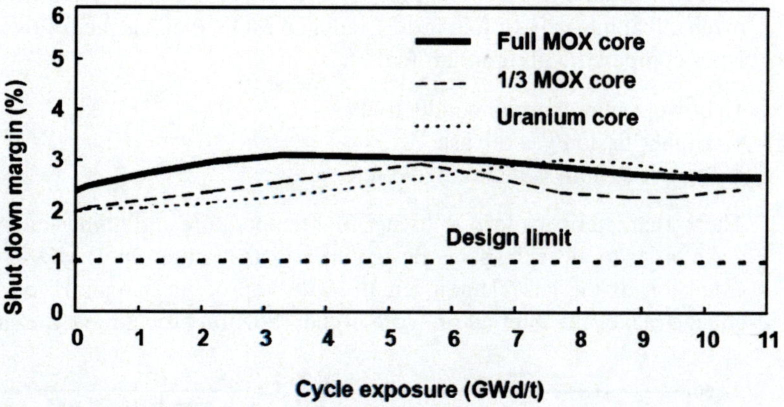

Fig. 6.9 Shutdown margins for different cores

6.5.4 Shutdown Margin

One of the most important design targets in reactor safety is the shutdown margin. The shutdown margin states that core *k*-effective should be less than 0.99 at the cold state, even if a control rod, or a pair of control rods connected to the same control unit, with maximum worth in the core, is withdrawn. The analysis results of shutdown margins show that they are enough during the whole cycle in the MOX cores as shown in Fig. 6.9. This means there is enough compensation for excess reactivity to allow the reduction of the control rod worth in MOX fuel.

6.5.5 Thermal Hydraulic Margins

In BWRs, the maximum linear heat generation rate (MLHGR) and minimum critical power ratio (MCPR) are representative values to measure the thermal hydraulic margins of the core. The MLHGR criterion is an operating limit of 44 kW/m, and the MCPR criterion is an operating limit defined by transient analysis. This MCPR limit generally consists of two stages, one is applied from the beginning of cycle to – 2,000 MWd/t to the end of cycle and another is for residual period of the cycle. The MLHGR and MCPR limits mainly come from getting a severe scram curve near the end of cycle.

The analysis results show both MLHGR and MCPR have enough margins through all cycles in the MOX cores as shown in Fig. 6.10.

6.5.6 Fuel Temperature and Internal Pressure

Fuel temperature and internal pressure are also essential characteristics for thermal mechanical integrity of fuel rods. Irradiated MOX fuel has the following characteristics compared with uranium fuel.

- Lower pellet thermal conductivity
- Higher fission gas release
- Higher helium generation and release

These characteristics lead to higher fuel temperature and higher internal pressure.

In order to meet the same design criteria as uranium fuel, MOX fuel adopts an extension of the gas plenum length at the top of fuel about 15 cm. This design change makes the internal pressure of the MOX fuel rod almost the same as that of

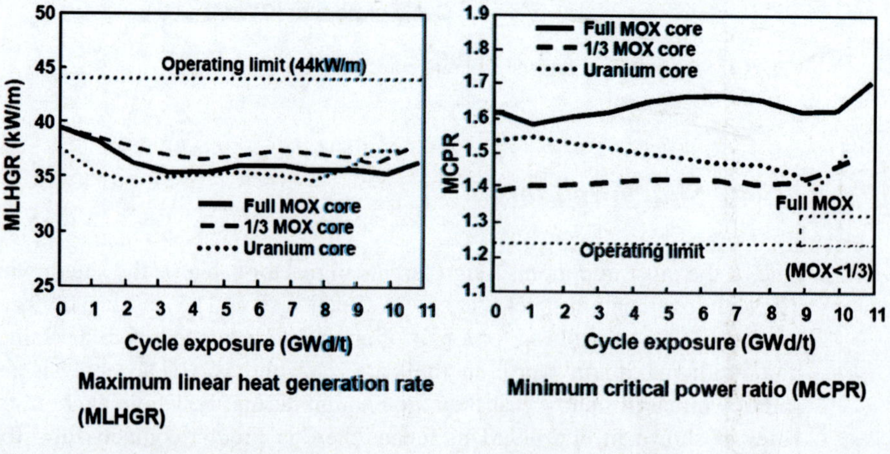

Fig. 6.10 Thermal hydraulic margins

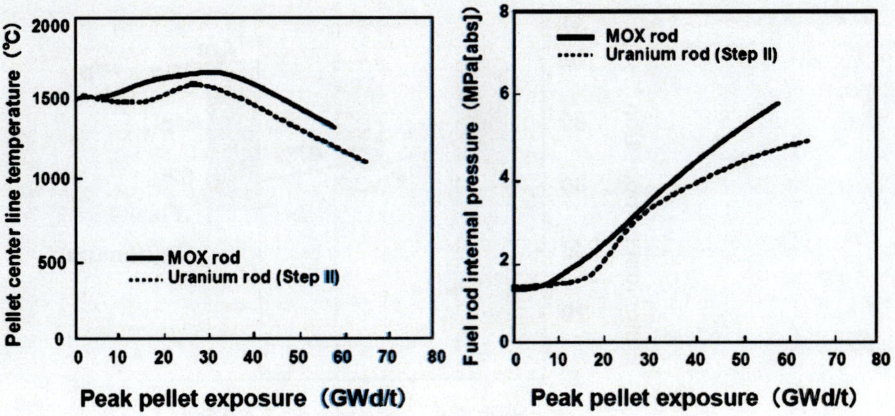

Fig. 6.11 Fuel temperature and internal pressure of MOX fuel

the uranium rod as shown in Fig. 6.11. The pellet centerline temperature is sufficiently lower than the fuel melting point which is approximately 2,600°C at the end of life.

6.5.7 *Initial Plutonium Composition*

Another unique aspect to take into consideration for the design of plutonium fuel is the variation of initial plutonium isotopic composition [5]. The isotopic composition depends on the reprocessed fuel burn-up, initial enrichment and neutron spectrum of the reactor, etc. In addition to that, Pu-241 will decay to Am-241 with a half-life of about 14.4 years. This means that there will be a loss of fissile material according to the timing of fuel loading after reprocessing. The MOX fuel design should compensate for the reactivity loss and the change of void coefficient and Doppler coefficient, etc., by the variation of plutonium composition. Based on a survey of reprocessing plant data, the variation of the initial plutonium composition was assumed to be from 62 to 75% of the Pu^f ratio. Detailed isotopic compositions are shown in Fig. 6.12.

Required bundle average plutonium enrichment and the Pu^f enrichment to get the same reactivity are shown in Fig. 6.13. The deterioration of Pu^f enrichment is shown in Fig. 6.13 according to fuel loading delay.

These variations of the initial isotopic composition of plutonium fuel affect some core characteristics. Figure 6.14 shows calculated thermal hydraulic characteristics for various composition cases. All cases allow operation within very small deviations, and there are sufficient margins to the operation limits of MLHGR and MCPR.

Void coefficient and Doppler coefficient were also analyzed among various plutonium isotopic compositions. Void coefficient and Doppler coefficient vary

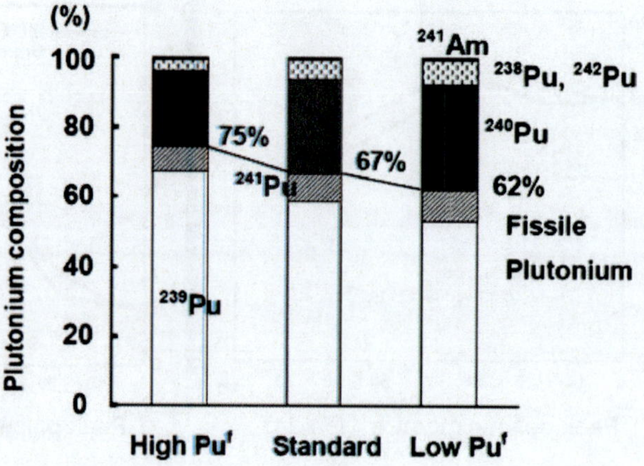

Fig. 6.12 Initial plutonium compositions for analysis

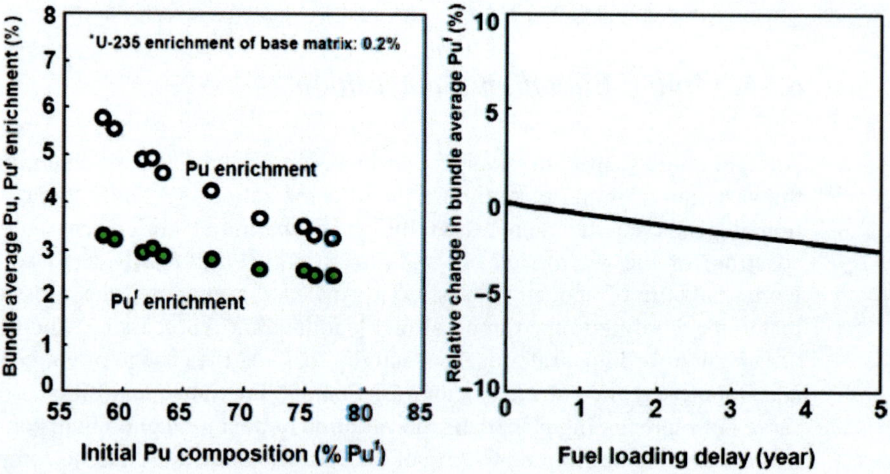

Fig. 6.13 Required bundle average Pu, Pu^f enrichment, and Pu^f enrichment deterioration versus fuel loading delay

according to the increase of initial Pu^f composition. These results show the appropriateness of allowances of safety analysis input used −4% for the void coefficient and +4% for the Doppler coefficient as shown in Fig. 6.15.

Shutdown margin and scram curve were also checked for various plutonium compositions. These parameters can be well controlled and have enough margins to the design target as shown in Fig. 6.16.

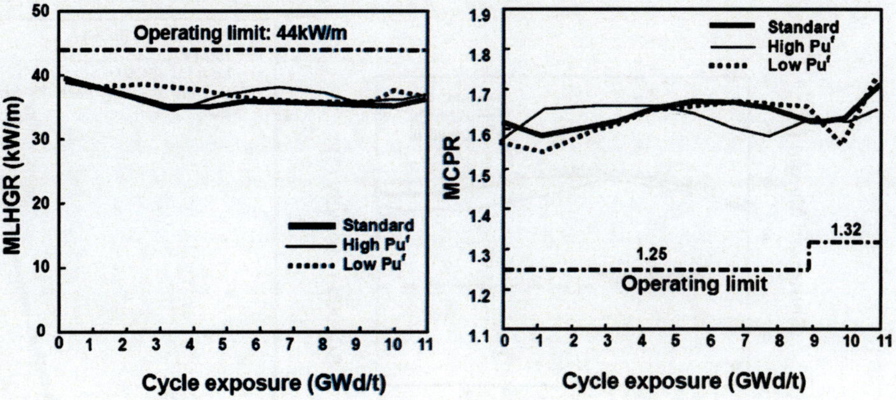

Fig. 6.14 Maximum linear heat generation rate (MLHGR) and minimum critical power ratio (MCPR) for various initial isotopic compositions of Pu

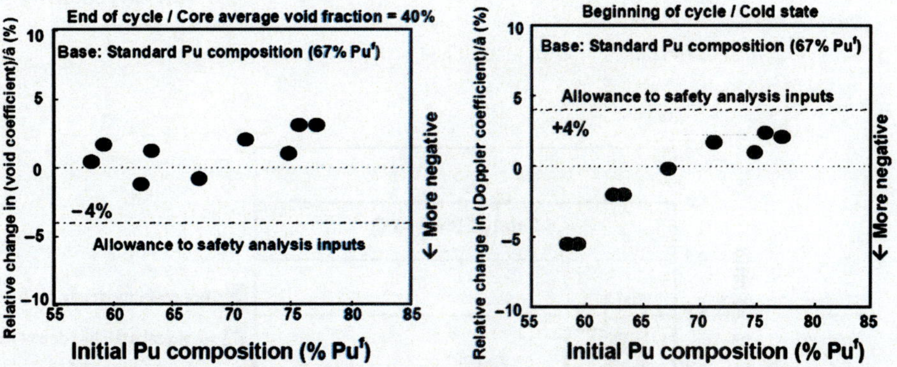

Fig. 6.15 Effect on void coefficient and Doppler coefficient

6.6 Core Dynamics and Safety Analyses

6.6.1 Stability Analysis

Stability analysis is also an important parameter for the safe operation of Nuclear Power Plants. Figure 6.17 shows the decay ratios for core stability and regional stability analysis [4]. In the full MOX core, the decay ratio increases slightly, but it is still under the criterion of 1.0.

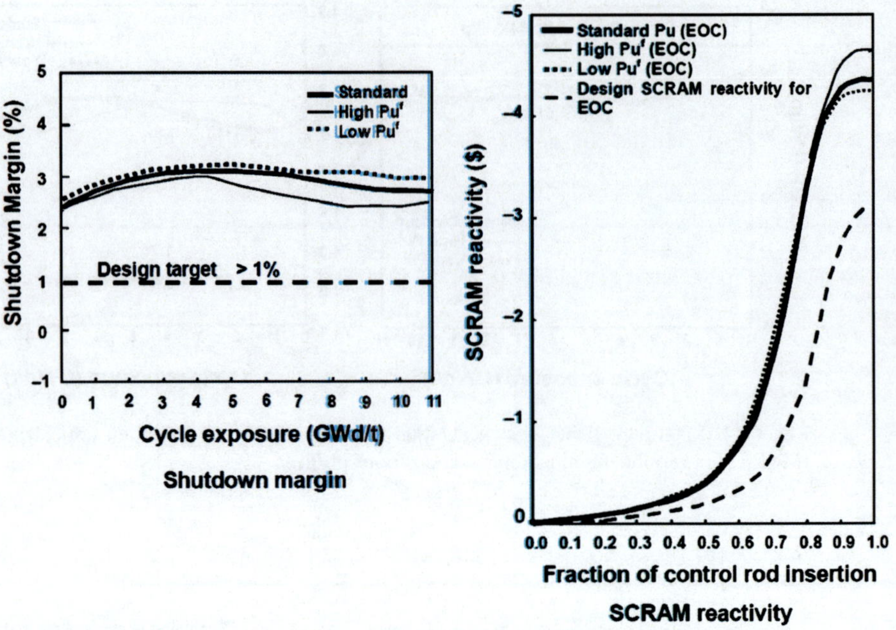

Fig. 6.16 Effect on shutdown margin and scram reactivity

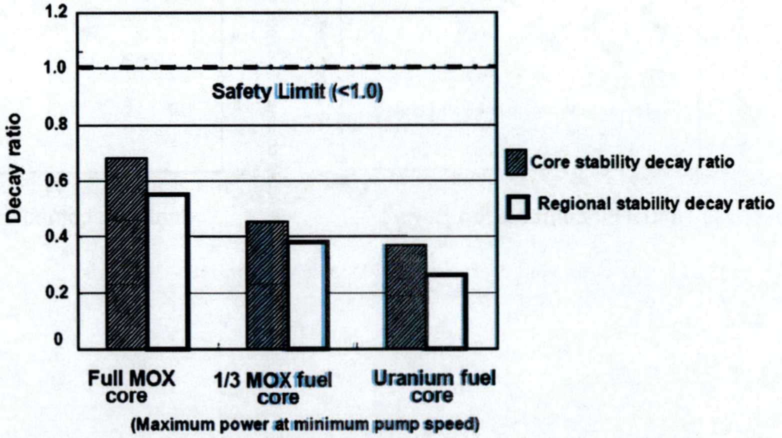

Fig. 6.17 Stability analysis for MOX cores

6.6.2 *Abnormal Transients During Operation*

Abnormal transients during operations were also analyzed for the full MOX core. As shown in Fig. 6.18, the most sever transient is the loss of feedwater heater, and the ΔMCPR is almost the same level as in the uranium core. The load rejection

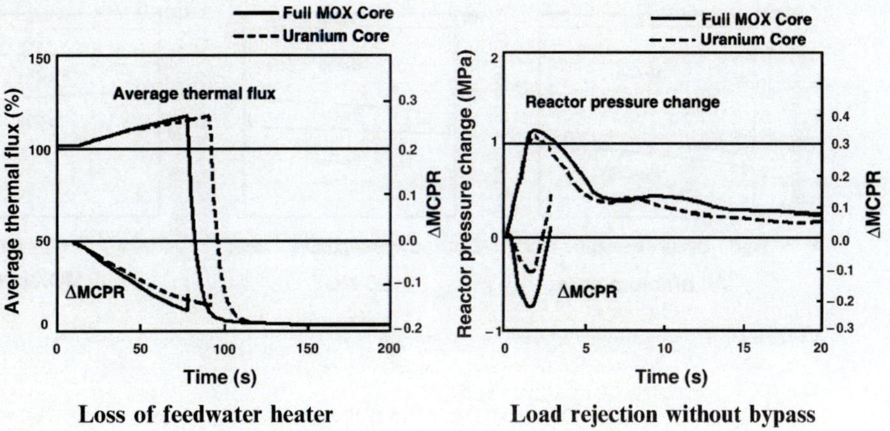

Fig. 6.18 Analyses of abnormal transients during operation

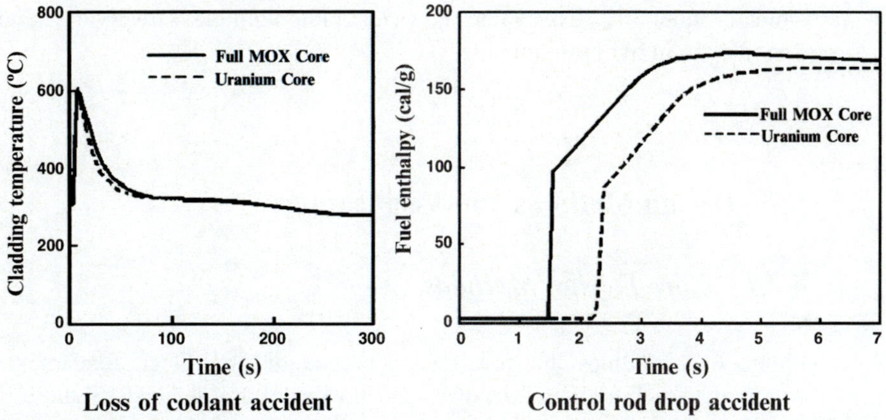

Fig. 6.19 Accident analyses

without bypass has a larger ΔMCPR; this requires a severe MCPR limit during operation. As shown in Fig. 6.18, there are no difficulties to operate the core under the severe limit of the MCPR.

6.6.3 Accident Analyses

Accident analyses also were performed as shown in Fig. 6.19. There is no significant difference in MOX and uranium cores for a loss of coolant accident. A control rod drop accident shows a slightly larger enthalpy for the MOX fuel core. But the difference is small enough to have no impact on radiation dose at the NPP boundary in the event of the control rod drop accident.

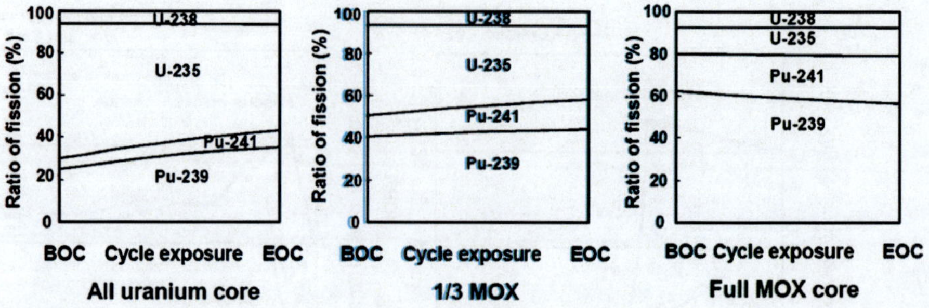

Fig. 6.20 Pu contributes to fission even in the U core

As shown in previous analyses, the impacts on various core characteristics are not significant in the full MOX fuel core. One of the reasons is the almost identical contribution of plutonium to fission even in the uranium core. The left-hand side graph of Fig. 6.20 shows the fission contribution of plutonium in the uranium core which is about 30%. This value shows that plutonium plays a very essential role in fission even in the uranium core.

6.7 Design Methods and Verifications

6.7.1 Core Design Methods

Figure 6.21 outlines the core design procedures [6]. Once fundamental plant conditions, such as thermal power and flow rate, are decided by the utility, the basic specifications, such as number of bundles and lattice configuration, are designed through repeated core simulations considering the burn-up history. Details of lattice specifications, such as enrichment, Pu^f content, and gadolinia parameters, are performed using the basic specifications. Core design features, such as number of reload fuel assemblies, loading patterns, and control rod patterns, are determined under the design limits by using the lattice design results.

HINES, TGBLA, and LANCR are example codes for preparing a lattice design (Fig. 6.22). These design codes generate *k*-infinity, macro cross-sections, delayed neutron fractions, and power distributions in a fuel assembly by using lattice configurations, isotopic composition and temperature, etc., as inputs.

PANACH, LOGOS, and AETNA are example codes for core design analysis (Fig. 6.23). These design codes generate *k*-effective, power distribution, operating parameters, such as MLHGR or MCPR, and burn-up for each fuel assembly, etc., based on the results of lattice design calculations.

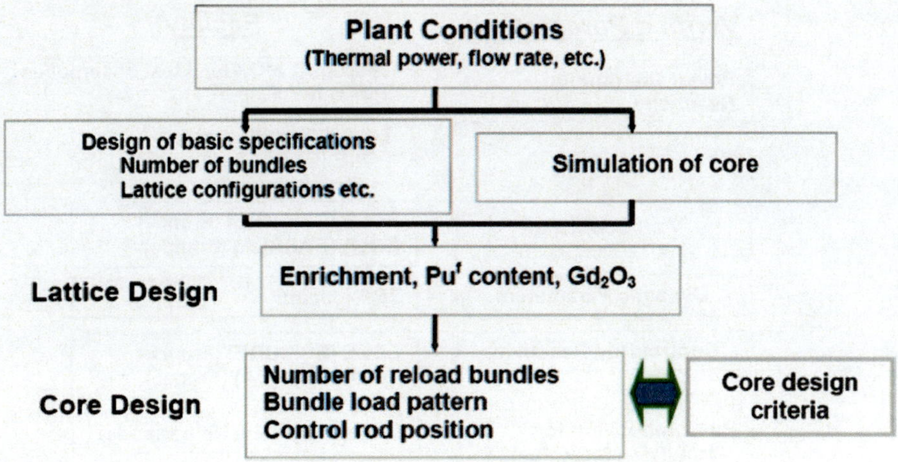

Fig. 6.21 Core design procedures

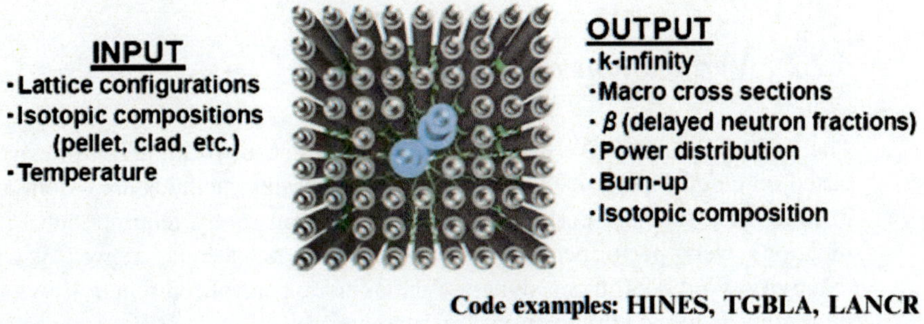

Fig. 6.22 Lattice design methods

Fig. 6.23 Core design methods

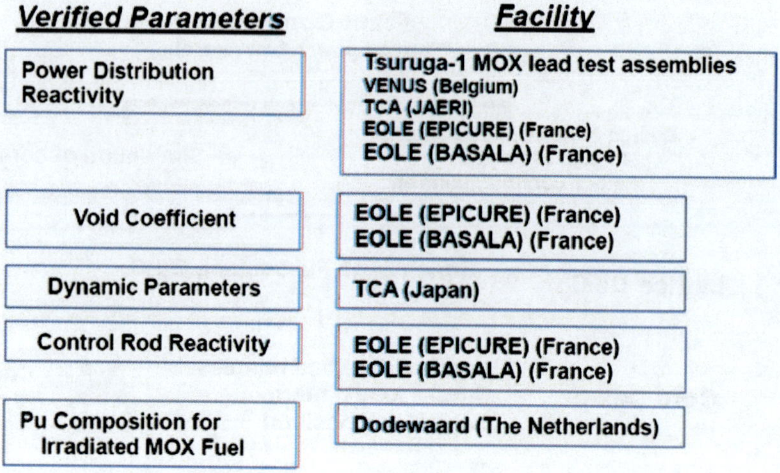

Fig. 6.24 Verifications of design methods

6.7.2 Verifications of Design Methods

The full MOX core ABWR represents the first in the world, and all designs were based on careful analysis. The verifications of design methods are essential points in order to assure the performance and to confirm safety requirements. The verifications were performed for various parameters, such as power distribution, reactivity, void coefficient, dynamic parameters, control rod reactivity worth, and plutonium isotopic composition, for irradiated MOX fuel. These parameters were verified by many experiments as shown in Fig. 6.24 such as for VENUS (Belgium), TCA (JAERI), EOLE (France), and Dodewaard (The Netherlands) as well as the results of MOX lead test assemblies in Tsuruga-1 NPP.

6.7.3 Verification for MOX Lead Test Assembly in Tsuruga-1

One example of verifications for power distribution and reactivity used data obtained from Tsuruga-1 NPP results [7]. In Tsuruga-1, two MOX assemblies, with the design shown in Fig. 6.25, were irradiated for three cycles during 1986–1990. They were 8 × 8 assemblies with two water rods and hollow MOX pellets. Discharged exposure reached 26.4 GWd/t. The assembly design and MOX fuel loading positions are shown in Fig. 6.26.

Figure 6.27 compares axial power shape for MOX and uranium bundles located at symmetrical positions using calculated and measured in-core monitor data. Results show sufficient analytical accuracy for MOX fuel.

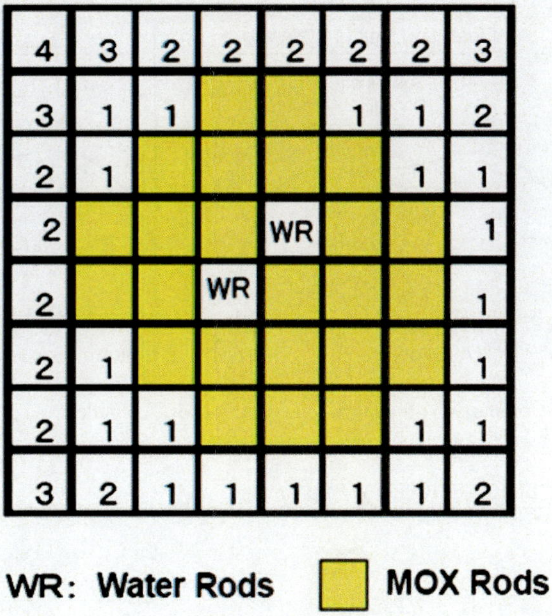

Fig. 6.25 Tsuruga-1 MOX fuel assembly

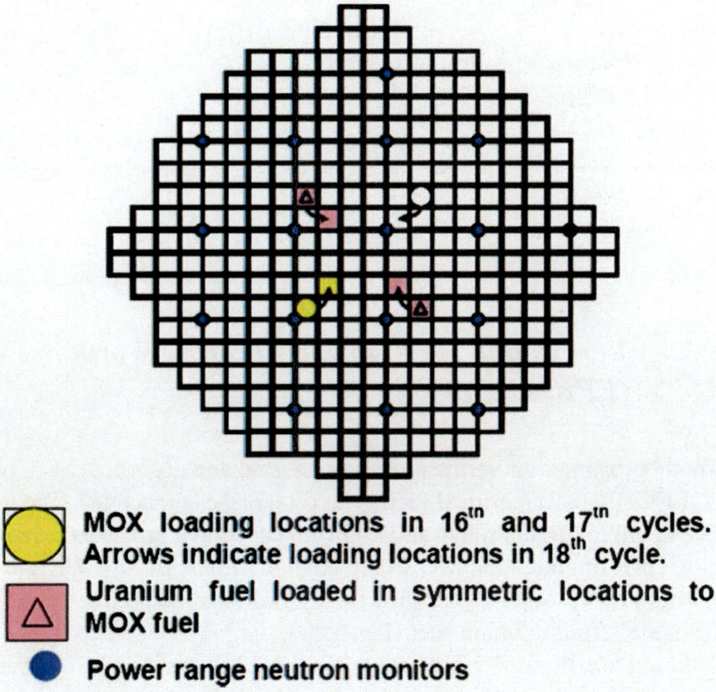

Fig. 6.26 MOX fuel loading position

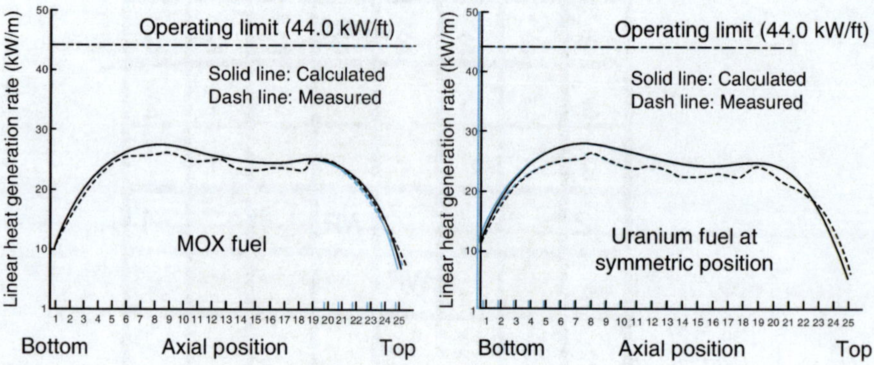

Fig. 6.27 Comparison of axial power shape for MOX bundle

Table 6.3 EOLE (EPICURE) test parameters

Fuel	Test types	Parameters	^{235}U enrichment (%)	Pu enrichment (%)	Pu^f fraction (%)	Fuel rod pitch (cm)	Moderator temperature
Uranium	Base	–	3.7	–	–	1.26	Room temperature
	Void reactivity	Void fraction					
	Absorber reactivity	Pyrex, B_4C					
MOX	Base	–	0.2(Tail U)	7.0	60–70	1.26	Room temperature
	Void reactivity	Void fraction					
	Absorber reactivity	Pyrex, B_4C, UO_2-Gd_2O_3, Hf					

6.7.4 *Verification for Void and Absorber Worth in EOLE (EPICURE)*

Another example of verification was for void and absorber worth performed in the EOLE (EPICURE) critical facility in Cadarache, France [8]. The tests simulated a LWR core for uranium fuel and MOX fuel as shown in Table 6.3. Figure 6.28 shows EPICURE uranium and MOX core configurations. In EPICURE tests, comparison of void and absorber worth shows very good agreement for the MOX fuel and no difference from uranium fuel (Fig. 6.29).

Based on these verification tests, the Japanese Nuclear Safety Commission reviewed and reported that thermal–mechanical and nuclear design methods, and safety analysis methods for uranium fuel were applicable to full MOX fuel core.

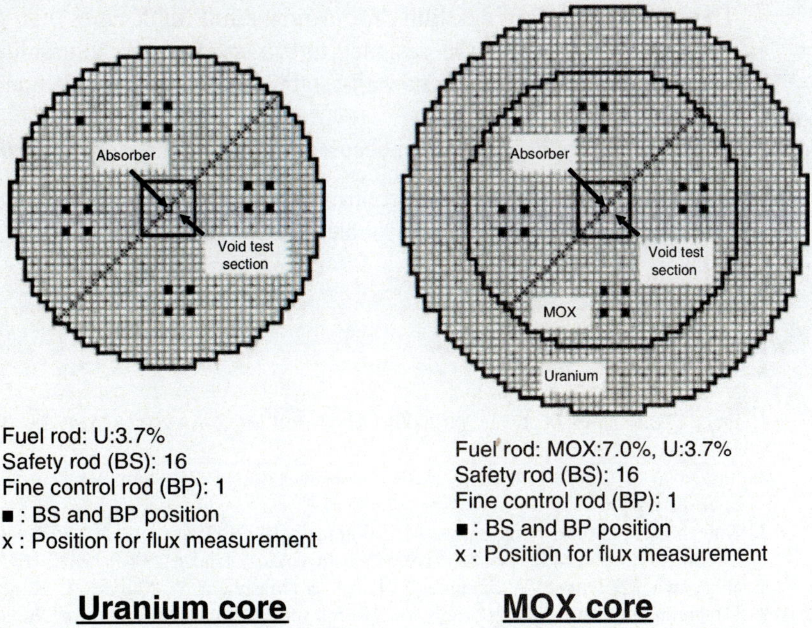

Fig. 6.28 EPICURE core configurations

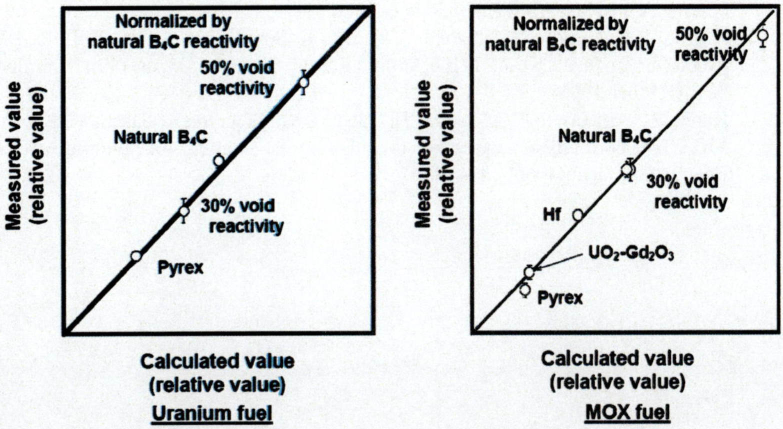

Fig. 6.29 Comparison of void/absorber worth in EPICURE tests

6.8 Summary

The design and safety analyses with verification experiments showed that it was possible to satisfy all design and safety limits and criteria with appropriate margins in the full MOX fuel core the same as in a conventional uranium core. This means:

- The ABWR has large flexibility for using the full MOX core.
- The full MOX core can be designed to be very close to the uranium fuel core without significant changes of well-established technology used and operating parameters.
- Design methods have sufficient accuracy for MOX fuel core analysis.

Full MOX capability will greatly contribute to conserve uranium resources and to enhance the flexibility of getting a stable energy supply.

References

1. Ihara T, Sasagawa M, Iwata Y (2009) OHMA full MOX-ABWR. ICAPP '09, Tokyo, Japan, 10–14 May 2009, Paper 9528
2. Sasagawa M (2006) Full MOX-ABWR core/fuel design. 21st Summer Seminar on Nuclear Fuel, July 2006
3. Kinoshita Y, Hirose T, Sasagawa M, Sakuma T (1999) Design of full MOX core in ABWR. Therm Nucl Power Eng Soc (Karyoku-Genshiryoku-Hatsuden) 50(2):194–201
4. Sasagawa M, Hirose T, Sakuma T, Izutsu S, Masuhara Y, Murata A, Kaneto K (1998) Development of full MOX core in ABWR. 6th International Conference on Nuclear Engineering, ICONE-6473, 10–14 May 1998
5. Hitachi Ltd. (2006) Variation of plutonium isotope composition of MOX fuel in full MOX-ABWR. HLR067 Rev.1
6. Hitachi GE Nuclear Energy Co., Ltd. (2008) Design analysis methods for MOX fuel loaded core in Full MOX-ABWR. HLR-066 Rev.2
7. Meguro S, Tsujimoto Y, Ishizaka Y, Kaneda K, Suzuki T, Nakakita T (1990) BWR-MOX fuel irradiation in Tsuruga unit-1 (II). Atomic Energy Society of Japan 1990 Autumn Meeting, D39, 225, October 1990
8. Kanda K, Yamamoto T, Matsuura H, Tatsumi M, Sakurada K, Sasaki M, Maruyama H (1998) MOX fuel core physics experiments and analysis – aiming for plutonium effective use. J At Energy Soc Japan 40(11):834–854

Chapter 7
CFD Analysis Applications in BWR Reactor System Design

Yuichiro Yoshimoto and Shiro Takahashi

Computational fluid dynamics (CFD) analysis has been used to evaluate phenomena related to the flow since the late twentieth century. Here, through some examples of its applications, the roles of CFD analysis in the actual design process are shown. The first example is an application to the design improvement for flow stabilization at a cross branch pipe in the recirculation loop of the jet pump-type BWR. The second example is an application to evaluations of the ABWR lower plenum flow characteristics and FIV stresses. The third example is an application to the development of a thicker reactor internal pump nozzle for seismic performance improvement. All of these applications were confirmed by tests before being applied to the design of actual reactor structures.

7.1 CFD Analysis Application to BWR-5 Recirculation System

The first example is a design improvement for flow stabilization at cross branch piping in the recirculation loop of a jet pump-type BWR. Although this engineering work was done in the 1980s, it remains as a good example for a practice exercise in present day CFD tool applications.

Some jet pump-type BWR plants have as many as 20 jet pumps in the RPV (reactor pressure vessel) system. This means ten jet pumps for each recirculation loop and five riser pipes from the header in a recirculation loop. This, in turn, means that there is a cross branch pipe at the center of the header. The example in this section deals with one phenomenon at this cross branch pipe.

Y. Yoshimoto (✉)
Hitachi-GE Nuclear Energy, Ltd, Tokyo, Japan
e-mail: yuichiro.yoshimoto.dv@hitachi.com

S. Takahashi
Hitachi, Ltd, Tokyo, Japan

T. Saito et al. (eds.), *Advances in Light Water Reactor Technologies*,
DOI 10.1007/978-1-4419-7101-2_7,

The stabilizing structure was investigated analytically and experimentally for the phenomenon whereby the flow rate of a loop with a cross pipe switches nonperiodically between a high level and a low level under certain flow conditions [1]. Numerical analysis showed that a pair of vortices appeared downstream from where the axes of the main pipe and the side branches crossed. As the upstream vortex grew larger, it drifted further upstream and resulted in swirls with a high pressure loss. Conversely, when the downstream vortex grew larger, the flow became nonswirling with a low pressure loss.

7.1.1 Introduction

One pipe element used for plant piping is the cross branch pipe. In piping systems with a cross branch pipe, in which side branches are connected to the main pipe at right angles and the axes of the side branches are at supplementary radial angles, the flow rate in each branch pipe and the entire system changes nonperiodically between high and low levels with certain flow condition change such as the flow distribution ratio to the branch pipe (Figs. 7.1 and 7.2).

Miura et al. [2] indicate that the presence of a swirling flow with vortices through the right and left branch pipes of a cross pipe is the cause for the changing flow rate phenomenon.

Fig. 7.1 Flow condition at cross pipe in H-pattern [$Q_0 = 0.121\ \mathrm{m^3/s}$] (Taken from [1] and used with permission from JSME)

Fig. 7.2 Flow condition at cross pipe in *L*-pattern [$Q_0 = 0.118\ \mathrm{m^3/s}$] (Taken from [1] and used with permission from JSME)

The mechanism and stability of this swirling flow that appears and disappears intermittently are closely interrelated with the pipe flow distribution ratio, branch region and neighboring flow channel structure, and flow conditions (drifting flow, pulsating flow, turbulence factor, etc.). Miura et al. [3] describe the results of an experimental study using an air flow test rig to explain the factors governing this flow alternation phenomenon including the possible effects of the branch flow distribution ratio and piping structural factors on the phenomenon.

This chapter describes the results of the CFD analysis and tests with a water flow test rig to select the cross pipe with a structure that does not generate the flow alternation phenomenon.

7.1.2 Symbols

d: Pipe diameter
H, L: High and low flow conditions
p: Pressure
Q_0: Main pipe flow rate
t: Time
u, v, w: x-, y-, and z-axial flow velocities at the central plane of the cross pipe
δ: Coefficient [as defined in (7.4) and (7.5)]
ν: Kinematic viscosity

7.1.3 *Analytical Study of Flow Stabilization*

7.1.3.1 Flow Conditions at Cross Branch Pipe Region

The results of the study on stabilization are summarized below based on the data obtained by the tests and analyses in Refs. [1] and [3].

1. When the main flow reaches the neck of the reducer after passing through the right and left cross branches, it starts departing from the pipe wall, which generates two vortices swirling in opposite directions (Fig. 7.3, vortices A and B).
2. When vortex A stabilizes, an *H*-pattern is observed.
3. When vortex B grows larger, extending to the inside of the branch pipes, an *L*-pattern is observed (Fig. 7.2).

7.1.3.2 Countermeasures for Flow Stabilization

To stabilize the flow in the nonswirling condition (*H*-pattern) with minimum loss at the branched pipe region, the following three methods were considered.

Method a: Vortex B must be stabilized. A straightening vanestraightening vane could be installed to prevent the generation of vortex vortex B inside the branched pipe region.

Method b: Vortex A must be stabilized. A deflector vanedeflector vane could be installed to prevent the generation of vortex A at the branched region of the cross pipe.

Method c: Generation of both vortices A and B must be prevented. Vortex A could be stabilized by eliminating the expanding flow region downstream from the branched region.

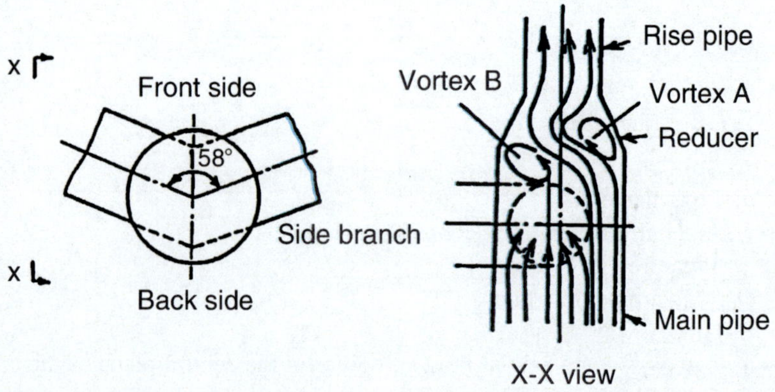

Fig. 7.3 Summary of cross branch pipe flowing conditions (Taken from [1] and used with permission from JSME)

7.1.3.3 Analytical Method

When analyzing the flow inside the branch pipe region, three-dimensional analysis is required in principle, but in the 1980s, it was impractical due to the considerable amount of necessary calculations. Therefore, attention was focused on the fact that when the flow at the main pipe inlet was symmetric at the border of the central cross section (Fig. 7.4 shows the analytical plane, [1]), the flows inside the branch pipes were also symmetric. In the central cross section, there was no flow perpendicular to the plane so the flow could be approximated as two dimensional. Because of the possible transfer (a three-dimensional effect) of momentum and mass at the branched region leading to the header pipe, the following basic equation was derived and analysis was made to take advantage of the three-dimensional effect in the branched pipe region.

Continuous equation:

$$\frac{\partial u}{\partial x} + \frac{\partial v}{\partial y} = -\delta \frac{\partial w}{\partial z} \tag{7.1}$$

Momentum equation:

$$\frac{\partial u}{\partial t} + \frac{\partial uu}{\partial x} + \frac{\partial vu}{\partial y} + \frac{\partial p}{\partial x} - v\Delta^2 u = -\delta \frac{\partial wu}{\partial z} \tag{7.2}$$

$$\frac{\partial v}{\partial t} + \frac{\partial uv}{\partial x} + \frac{\partial vv}{\partial y} + \frac{\partial p}{\partial y} - v\Delta^2 v = -\delta \frac{\partial wv}{\partial z} \tag{7.3}$$

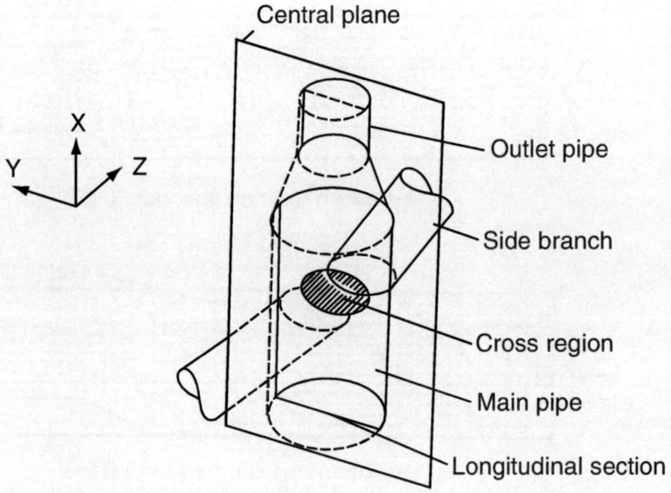

Fig. 7.4 Analytical plane (Taken from [1] and used with permission from JSME)

Where

$$\delta = 1.0 \text{ (branched region)} \quad (7.4)$$

$$\delta = 0.0 \text{ (regions other than above)} \quad (7.5)$$

7.1.3.4 Analytical Results

The analytical results of the flow velocity distributions in the *H*- and *L*-patterns of the standard cross branch pipe with a reducer are shown in Fig. 7.5. The analysis simulated both patterns by changing the branch flow ratio (flow ratio to central branch pipe). It was then confirmed that the analytical method enabled the examination of vortex formation status in the cross branch pipe.

In the above stabilization methods, it was clear that the straightening vane (method a) prevented vortex B (swirl flow). Then, methods b and c were examined here. From Fig. 7.6a, it was seen that the *H*-pattern was maintained with vortex A being stabilized because the stagnant portion of the flow was kept at a constant position by the diverter plane at the side wall of main pipe.

Figure 7.6b shows results for method c, which focused on preventing the formation of vortices A and B. Installing a branched pipe immediately before the main flow outlet in the center eliminated the expanding region of the flow at the back of the branched region. As a result, vortices A and B could be prevented and the flow was stabilized in the *H*-pattern nonswirling flow state. Analysis proved that both methods b and c were realistic from a structural viewpoint.

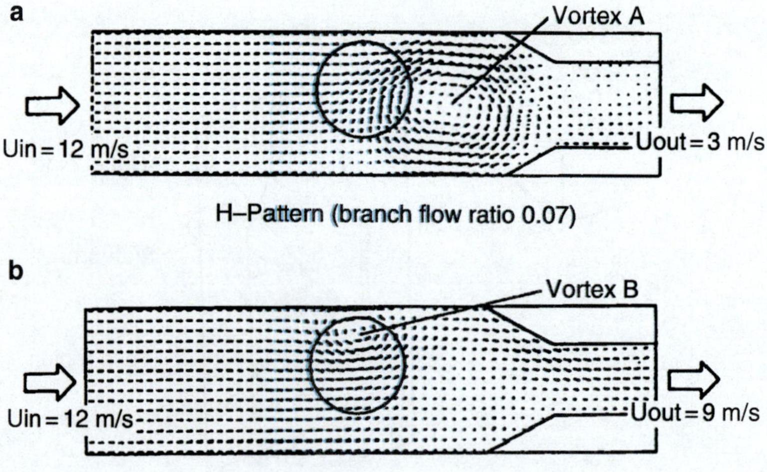

H–Pattern (branch flow ratio 0.07)

L–Pattern (branch flow ratio 0.2)

Fig. 7.5 Analytical results of standard cross branch pipe (Taken from [1] and used with permission from JSME)

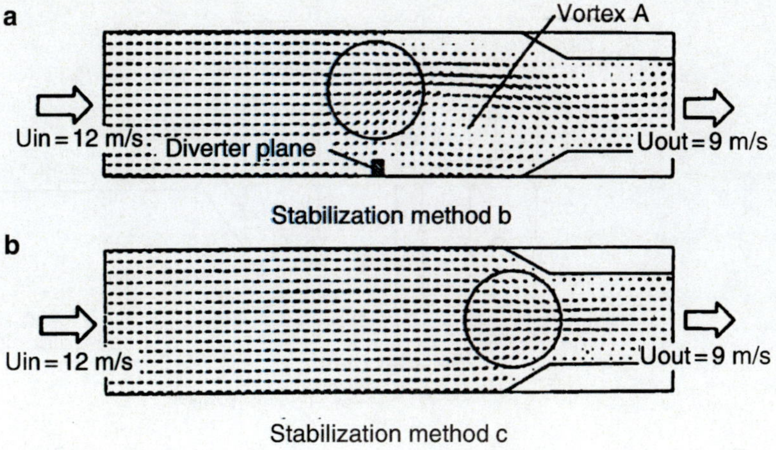

Fig. 7.6 Analytical results of stabilization method (Taken from [1] and used with permission from JSME)

7.1.4 Verification of Flow Stabilization by Tests

7.1.4.1 Test Rig and Method

Figure 7.7 shows a schematic flow diagram of the test rig. The cross pipe and branched pipe were made of transparent acryl through which the water flow could be seen. The test results of the improved cross branch pipe for method c are shown here. Method c gave the lowest pressure loss at the branch regions among the above three stabilization countermeasures.

The flow alternation phenomenon for the hetero-diameter T-type cross branch pipe was observed with the normal shape shown in Fig. 7.8a, while the flow stabilization effect was verified using the hetero-diameter T-type cross branch pipe with a sleeve shown in Fig. 7.6b. The latter cross branch pipe was made by simply adding a sleeve to the normal shape shown in Fig. 7.8a.

It was found in the preliminary test (Fig. 7.9) that if a 1/4 d_0 size (d_0: main pipe inside diameter) diverter was provided at the $1.5d_0$ position in the upper stream of the cross branch pipe, the formation frequency of the *H*- and *L*-patterns changed as the installation angle changed. This was attributed to the fact that by positioning this diverter at $1.5d_0$ in the upper stream of the branched region, the main stream was deflected according to the angle of the diverter in the branched region. In this way, the *L*-pattern was created with the 180° direction diverter and the *H*-pattern was created with the 0° direction diverter.

To examine the possible impact of this drifting flow on the flow alternation pattern, the effect on the vortex formation status by the flow-velocity distribution at the main flow inlet was assessed by the analysis; these results are shown in Fig. 7.10. Analysis

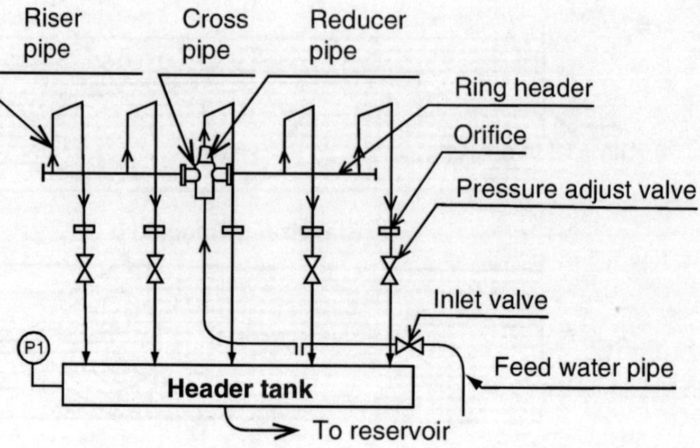

Fig. 7.7 Schematic flow diagram of the test rig (Taken from [1] and used with permission from JSME)

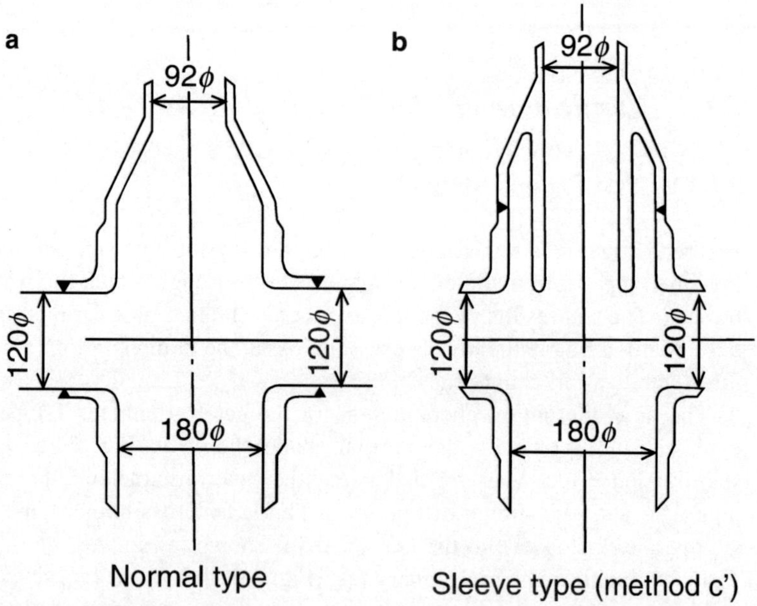

Fig. 7.8 Tested cross branch pipes (Taken from [1] and used with permission from JSME)

confirmed that the vortex B (*L*-pattern) was created by the flow drifting in the 180° direction. But vortex A (*H*-pattern) was created by the flow drifting in the 0° direction.

By adding this upper stream effect to the cross branch pipe by using the diverter, the flow stabilization effect of the hetero-diameter T-type cross branch pipe was confirmed for the flow fluctuation in the upper stream.

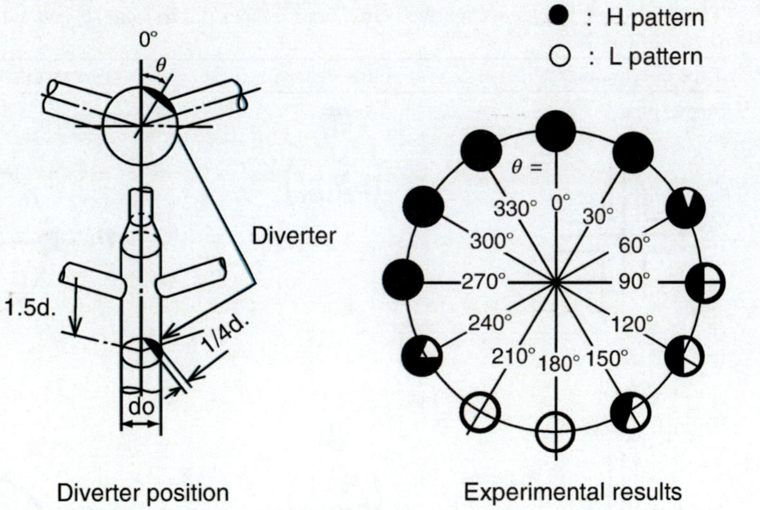

Fig. 7.9 Impact of drifting flow in upper stream of cross branch pipe (Taken from [1] and used with permission from JSME)

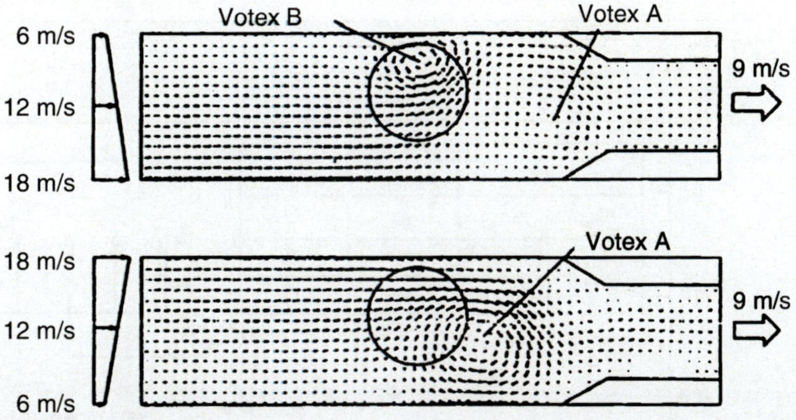

Fig. 7.10 Analytical results of drifting flow impact

7.1.4.2 Test Results

Table 7.1 summarizes the test results for the improved cross branch pipes. Both the sleeved hetero-diameter T-type cross branch pipe and the normal hetero-diameter T-type cross branch pipe had stabilized flow in the H-pattern without or with a diverter, regardless of the installation angle.

Table 7.1 Test results of improved cross branch pipes (taken from [1] and used with permission from JSME)

<table>
<tr><th>Type of cross pipe</th><th>Flow pattern</th><th>Test results with diverter</th></tr>
<tr><td>Sleeve type
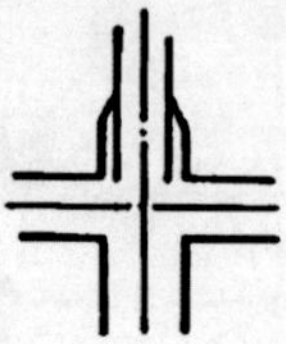</td><td>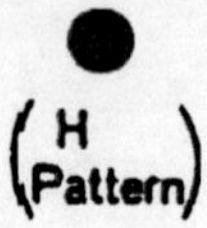
</td><td>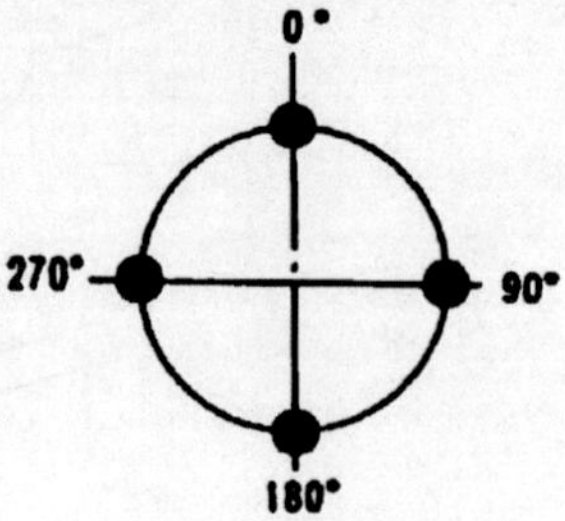
</td></tr>
<tr><td>Normal
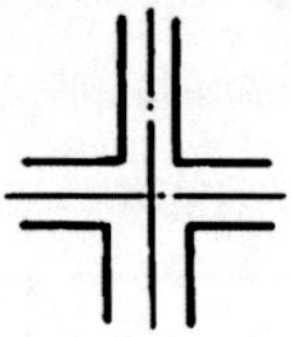</td><td>
</td><td>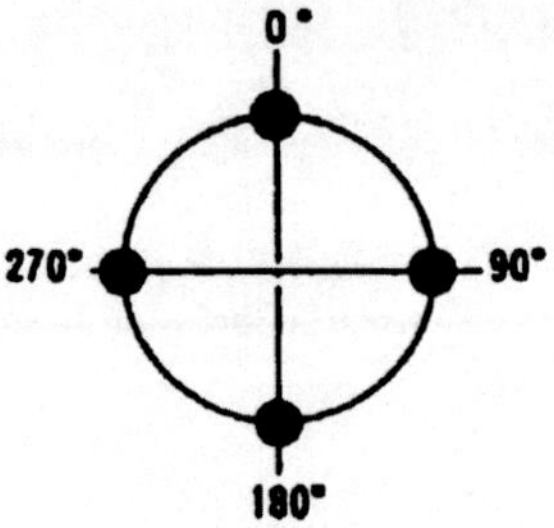
</td></tr>
</table>

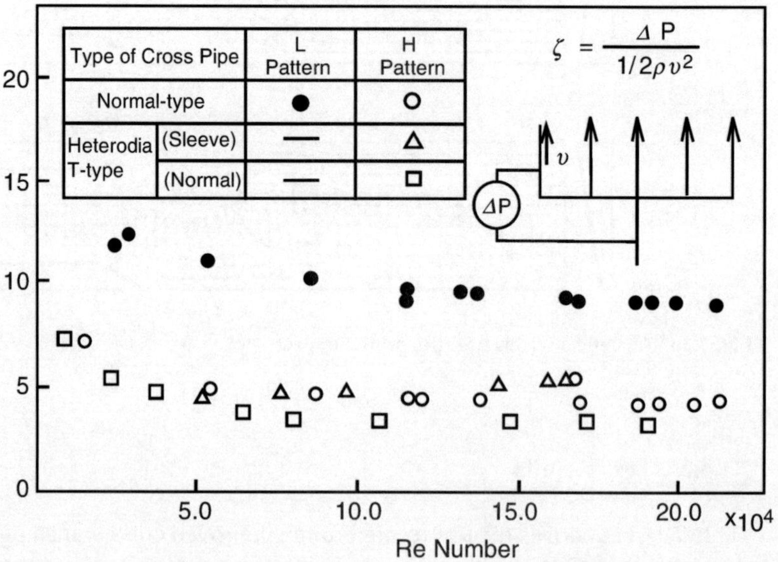

Fig. 7.11 Pressure loss coefficients of various cross branch pipes (branch flow ratio 0.2) (Taken from [1] and used with permission from JSME)

Further, a check was made without the diverters to see if the flow rate alternation phenomenon occurred; this was done by changing the main pipe flow rate and branch flow ratio (center pipe flow rate/main pipe flow rate). As shown in Fig. 7.11, the improved cross branch pipes had nearly the same pressure loss coefficient as for the *H*-pattern of the normal cross pipe and flow was stabilized as the *H*-pattern became independent of the branch flow ratio. It was therefore confirmed that the improved cross branch pipes showed stabilized characteristics with small pressure loss against upper stream flow fluctuation, changing flow rate, and branch flow ratio.

7.1.5 Conclusions

Based on these analytical results and test results, the hetero-diameter T-type design has been applied to cross branch pipes in jet pump-type BWR plants with 20 jet pumps.

In the actual engineering work, CFD analysis was only used to reduce the number of necessary test cases and to investigate the potential countermeasures before tests. Even for this purpose, CDF analysis applicability needed to be confirmed by comparison with the tests. CFD analysis could not replace the confirmation tests.

Although this study was done in the 1980s, it remains as a good example for a practice exercise applying new CFD tools.

7.2 CFD Analysis Application in an ABWR

7.2.1 ABWR Lower Plenum CFD Analysis and Reactor Internals FIV Stress Evaluation

The ABWR uses reactor internal pumps (RIPs) to drive core flow rate in the RPV. From the early stage of its development, flow in the lower plenum of the ABWR and flow-induced vibration (FIV) of lower plenum internals were recognized as important confirmation items before use based on US Regulatory Guide 1.20. Consequently, many small-scale tests, actual size tests, and a preoperational test at the first ABWR plant have been done that provide extensive data. These data have been used to confirm the capability of current CFD analysis tools and to study procedures for their suitable application for engineering purposes.

7.2.1.1 Application to 60° Sector 1/5-Scaled Test

The experimental apparatus (Fig. 7.12) consisted of two 1/5-scaled model RIPs and an RPV which had various structures in the lower plenum. The impellers, diffusers, nozzles, and RPV of the 1/5-scaled model imitated those of the ABWR.

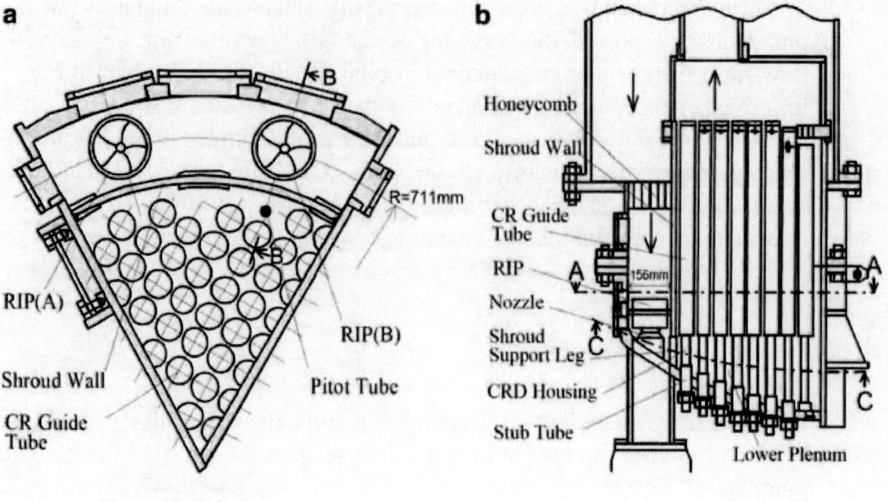

Fig. 7.12 Schematic diagrams of experimental apparatus

Table 7.2 Calculation conditions for experimental analysis (taken from [4] and used with permission from JSME)

Code name	STAR-CD
Number of grids	4,600,000
Algorithm	SIMPLE method
Scheme of advection term	First-order upwind difference scheme
Turbulence model	$k - \varepsilon$ model

Calculation conditions are shown in Table 7.2. The calculation was continued to steady-state convergence using the SIMPLE method and the first-order upwind difference scheme. Velocity of the inlet boundary was uniform and had no prerotation. The exit region was modeled as a zero gradient outlet. Fluid temperature and pressure were room conditions.

Analytical results at the pitot tube location are shown in Fig. 7.13. Comparison with the test results for flow velocity distribution at the shroud support leg opening and the analytical results showed good agreement.

7.2.1.2 Evaluation of FIV Performance of Lower Plenum Structure

In order to estimate the FIV stress level of lower plenum internals, an analysis was done for the actual size ABWR and in the reactor-rated condition. Calculation conditions were same as the conditions applied to the 1/5-scaled model except that

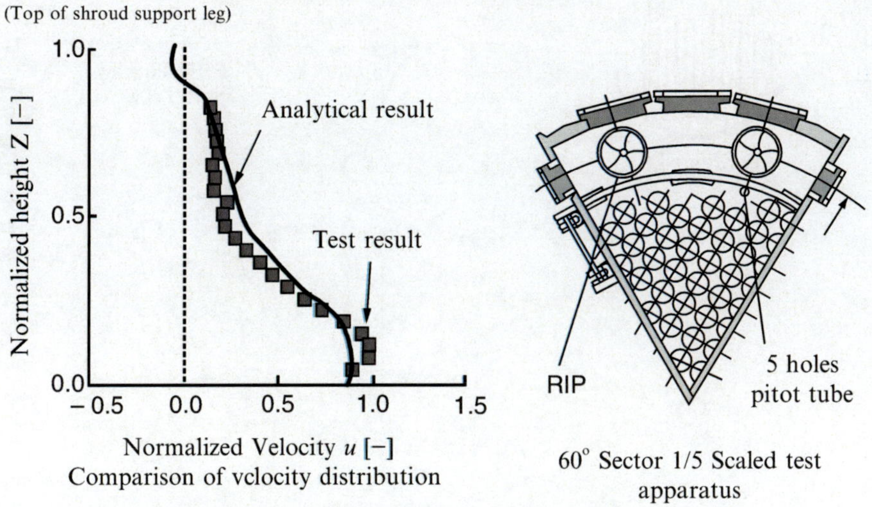

Fig. 7.13 Velocity distribution comparison between analysis and test results

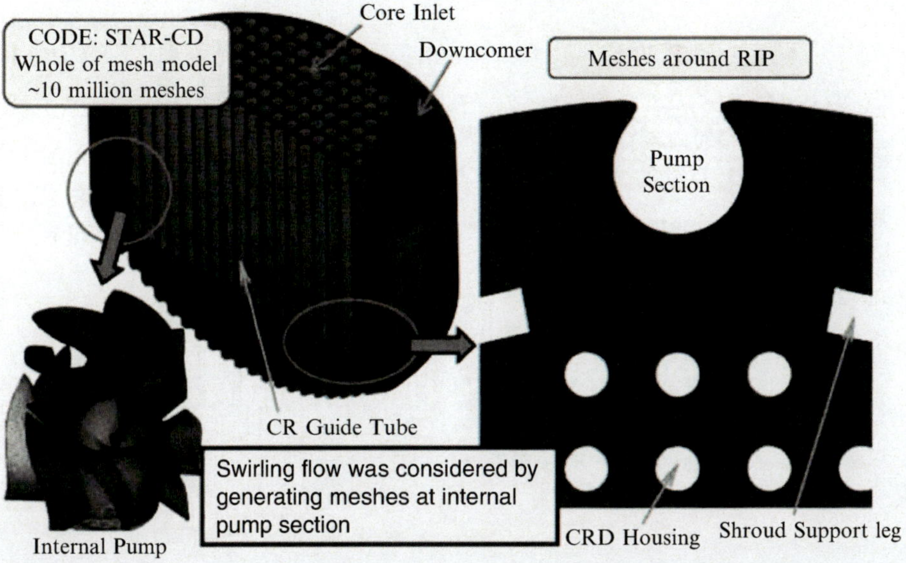

Fig. 7.14 ABWR lower plenum CFD model (5 RIP sector, 10,000,000 grids)

fluid temperature and pressure were at the rated reactor conditions. Computational grids used for the calculations are shown in Fig. 7.14.

Two typical flow patterns of the ABWR lower plenum are shown in Figs. 7.15 and 7.16. Along the RPV bottom, cross flow was observed around the peripheral control rod drive (CRD) housing. In the vertical plane, a high flow rate was

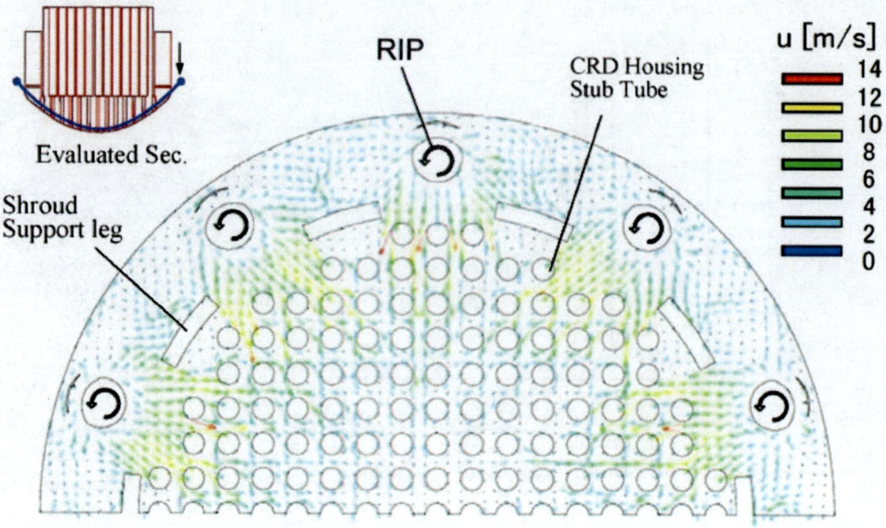

Fig. 7.15 Flow pattern of ABWR lower plenum along the RPV bottom

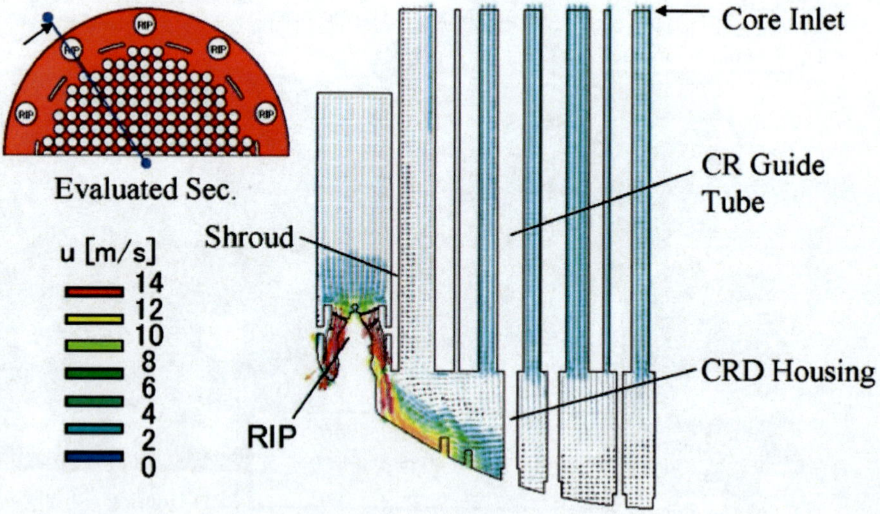

Fig. 7.16 Flow pattern of ABWR lower plenum in the vertical plane

observed at the bottom of shroud support leg opening and parallel flow was observed around control rod guide tubes and the core inlet.

Based on the fluid force equation for FIV (Fig. 7.17), fluid force was evaluated using CFD analytical results. Evaluated FIV stresses at the bottom of the CRD housing were within the allowable level (Fig. 7.18).

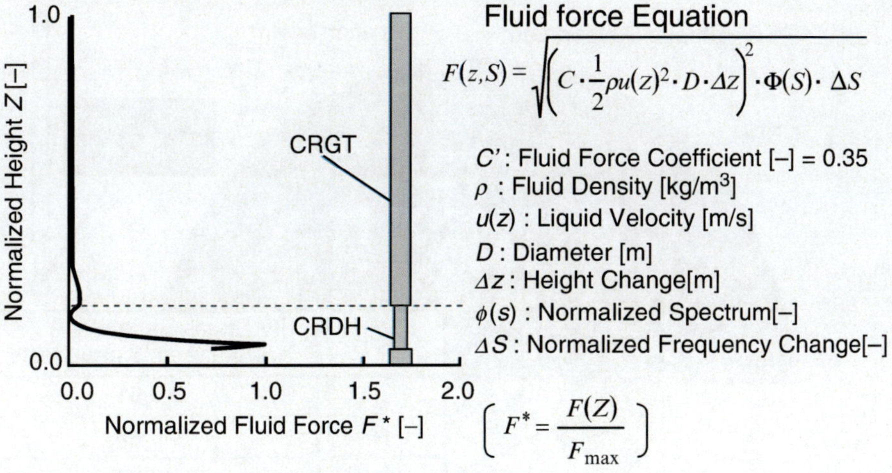

$$F(z,S) = \sqrt{\left(C \cdot \frac{1}{2}\rho u(z)^2 \cdot D \cdot \Delta z\right)^2 \cdot \Phi(S) \cdot \Delta S}$$

$$\left(F^* = \frac{F(Z)}{F_{\max}} \right)$$

Fig. 7.17 Fluid force equation based on CFD analysis result (CRGT control rod guide tube, CRDH CRD housing)

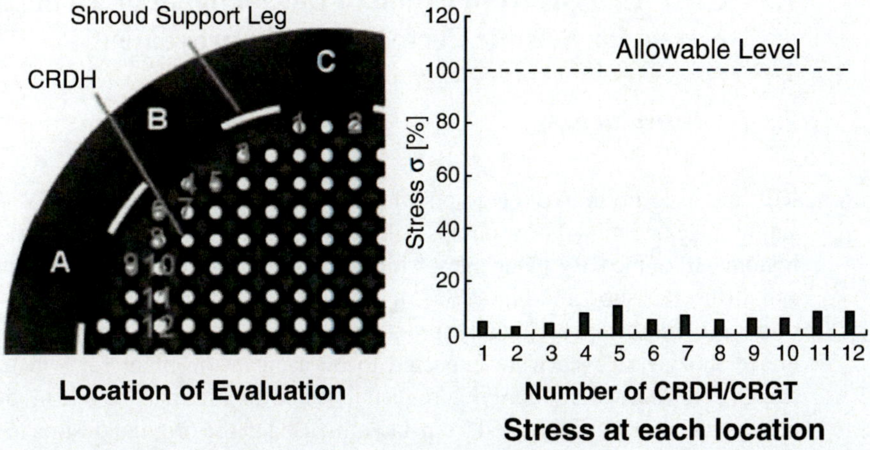

Fig. 7.18 Evaluation results of FIV stresses for lower plenum structures in RPV

In the first ABWR, a flow-induced vibration test was conducted during the start-up test. Figure 7.19 compares start-up test results and analytical results; agreement was good between them.

7.2.2 Conclusion

Based on these evaluations, the CFD analysis model was judged useful to evaluate phenomena related to the flow (especially FIV) and the evaluation could be done at lower cost than large-scale experiments.

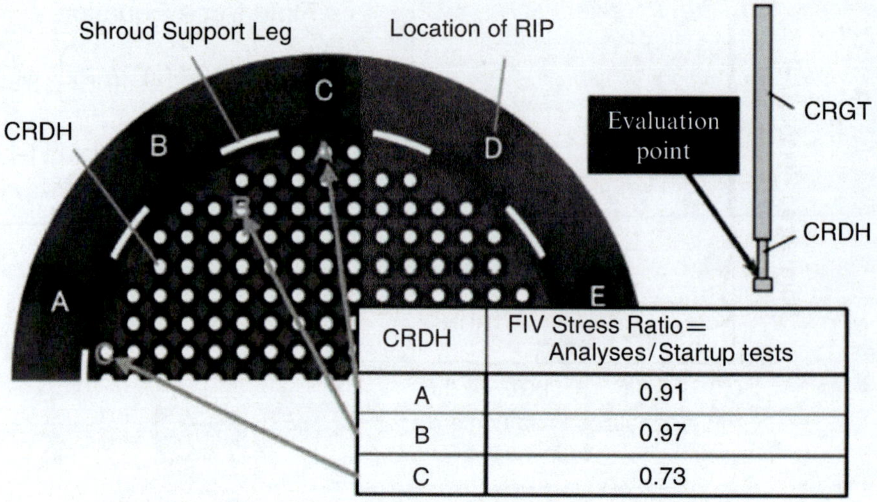

Fig. 7.19 Comparison of the FIV stresses between analysis and start-up test results

7.3 CFD Analysis Application to Development of a Thicker RIP Nozzle for Seismic Performance Improvement

7.3.1 Introduction

RIPs are used in the ABWR to circulate the reactor coolant in the RPV. RIPs are single stage and mixed flow pumps with wet motors, and these are supported on the bottom part of the RPV by nozzles, which are located downstream from the impeller and diffuser as shown in Fig. 7.20 [5].

To provide better earthquake-proof performance, larger-diameter nozzles with thicker nozzle sleeves were expected to be used in the plant for which seismic design conditions were more severe than those of the original ABWR [4, 5]; hence, nozzle diameter was increased from 445 mm, used in the original design, to 492 mm (Fig. 7.20).

However, as the RIP nozzle was located just downstream from the diffuser, too large nozzle diameter might change flow characteristics which affect the FIV of the structures in the lower plenum. It was necessary to select the maximum allowable nozzle diameter that did not affect flow characteristics and to clarify that the influence of the large diameter nozzle on any performance parameters was negligible.

7.3.2 Method of CFD Analysis

CFD analysis is useful to evaluate phenomena related to the flow in the reactor lower plenum at lower cost than for large-scale experiments. The parameter survey is also easy to conduct for CFD analysis.

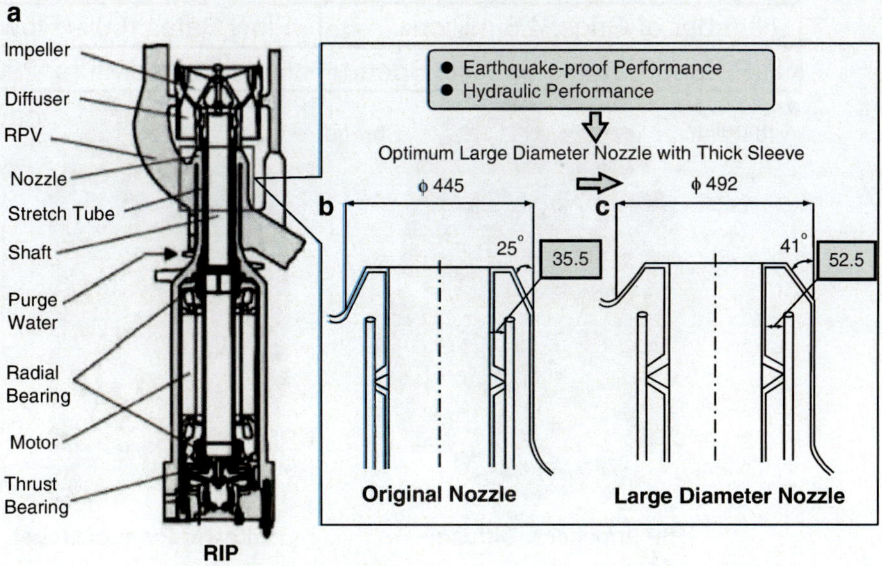

Fig. 7.20 Reactor internal pump and its nozzles (Taken from [5] and used with permission from JSME)

Table 7.3 Calculation conditions

Code name	STAR-CD
Algorithm	SIMPLE method
Scheme of advection term	UD and QUICK
Turbulence model	Standard $k - \varepsilon$ model
RIP nozzle	445 mm (original nozzle)
	492 mm (optimum nozzle)
	525 and 550 mm (reference)

Calculation conditions are shown in Table 7.3. Boundary conditions used in the analysis were the same as the method used in Sect. 7.2.1. Rotation of the RIP was simulated in all calculations by adding the Coriolis and centrifugal forces to the basic equations. RIP impeller and diffuser were modeled in detail as shown in Figs. 7.14 and 7.21.

Influence of larger nozzle diameter of the RIP was investigated with respect to flow characteristics. Four RIP nozzle sizes, 445, 492, 525, and 550 mm, were selected for the calculations.

7.3.3 *CFD Analysis Qualification with 1/5-Scale Tests*

To investigate the velocity distribution for the large-diameter nozzles and the validity of CFD analysis, velocity measurements with a five-hole pitot tube were made for

Number of Grids: 4.6 million	Flow Rate: 100% Flow
RIP Speed: Same Specific Speed	Fluid Temperature: 293K

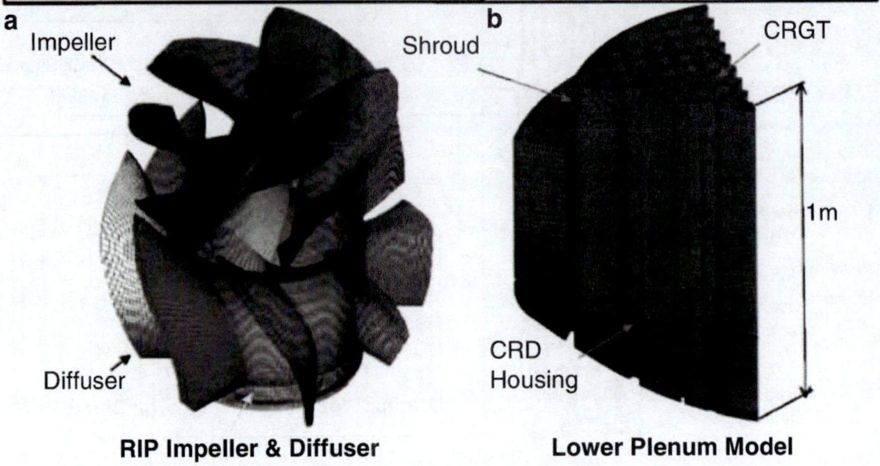

Fig. 7.21 Computational grids for 1/5-scale test model

445, 492, and 550 mm nozzles. The pitot tube was located at the center of the exit opening of the shroud support leg. Fluid temperature and pressure were ambient conditions. RIP speed and flow rate were design conditions of the 1/5-scale RIP.

Figure 7.22 shows both test and calculation results of 1/5-scale tests. Velocity *u* was velocity-normalized by the mean velocity of the opening of the shroud support leg. *H* was length-normalized by the height of the opening. As for test results, the velocity distribution of the 492 mm nozzle was almost the same as that of the 445 mm nozzle. Both flows went along the RPV bottom head. However, the flow pattern of the 550 mm nozzle changed from the patterns of the 445 and 492 mm nozzles. Velocity was also high at *H* of 0.5 for the 550 mm nozzle.

Comparing test and calculation results showed that the trends of the two were almost the same. The velocity distribution of the 550 mm nozzle obtained by CFD analysis was also changed from that with the 445 mm nozzle.

This change was unfavorable for FIV characteristics of the structures in the ABWR lower plenum. It would be desirable to have a high velocity fluid flow along the bottom of the RPV because of the lower moment of fluid force acting on the structures. It was confirmed that the 492 mm nozzle maintained the flow along the RPV bottom head and CFD analysis could simulate the influence of nozzle diameter change.

7.3.4 Evaluation of Influence in the Actual ABWR

To estimate the design margin for the nozzle diameter and investigate the difference of structures and conditions between the test model and the actual ABWR, the

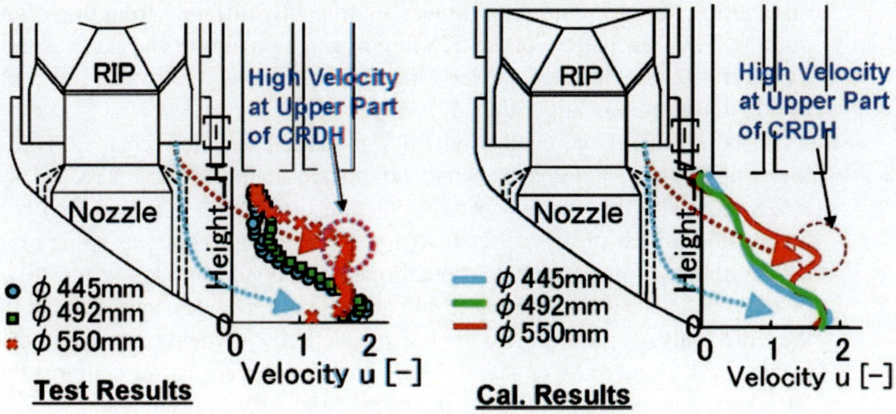

Fig. 7.22 Comparison of test and calculation results of 1/5-scale tests

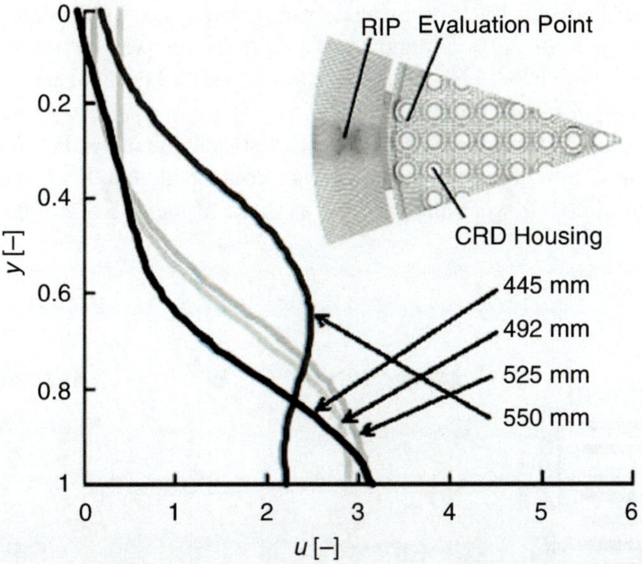

Fig. 7.23 Velocity distribution of calculation results under actual plant conditions (Taken from [5] and used with permission from JSME)

full-scale ABWR lower plenum model analysis was done with the 445, 492, 525, and 550 mm nozzles under actual plant operating conditions.

Velocity distributions for the nozzles between the CRD housings are shown in Fig. 7.23. Significant influence from the nozzle diameter could be seen at these evaluated points. Velocity was increased between CRD housings due to contraction of the flow channel compared with the mean velocity of the opening. The velocity distribution of the 550 mm nozzle was significantly different from that of the other nozzles. The flow pattern of the 525 mm nozzle was almost the same as that of the 445 and 492 mm nozzles, so the change of the flow pattern occurred when the nozzle diameter was larger than 525 mm.

Based on the above evaluations, it was judged that flow characteristics in the lower plenum did not change when the nozzle diameter was increased, but kept <525 mm. The flow pattern of the 492 mm nozzle was confirmed to be almost identical with that of the original 445 mm nozzle and there was some margin for nozzle diameter under the actual plant operating conditions.

Flow characteristics around the 445 and 492 mm nozzles were also compared in the CFD analysis. Flow around the RIP nozzle in the vertical cross section is shown in Fig. 7.24 for these two nozzles. Their flow characteristics were almost the same. RIP discharge fluid flowed along the nozzle and RPV bottom head.

Flow characteristics in the lower plenum of the RPV for the 445 and 492 mm nozzles are also shown in Fig. 7.25. The velocity vectors are shown in the cross section of the lower plenum along the RPV bottom head in which velocity was comparatively high. All RIPs were rotating in the counterclockwise direction. Flows between CRD housings went toward the RPV center and were almost uniform. Cross flow around the CRD housings was decreased and small in the center of the RPV. Flow characteristics of the 492 mm nozzle were almost the same as those of the 445 mm nozzle.

Figure 7.26 compares flow characteristics in the one RIP-tripped case for the 445 and 492 mm nozzles. Again it was confirmed that flow characteristics of the 492 mm nozzle were almost same as those of the 445 mm nozzle.

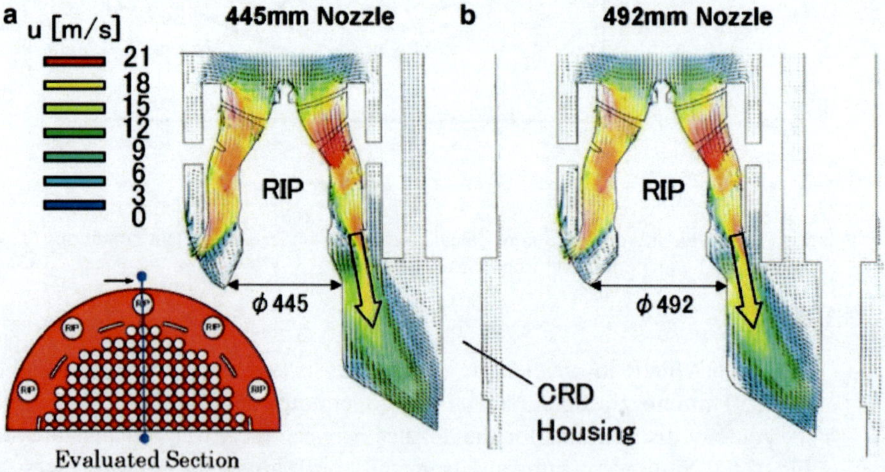

Fig. 7.24 Flow pattern comparison around nozzle

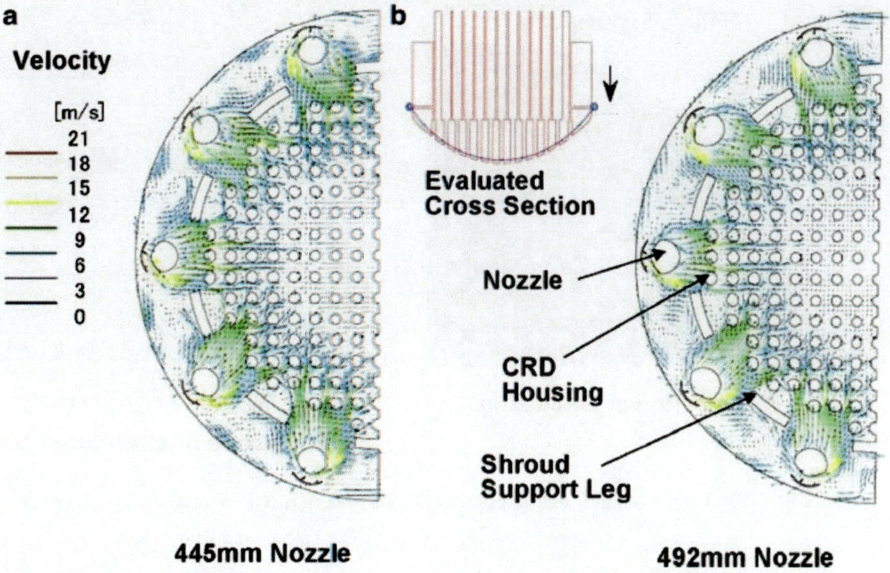

Fig. 7.25 Flow pattern comparison in lower plenum (all RIPs in6 operation case)

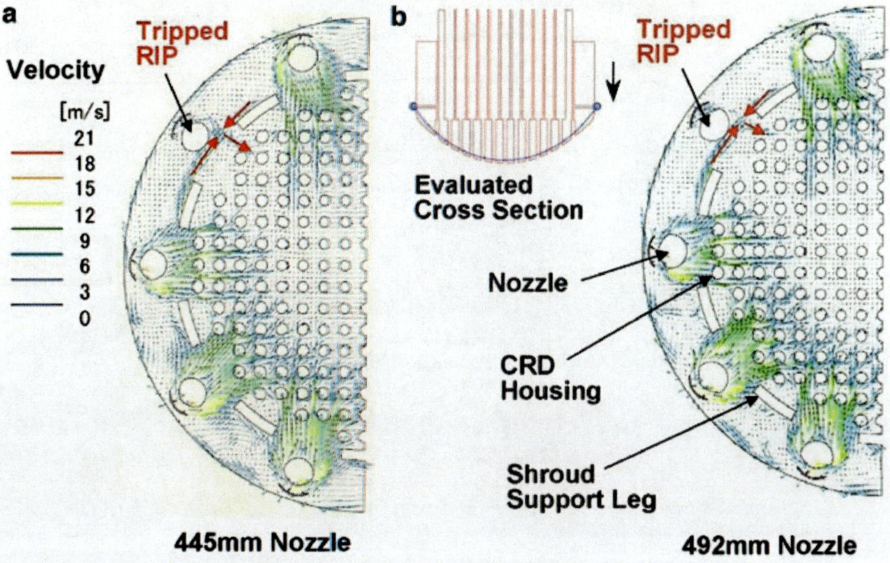

Fig. 7.26 Flow pattern comparison in lower plenum (one RIP-tripped case)

The FIV stress analysis was conducted for ABWR actual operating conditions using the above CFD results. FIV stresses at the bottom of the same CRD housings as shown in Fig. 7.18 were evaluated with the 492 mm nozzle. Figure 7.27

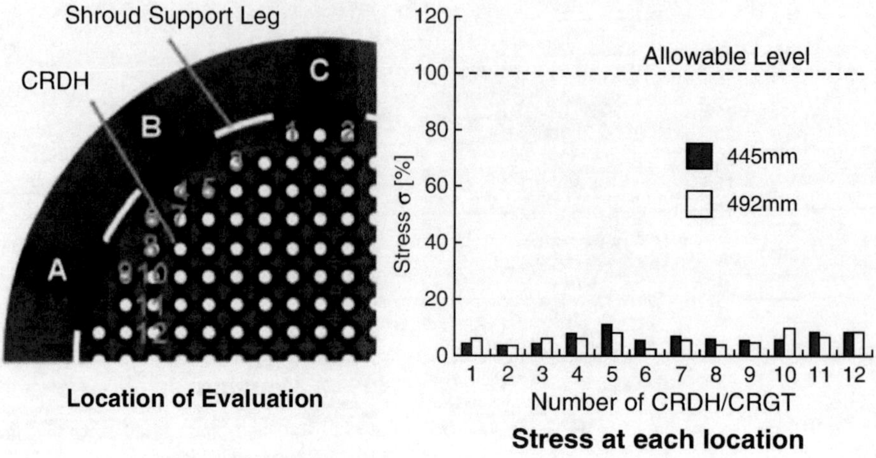

Fig. 7.27 Lower plenum FIV stress analysis for 492 mm RIP nozzle

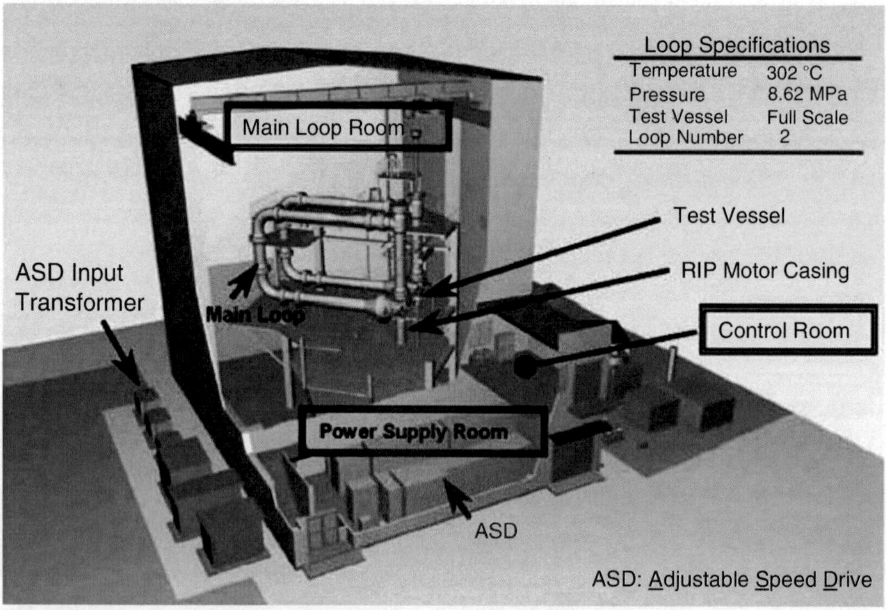

Fig. 7.28 Hitachi-GE Nuclear Energy ABWR RIP Test Center (Taken from [6] and used with permission from JSME)

compares results with 445 and 492 mm nozzles. It was confirmed that FIV stresses of CRD housings with the 492 mm nozzle were almost the same level as those with the 445 mm nozzle and stresses of both nozzles were within the allowable level.

7.3.5 Conclusion

Based on these tests and CFD analysis, 492 mm nozzle design was developed. Before use in the actual plant, the 492 mm nozzle was verified through the RIP full-scale tests at the ABWR RIP Test Center (Fig. 7.28). The RIP with the 492 mm nozzle could maintain good performance regarding hydraulics, coast down and vibration characteristics; it compared well with the original nozzle (445 mm) and use of the 492 mm nozzle was judged appropriate for the RIP in the ABWR. [6] This design was subsequently applied in the fourth and fifth ABWRs manufactured in Japan.

7.4 CFD Analysis Application to the Next Generation Reactors: Some Concluding Remarks

1. "Test before use" has been the dominant engineering culture. CFD analysis has steadily increased its roles in new design development as a replacement for tests. But applications of CFD still require specific verification by tests.
2. CFD analysis has been applied to the BWR-5 and ABWR in the early stage of development mainly for the single-phase flow region.
3. For the next generation BWR, wider applications of CFD (including two-phase flow analysis) can be expected.

Acknowledgments This chapter used papers originally published by JSME and ASME. The authors express many thanks to JSME and ASME for their permission to reproduce some of the figures and tables included here.

References

1. Ohki A, Miura S, Yoshimoto Y et al (1990) Unstable Flow Phenomena through a Pipe System with Cross Branch Pipe (Investigation of Flow Stabilizing Structure for Unstable Discharge Phenomena). JSME Intl J 33(4):680
2. Miura S, Yoshinaga Y et al (1987) Unstable phenomenon in flow through a pipe system with a cross pipe (1st report, Generation of flow instability and its condition). Trans JSME 53(485, B):35, in Japanese
3. Miura S, Yoshinaga Y et al (1988) Unstable phenomenon in flow through a pipe system with a cross pipe (2nd report, The influence of the branching discharge ratio and structural factor of a pipe upon the unstable discharge phenomenon). Trans JSME 54(503, B):1607, in Japanese
4. Takahashi S et al (2000) Influence of nozzle diameter on reactor internal pump performance in reactor pressure vessel, ICONE-8102
5. Takahashi S et al (2003) Evaluation of flow characteristics in the lower plenum of the ABWR by using CFD analysis, ICONE11-36393
6. Takahashi S et al (2001) Full scale test results as part of the development of a large diameter nozzle with thicker sleeve for the ABWR internal pump, ICONE-9- No. 680

Chapter 8
Next Generation Technologies in the Digital I&C Systems for Nuclear Power Plants

Tatsuyuki Maekawa and Toshifumi Hayashi

In this chapter, overviews of digital technologies for instrumentation and control (I&C) systems and the main control room (MCR) of boiling water reactor nuclear power plants (BWR NPPs) are explained. Then, cutting edge fundamental technologies and future possibilities for the next generation I&C systems are described.

8.1 Overview of I&C Systems for NPPs

Figure 8.1 shows a simplified overview of a BWR NPP. The NPP consists of a nuclear reactor and a turbine/generator. The instrumentation system for the turbine/generator is almost the same as that of a conventional thermal power plant, and most of the dissimilar items are related to the nuclear reactor part. The reactor I&C systems are composed of nuclear instrumentations, reactor control systems, and safety and reactor protection systems.

For nuclear instrumentations, reactor thermal power, ranging from less than a watt to over a thousand megawatts, must be measured by neutron flux using different types of neutron detectors in various measurement ranges. A large number of neutron detectors must be installed in in-core locations under a harsh environment (300°C, 7 MPa, irradiated by neutrons and gamma-rays). Reactor control systems are composed of reactor power control systems and a reactor water level control system. The reactor power control systems manage the recirculation flow and the control rods. The reactor water level is measured as water mass by differential pressure measurements, because the BWR water surface is not distinctly seen due to the two-phase flow. Steam flow to the turbine is controlled by the turbine I&C system consisting of a reactor pressure control system and a turbine control system.

T. Maekawa (✉) and T. Hayashi
Toshiba Corporation, Tokyo, Japan
e-mail: tatsuyuki.maekawa@toshiba.co.jp

T. Saito et al. (eds.), *Advances in Light Water Reactor Technologies*,
DOI 10.1007/978-1-4419-7101-2_8, © Springer Science+Business Media, LLC 2011

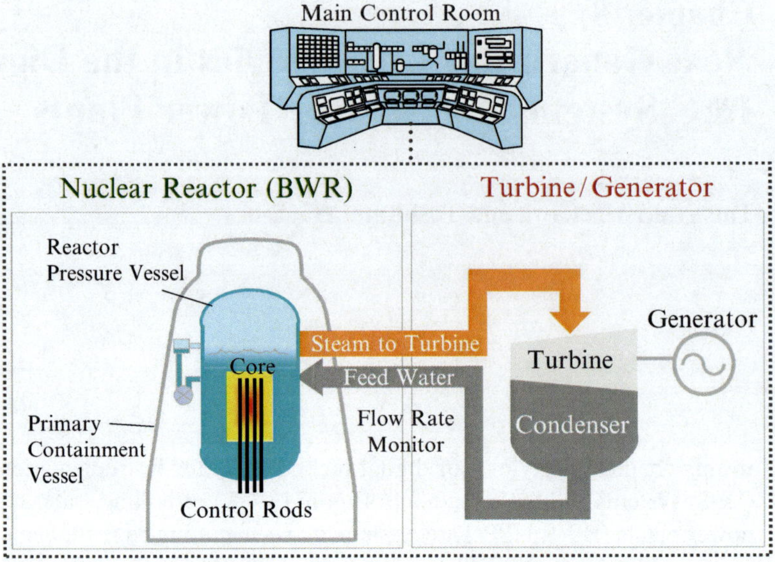

Fig. 8.1 Overview of a BWR NPP

The most unique item in the reactor I&C system is the safety and reactor protection system. The purpose of this system is to prevent reactor conditions from deviating beyond safe limits, but if safe limits are exceeded, to mitigate the consequences. If an abnormal condition occurs, the reactor protection system activates reactor shutdown and engineered safety features (ESFs). Then, core isolation and cooling, pressure reduction, emergency power startup, containment, and air filtration for radioactive materials are activated. In this protection system, the consequence of greatest concern is the release of radioactive materials.

There are some design criteria for safety systems of the NPP I&C systems such as redundancy, separation, reliability, testability, diversity, qualification and quality assurance, and appropriate compliance with codes and standards. The safety systems must incorporate sufficient redundancy and electrical and physical separation independence to ensure that no single failure or removal of any component results in loss of the safety and protection functions. The systems must be separated from other I&C systems. If common subsystems are used, each safety system should not be affected by a failure of any nonsafety instrument or control system. The safety systems must be highly reliable and designed to fall into an acceptable safety state (fail-safe or fail-as-is state) if they are disconnected or lose power. The safety systems must permit periodic testing of their functions during normal operation. In some cases, diversity of methods or equipment should be considered. All information about the I&C systems, including the turbine/generator I&C systems, are collected by the MCR. The monitoring and control console panels are installed in the MCR which has both main and subpanels. Plant operators work in the MCR as they operate the plant.

8.2 Chronicle of Digitalization

8.2.1 The First Generation

Construction of commercial NPPs started in the 1950s. In those days, analog components (e.g., relays, switches, and meters) were adopted to configure I&C systems. In Japan, the first commercial plant Tokai-1 was turned over in to the utility in 1966. Since then, the improvement and development of I&C systems and the MCR have gone through three generations. Simultaneously, digitalization has also been carried out step-by-step through three generations as shown Fig. 8.2.

In this chapter, "digitalization" includes all of the following meanings: digitized signals by ADCs, digital signal and digital data processing by microcomputers, digital signal transferring by networking, and sophisticated human machine interfaces (HMIs) using computer technologies. These digitalizations have taken place along with the evolution of computer technologies. At first, the radioactive waste disposal system was digitalized in the mid-1980s. Next, based on these results and experiences, the nonsafety systems were digitalized. For the ABWR (advanced BWR) I&C in the mid-1990s, the safety systems were digitalized finally as the SSLCs (software safety logic circuits).

The rest of this section gives a brief chronicle of digitalization in Japanese NPPs, focusing on the MCR. In their infancy, the control panel used for nuclear reactors was very simple. Then, the first design concept for commercial plants was established. In the case of Japan, the design of the MCR can be divided into three generations since the end of the 1960–1970s.

The first generation of the MCR design was based on that of contemporary thermal power plants. It consisted of a bench-type panel with hard-wired

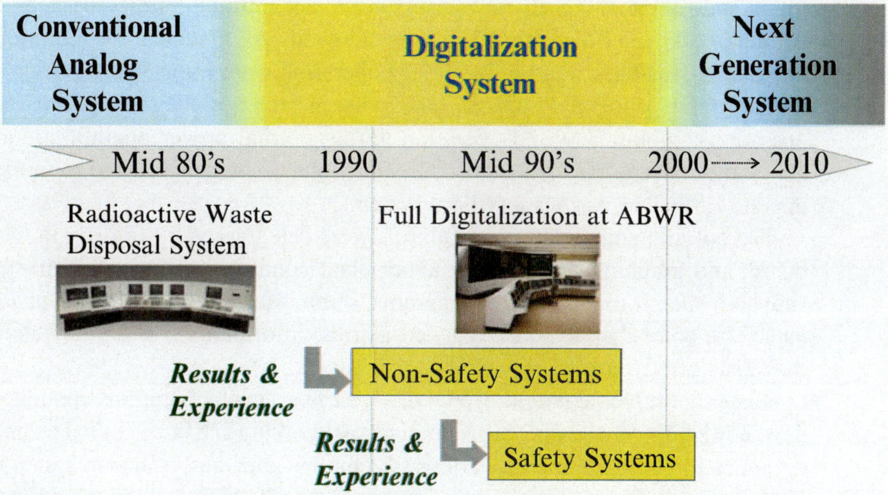

Fig. 8.2 Digitalization process

HMI devices, such as switches, meters, and lamps. During this generation, the Three-Mile Island (TMI-2) accident occurred on March 28, 1979 near Harrisburg, Pennsylvania [1]. In this accident, a gradual loss of cooling water led to partial melting of the fuel rod cladding and the uranium fuel, and the release of a small amount of radioactive material. The accident was caused by a combination of personnel errors, design deficiencies, and component failures. As a result, the accident had a great impact on various fields. It brought about sweeping changes involving emergency response planning, reactor operator training, human factor engineering, radiation protection, and many other areas of NPP operation. Human performance was identified as a critical part of plant safety after this accident, and the requirements for plant design and equipment were drastically changed to deal with this. The following improvements in the system design were carried out.

- Clarification of operating range of instruments.
- Classification of important operations and annunciators.
- Reconsideration of panel layout of information displays and switches.
- Expansion of computer technology utilization. In particular, grouping and coordination of related information from plural system outputs were utilized for accurate information supplementation.

8.2.2 The Second Generation

Based on lessons learned at TMI-2, the development of second generation systems was started. In the mid-1980s, Toshiba designed and released the second generation MCR, called Plant Operation by Display Information and Automation (PODIA™) [2]. A new operator interface, which consisted of two kinds of separated panels (main and subpanels) with color displays and a simplified mimic board, and partial automation of auxiliary systems were adopted. As a result, the monitorability, operability, visibility, and reliability of operation were improved.

The main panel of PODIA™ had functions for normal startup and shutdown operations, monitoring and operation during normal power operation, and emergency monitoring of plant status. The functions of subpanels were to monitor and operate ESFs and various auxiliary systems.

The colored annunciators and alarms were categorized by priority in the mimic board, and graphical information about plant conditions using color displays was shown in such forms as system diagrams, summarized plant information, and trend data. The color displays also showed alarms, information on standby state, early diagnosis of errors, and surveillance test guides.

Since the implementation of PODIA™, further development has continued. With the many significant advances in computer, networking and other digital technologies, digitalization was extended to various systems and apparatuses in plants, step-by-step. Finally, the plant-wide digitalization system for the ABWR was completed as the third generation system and operation was started in 1996.

8.2.3 The Third Generation System for the ABWR

Toshiba named the MCR in this generation as the "advanced-PODIA™" (A-PODIA™) [3]. Figure 8.3 shows an experimental mockup of A-PODIA™. In the third generation, a sophisticated HMI was developed and designed based on workload analyses.

The HMI of A-PODIA™ consists of a large display board and a compact console desk as shown in Fig. 8.3. Large panels provide the operating crew with common recognition of plant status and they include an essential annunciator panel, a large mimic panel, and a large display. System integrated alarms are installed in the upper part of the large display panel, and major failure, minor failure, and the actuation of mitigating functions are identified using three colors. The compact main console promotes high operability and provides controls and monitoring information except during periodic inspections. It typically has seven CRTs (LCDs are adopted in the latest systems), 17 flat display panels with color LCDs and emergency hard-wired switches to facilitate annual inspections, an auxiliary console with 31 flat display panels, and about 100 hard-wired switches are provided at the lower part of the large display panel. These flat display panels are interfaced with touch screen to minimize operator burden.

A power plant involves a vast amount of control and operating information, and the basic design concept of this HMI system is to structure control actions and their related information according to the hierarchy of the integrated digital control system. In this new system, monitoring information is classified into three levels: plant, system, and equipment levels. It is most important in assigning information to consider the quality and quantity of the information. Furthermore, compared with the second generation, the automated operation scope was extended, as well.

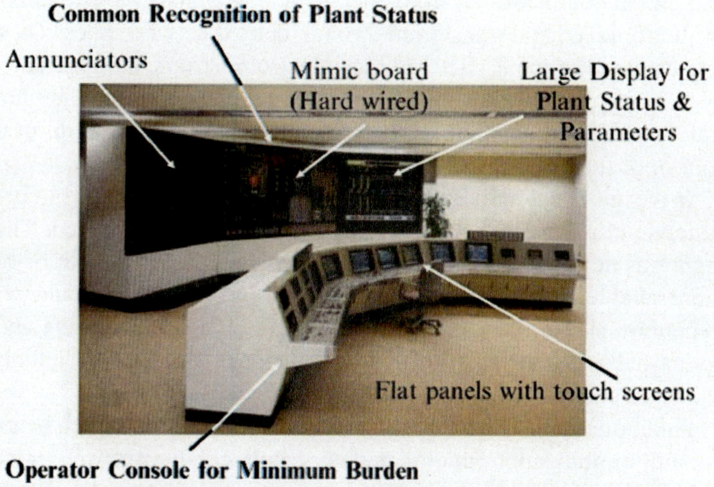

Fig. 8.3 An experimental mockup of the A-PODIA™

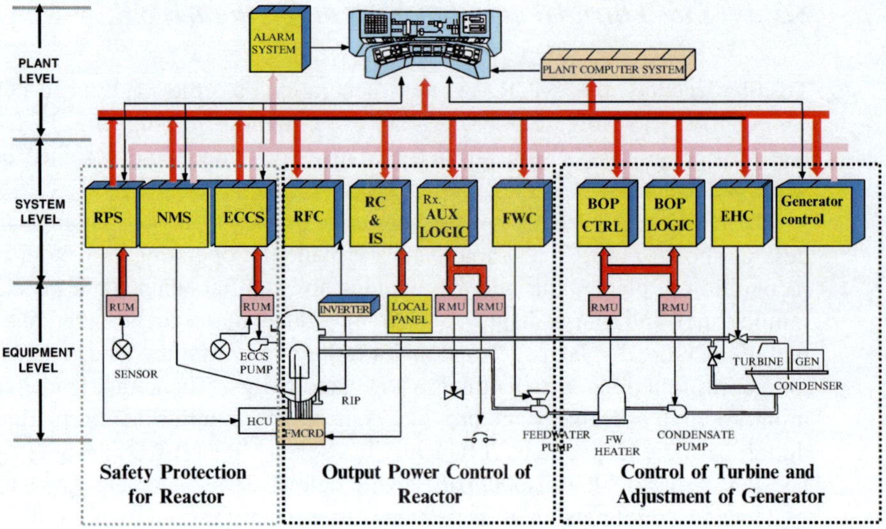

Fig. 8.4 Schematic diagram of the plant-wide digitalization system

Figure 8.4 is a schematic diagram of the plant-wide digitalization system of the ABWR. The system consists of three physical levels as mentioned above. The first level is designed for managing plant level information and functions. The MCR with the plant computer system is in this level. The second level is designed for managing each control system of the plant. Each control system consists of a number of highly reliable and intelligent digital controllers. In the third level, various sensors, actuators, local control instrumentations, and remote multiplexing units (RMU) are included. Analog signals of sensors are converted to digital signals and digital commands are converted to analog signals for actuators. Digital signals are multiplexed and transferred through optical network lines. These functions are implemented in the RMUs and local control instrumentations. The second and the third levels can be categorized into three groups regarding the purpose. These are follows: safety protection for the reactor, output power control of the reactor, and control of the turbine and adjustment of the generator.

In the third generation, not only nonsafety systems, but also safety-related systems (safety and reactor protection system) were wholly digitalized. "Two-out-of-four" logic was newly adopted rather than the conventional "one-out-of-two" to establish more reliable plant operation. Self-diagnosis by the microcomputer was implemented to confirm system validity in operation. The overall monitoring and control via an optical networking system are done in the control room where digitalized information is integrated.

Additionally in the safety-related systems, attention had to be paid to software. For full digitalization, microcomputers with suitable software were implemented. To qualify the software logic and products as safety-related systems, a "verification and validation (V&V)" method was newly adopted. Subsequently, application guidelines for programmable digital computers in the safety and protection system were established as JEAG-4609 [4] in Japan.

8.3 Advantages and Evaluation of the Third Generation Digital System

8.3.1 Advantages of Digitalization

In the third generation plant-wide digital system, the operational safety and reliability and the reliability of instrumentations are improved. The system reliability is enhanced by the "two-out-of-four" logic. Operator workload is reduced by the expansion of automated operation scope and information sharing further supports the improvements.

In this section, some evaluation results obtained from operators are described for the third generation digital system, especially the HMI. Figure 8.5 shows the effect of automated operation. An operator workload comparison is shown during a plant trip sequence and a startup sequence. The workload of the third generation is clearly lower and flatter compared with that of the second generation HMI and, as a result, stable and reliable operation can be established.

Figure 8.6 shows operators' impressions of the third generation HMI of the MCR. This HMI was well received by operators after simulator training.

Next, Fig. 8.7 shows an example of automated operation. In this case, automatic pressurization to keep a constant heat-up rate was done. Compared with the case of manual operation, the automated operation function offered faster startup. To get

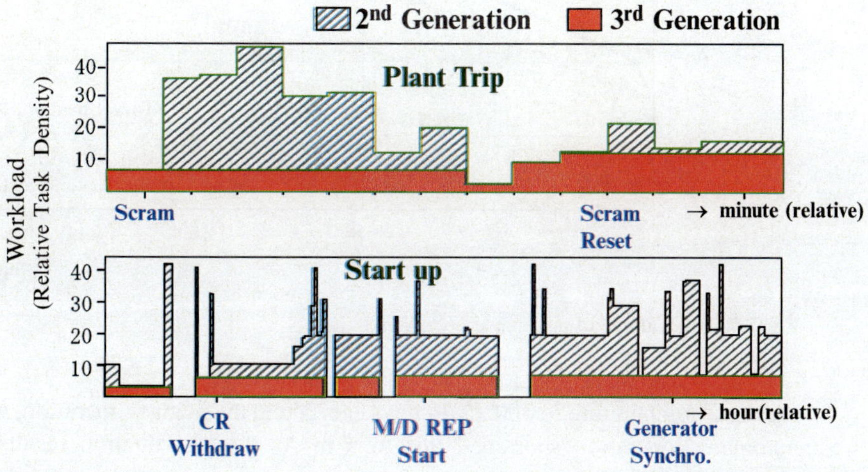

Fig. 8.5 Comparison of operator workload during a plant trip sequence and a startup sequence

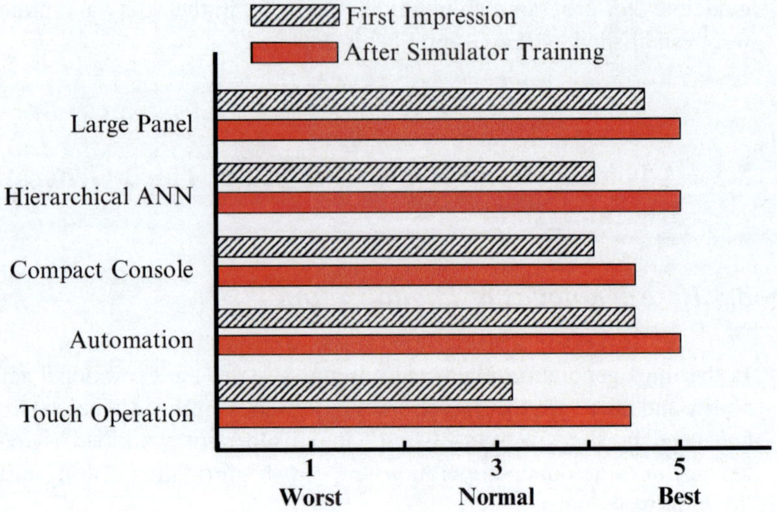

Fig. 8.6 Operators' impressions of the third generation HMI

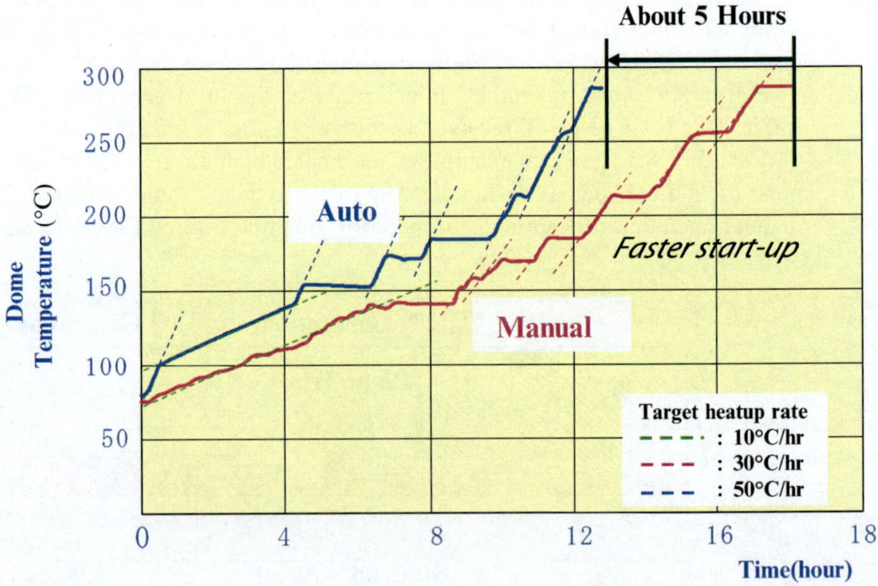

Fig. 8.7 An example of automated operation

the dome temperature to 300°C, 17 h was needed in the manual operation, while the automated operation shortened this to 5 h. As these evaluation results clearly showed that operability, operational safety, and operational reliability were improved.

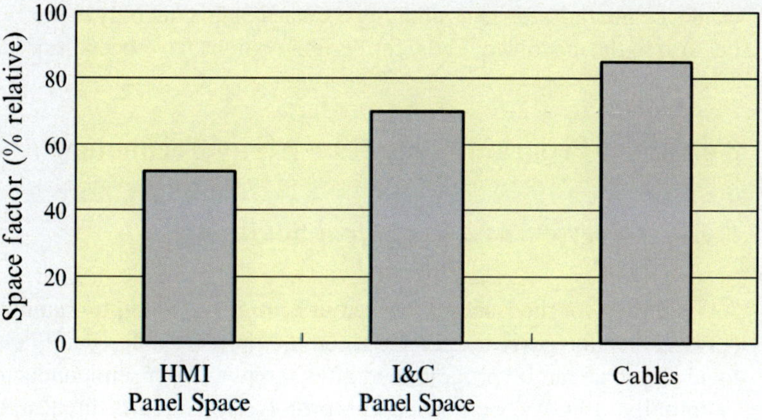

Fig. 8.8 Reduction of space and amount of cables

Figure 8.8 shows the merits of physical quantities in comparison to the second generation system. The HMI panel space was reduced to 52%, the I&C system panel space was reduced to 70%, and the length of cables was reduced to 85%. By adopting the digitalization and the multiplexed signal transmission, the total amount of space needed and the number of items were drastically reduced.

After the first implementation of the third generation system at Kashiwazaki-Kariwa NPP-Units 6 and 7, the digitalization trend continued to other BWR NPPs at Hamaoka Unit 5 and Shika Unit 2. These NPPs are operating. The digitalization trend also has proceeded in PWRs. Ikata NPP Units1 and 2 are being renovated and Tomari NPP Unit 3 is under construction at this time (2009). Newly planned NPPs in the United States are also expected to adopt these fully digitalized I&C systems and the MCR, as well.

8.3.2 Issues and Solutions

As described in the former section, there are many advantages in digitalization. For the MCR, the HMI has been well received. For the instrumentations themselves, intelligent CPU systems can pack in as many and various functions as needed. As a result, problems caused by human error are decreased, and operational safety and reliability relating to the operation and the instrumentations are improved. However, in the years since the first ABWR I&C systems were realized, some issues related to the digitalization have been recognized.

The first and most important issue is that the commercial lifetime of CPUs is too short. Secondly, the V&V man power for redesign using new alternative CPUs would be massive. Redesigning of the system due to the obsolete CPUs would require a long time to obtain the approval of the regulatory agency. In these situations, the sustainability of product supply and design for long-term maintenance cannot be guaranteed. This situation is unacceptable not only to the system supplier, but also to the customer. Thus, further innovations have been desired.

8.4 Field Programmable Gate Array Technology

8.4.1 Overview as a Technical Solution

As a solution for the issues described in Sect. 8.3.2, field programmable gate array (FPGA) technology is considered to be the most feasible. The FPGA is a type of logic chip that can be programmed after shipping from semiconductor factories.

Initially, FPGAs were used for prototyping circuits in design procedures. However, they are also suitable for low volume applications, such as those in the nuclear power industry. Nowadays, FPGAs are used not only for prototyping circuit designs, but also for assembling commercial chips in various applications. Applications of FPGAs have been rapidly increasing and they are now applied in cell phones, electronic components for automobiles, computers, data storage devices, industrial electronics, and aerospace electronics. In order to broaden their application range further, various workshops for FPGA application technology have been held worldwide. The first workshop in the field of NPP I&C was held by IAEA in October 2008 [5]. Participants at the workshop included IAEA members, researchers from a national laboratory in the USA, and persons from FPGA suppliers, instrumentation and system companies, plant operation companies, and electric power companies. They reported on the recent activities and technological topics, and discussed various issues of FPGA application in the field of NPP I&C.

It is expected that by using FPGAs, the most important feature will be that a product life cycle does not stick to the CPU market. The long-term availabilities of FPGAs are based on a limited set of components. However, they can be used to implement a large variety of electronic functions. FPGAs have the proven reliability of integrated circuits for design and manufacturing processes. Furthermore, FPGAs have the ability to transfer functions to other technology when necessary, with limited effort using a logic scripting language, such as VHDL (very high speed integrated circuits hardware description language).

Figure 8.9 compares system design schemes for an analog system, a CPU-based digital system, and a FPGA-based digital system. The scheme of the CPU-based digital system is rather complicated, compared with the other two schemes. The FPGA system uses hardware-based dedicated logics without the CPU and operating system. Then, only necessary functions have to be implemented in the FPGA hardware component. As shown in Fig. 8.9, the design scheme of the FPGA-based system is simple and similar to an analog system. Then, the amount of logic to be qualified is lower than for a CPU-based system, and the activity demanded for the V&V is minimized and affordable. As a result, the FPGA technology has a good possibility for sustainability of maintenance, supply, and design.

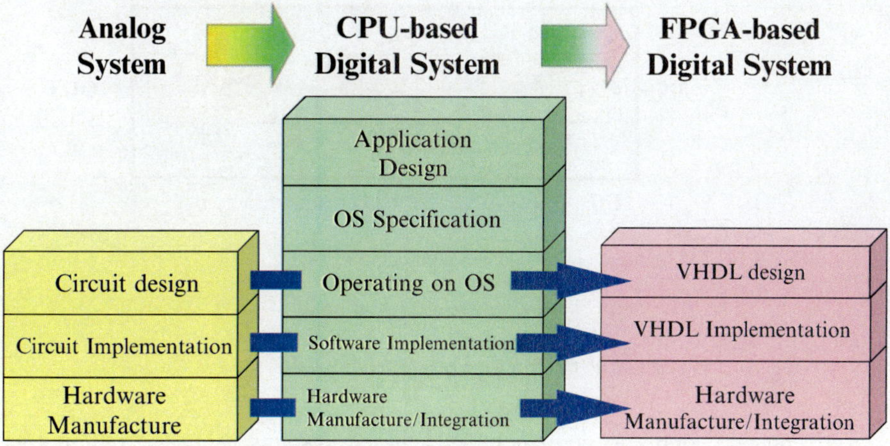

Fig. 8.9 Comparison of system design schemes

8.4.2 Design Concept of FPGA-Based Systems

In the design procedures of FPGA-based systems, the primary concern is the reliability of FPGA logic. A simple method to ensure the FPGA logic is full pattern testing, in which every possible combination of input signals is entered into the FPGA, and its outputs are checked. However, full pattern testing of whole circuits is not practical for large-scale digital circuits. In the case of safety-related I&C systems of NPPs, the number of input signal combinations may become so great that testing cannot be finished in a practically acceptable time. Instead of full pattern testing, the functional element (FE) approach has been proposed by Toshiba [6].

In this proposed method, there are two important points. The first point is that an FE is defined as the minimum logical element that performs a certain function in an FPGA, and it is so simple that its function can be verified through full pattern testing. The size of an FE is limited by the time needed to complete full pattern testing. Then, FPGA logic circuits are constructed using combinations of verified FEs. Figure 8.10 shows this concept.

The second point is to ensure that FPGAs are correctly built from FEs, i.e., that all connections between FEs are correctly installed and they operate correctly. For this purpose, test indices, which can determine whether the test cases are sufficient to ensure the connections between FEs, should be surveyed. Toggle coverage has been selected as one index. In FPGA testing, a change in input signals is examined and verified by some connections between FEs from logic zero to logic one, and by other connections from logic one to logic zero. The toggle coverage is the ratio of the number of examined connections to the number of operable connections in the testing. It should be noted that some connections are directly linked to a ground line or a power line in the FPGA. They should be excluded from the toggle coverage calculation. After finishing all the tests, the FPGA becomes a qualified integrated circuit (QIC).

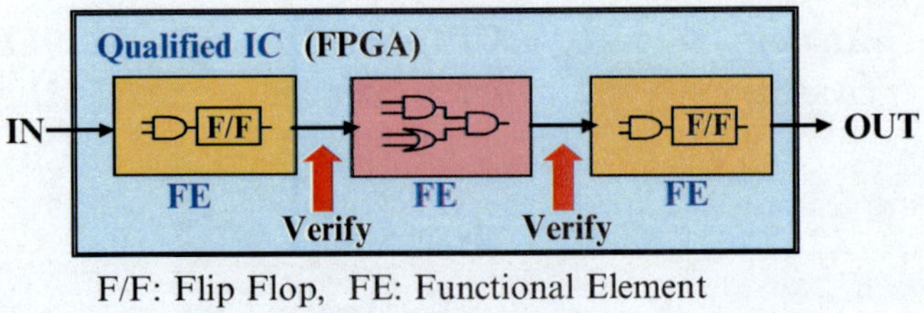

Fig. 8.10 FPGA logic circuits constructed using FEs

8.4.3 Overview of FPGA-Based Systems

The procedures to construct FPGA-based systems using the QIC are explained below. Figure 8.11 shows the concept of the hierarchical structure of a FPGA-based system. The system consists of some units. Each unit contains some modules, and a module is built using some printed circuit boards, including some FPGAs (QICs). This hierarchical structure is explained in detail using an example of a FPGA-based power range neutron monitor (PRNM) that Toshiba developed. The PRNM measures the neutron level in the BWR core in the power range (i.e., above about 10% rated power).

Figure 8.12 shows the PRNM configuration using neutron detectors. The PRNM is placed in the MCR. Electrical signals are obtained from the neutron detectors installed in the core and from differential pressure transmitters placed at each recirculation loop. The number of neutron detectors is 172 for a typical BWR-5 plant. The PRNM consists of two equivalent divisions. Each division processes 86 detector signals with four channels to which the signals are divided and assigned.

The units of the PRNM system consist of a chassis that has front slots and back slots to house modules. Each unit consists of several modules. There is a vertical middle plane between the front and back slots in each unit. This plane consists of two circuit boards. These circuit boards provide backplanes for the front and rear modules. Modules plug into the backplanes using connectors. When a module is plugged into the appropriate connector, the module is powered, and it exchanges data with other modules in the same unit. For inter-unit data transmission, the units are equipped with two types of communication modules, the transmission module and the receiver module. These modules link two units with one-way point-to-point data transmission through fiber optic cables. These modules are used for data transmission from the PRNM system to external devices, and from external devices to the PRNM system. Each module consists of one or more printed circuit boards and a front panel. The printed circuit boards have FPGAs for signal processing and for the HMI. The front panel is connected to the HMI FPGAs, and the HMI allows plant operators or maintenance personnel to enter appropriate set points.

The PRNM contains the local power range monitor (LPRM) modules that correspond to individual neutron detectors. Figure 8.13 shows the configuration of an LPRM module.

Fig. 8.11 Hierarchical structure of FPGA-based system

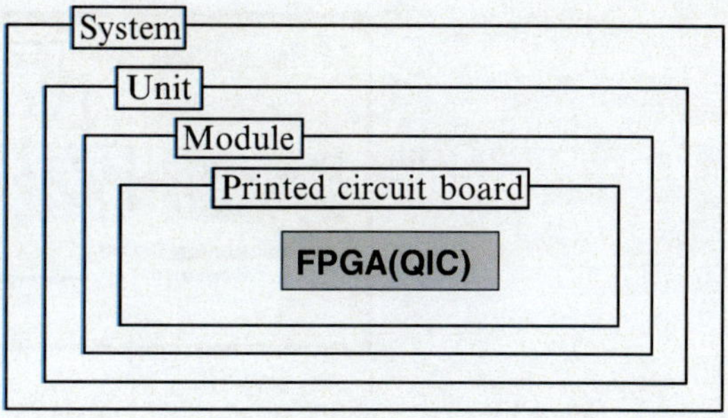

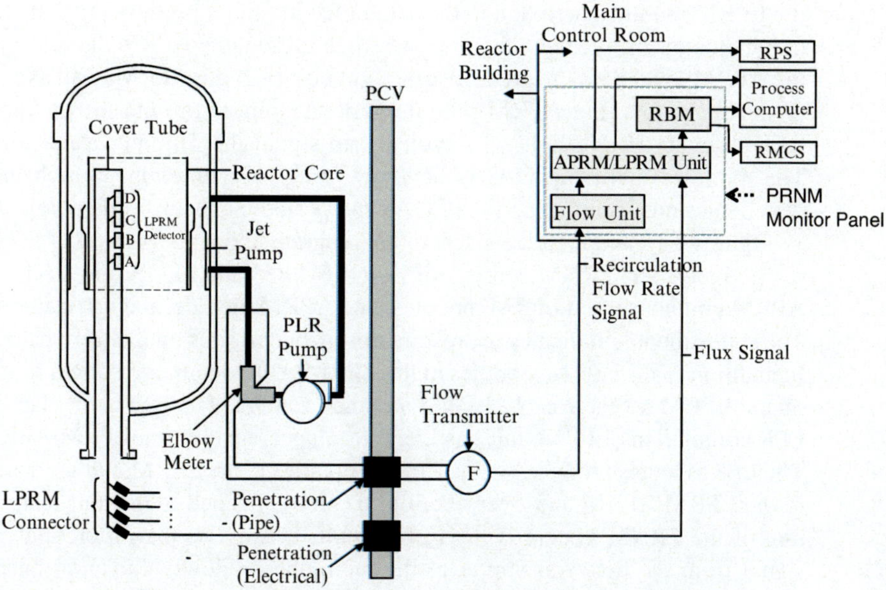

Fig. 8.12 Configuration of the PRNM using neutron detectors for BWR NPPs

Each LPRM module obtains an electrical signal from the detector, amplify the signal, and convert the analog electrical signal to a digital signal. The filter FPGA utilizes a digital low pass filter featuring power supply noise reduction. The LPRM modules multiply a gain value on the filtered signal, and produce the LPRM level, which is transmitted to an average power range monitor (APRM) module included in the PRNM and to external devices through analog output modules. The LPRM modules compare the LPRM level with a predetermined set point, and if the LPRM level exceeds the set point, the LPRM modules generate an alarm. In addition, each LPRM module has an input interface (I/F) and parameter FPGA, which allow the plant maintenance personnel to make calibration and set point changes.

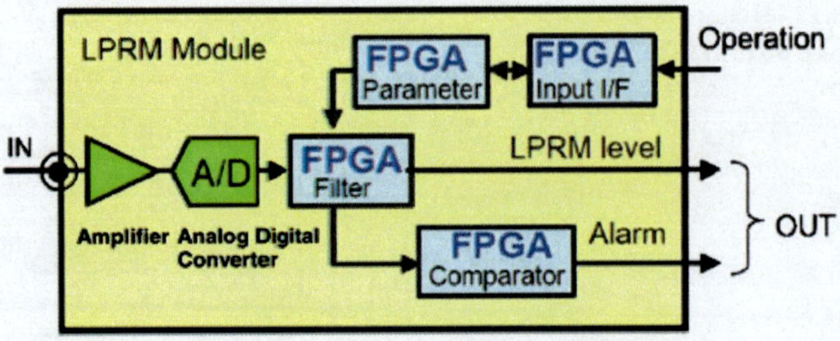

Fig. 8.13 Configuration of an LPRM module

An electrically erasable programmable read-only memory (EEPROM) is attached to the parameter FPGA to retain the set points.

The APRM module has several FPGAs to average up to 22 LPRM levels from the LPRM modules included in the same PRNM, and it produces the APRM level that indicates the average reactor power. In addition, the APRM module compares the APRM level with the predetermined set points. If the APRM level exceeds one of the set points, the APRM module generates an alarm and a trip signal. The APRM level "High–High" is a typical trip signal that initiates a reactor scram. The modules including FPGAs have self-diagnosis functions, which include a watchdog timer monitoring the FPGAs and periodic checks for data transmission.

Figures 8.14 and 8.15 are a schematic diagram and a photograph of the PRNM, respectively. The PRNM consists of LPRM/APRM and FLOW units. The LPRM/APRM unit houses ten LPRM modules, one APRM module, and one status module. The status module indicates the results of self-diagnostics on the front panel LEDs. In addition to the LPRM modules in the LPRM/APRM unit, the APRM module can obtain LPRM levels from 12 LPRM modules which are housed in another unit, the LPRM unit. This LPRM unit has a similar appearance to the LPRM/APRM unit. The LPRM levels produced in the LPRM modules in the LPRM unit are transmitted to the LPRM/APRM unit over fiber optic cables attached to the unit rear. The last unit of the PRNM system is the FLOW unit. It converts the differential pressure signal from the flow transmitter to the recirculation flow value and transmits the value to the APRM module which uses the flow value to calculate a flow-biased trip set point.

8.5 Development Process of FPGA-Based System

FPGAs themselves are hardware. However, the logic for these FPGA-based components is designed and manufactured by a process which is similar to that for generating software products. The logic to be embedded into an FPGA is written in a hardware description language, such as VHDL, and the code is converted into a fuse map that determines the circuit in the FPGA. To implement this process for the safety-related I&C systems, a high quality design and manufacturing processes

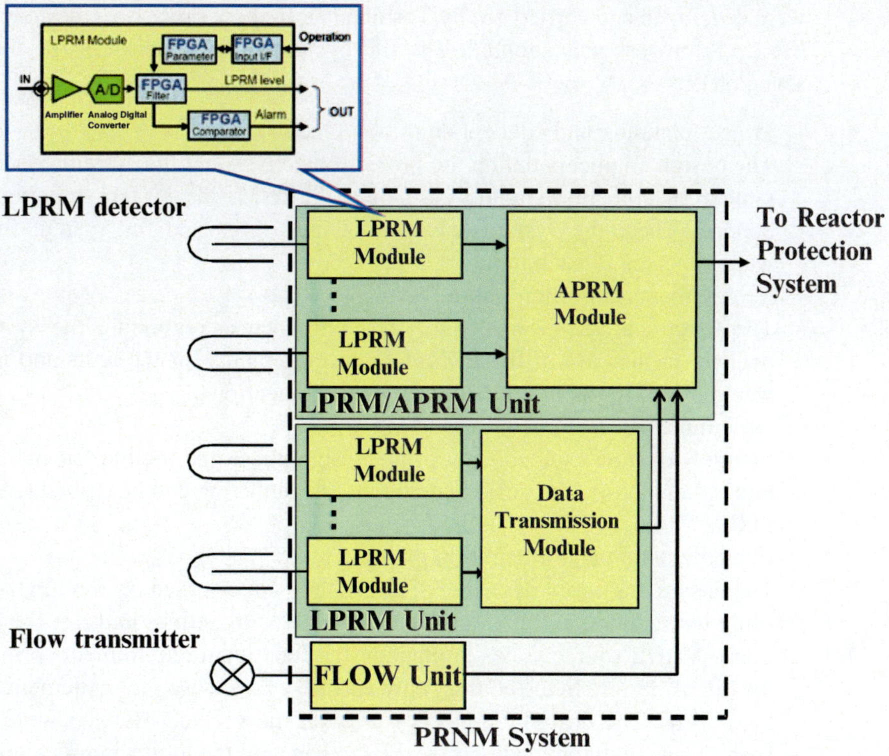

Fig. 8.14 Schematic diagram of PRNM

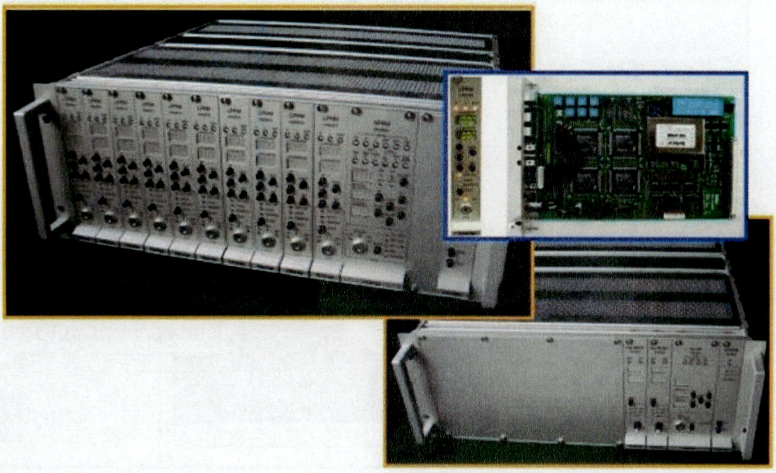

Fig. 8.15 Photograph of PRNM

were developed and carried out by Toshiba [7]. In these processes, the concept of a life cycle process was adopted. The life cycle process consists of the following seven phases.

- Project planning and concept definition phase
  The design engineers define the project objectives and the system concept as a system specification. In the case of the PRNM, the system functions were defined. Then, the system was divided into units and the system functions were allocated to each unit.
- Requirements definition phase
  The functional requirements for the components comprising the system are defined. In the case of the PRNM, the requirements for the units and modules were defined in the unit/module design specifications.
- Design phase
  Figure 8.16 shows the concept of the design phase and the implementation and integration phase. The design engineers design the logic to be embedded in each FPGA.
- Implementation and integration phase
  The design engineers describe VHDL source codes based on the FPGA design using text editors, and then convert the codes into netlists and test the netlists. These VHDL source codes implement the functional requirements provided in the FPGA design. In the coding, only verified FEs are used to implement specific logic steps. The design engineers start with the verified FEs and write VHDL source code interconnecting those FEs to generate the logic circuits required for the FPGA.

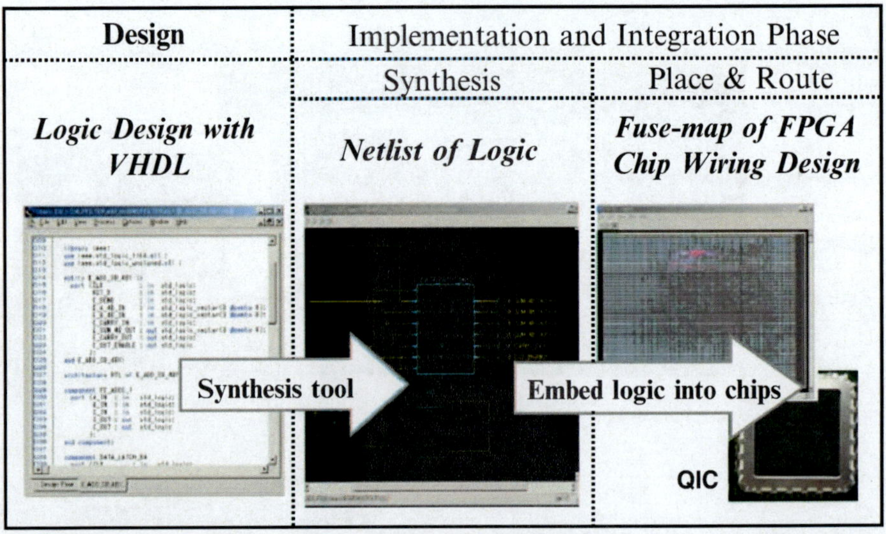

Fig. 8.16 Concept of design phase and implementation and integration phase

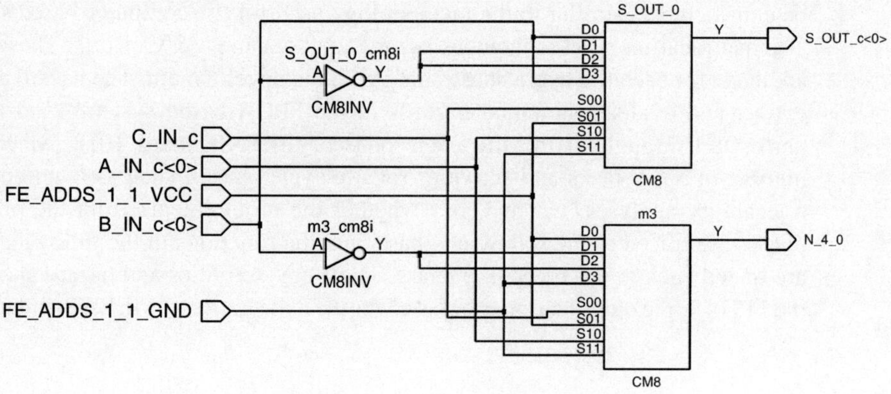

Fig. 8.17 Block diagram generated from a one-bit full-adder netlist

The design engineers convert the VHDL source codes into netlists. All FEs used in the design are individually converted into FE netlists, and merged into an integrated netlist that retains the verified FE structures. As an example, Fig. 8.17 shows the block diagram generated from a one-bit full-adder netlist. A netlist contains information representing how gates or cells are connected. After the integrated netlist is produced, the next step is producing a fuse map, which defines the connection among cells in the FPGA. The engineers embed the fuse map in the FPGA using an FPGA programming tool and perform FPGA validation testing. After FPGA validation testing is completely finished with a satisfactory result, the engineers register the FPGA fuse map.

– Unit/module validation testing phase
  Before this phase, the FPGAs embedding the logic are soldered on printed circuit boards and fabricated as modules. In this phase, unit/module validation testing is performed to validate the modules containing the FPGAs, and the units containing modules.
– System validation testing phase
  The units are integrated into a system, and system validation testing is performed.
– Operation and maintenance phase
  The system is finally installed in the plant, and operated. Appropriate maintenance is performed as needed.

8.6 Logic Qualification of FPGA-Based Systems

8.6.1 Qualification Process

Because the development process of the FPGA-based systems is similar to that of computer-based systems, the qualification process for the FPGA-based systems can be considered as similar to the methodology used for the computer-based systems. The qualification process includes hazard analyses and V&V efforts. These efforts are made for each life cycle phase along with the development. The hazard analyses extract and evaluate potential hazards in the FPGA-based systems, and are also performed throughout the life cycle phases. IEEE Standard 1012 [8] defines a number of V&V tasks and activities for life cycle phases, such as requirements of traceability analyses, which verify whether the requirements from the preceding phases are traced to the following phase, and the functions in the following phases are traced back to the preceding phase. The V&V activities and hazard analyses in the FPGA logic qualification process cover the expectations of IEEE Standard 1012.

8.6.2 FPGA-Specific Issues

Although the FPGA development process is similar to the software development process, FPGAs are not the same as computers. Basically, FPGA-based systems have many advantages over computer-based systems, such as that the logic design is rigorous, simple, deterministic, and verifiable. As mentioned above, it is a kind of hardware-based dedicated logics. However, there are some issues to be addressed in the qualification of FPGA-based systems. Assumptions that the following hazards may occur in each development phase should be considered.

- Project planning and concept definition phase
  System specification errors are possible hazards because the system specifications are established in this phase.
- Requirements definition phase
  Errors in these specifications are possible hazards because the unit/module design specifications are established in this phase.
- Design phase
  FPGA design errors are possible hazards because the FPGA design is established in this phase.
- Implementation and integration phase
- VHDL source codes are produced as combinations of FEs and converted into fuse maps. The fuse maps are programmed into the FPGAs, and FPGA testing is performed. The possible hazards in this phase are coding errors, small logic errors (FE errors), timing errors, logic synthesis errors, place and route errors, and logic embedding errors.
- Unit/module validation testing phase
  Unit/module validation testing errors are possible hazards in this phase.
- System validation testing phase
  System validation testing errors are possible hazards in this phase. The FPGA, unit/module, and system validation testing errors cause no new defects, instead insufficient testing is likely to fail to detect errors in the system. There are three causes leading to insufficient testing. The first one is test procedure errors

including insufficient or inappropriate test cases, test designs, or test methods. The second is errors in test equipment or test equipment software. The last is inappropriate performance of the testing, i.e., test personnel do not perform the testing in accordance with the test procedures that are the products of the test design. These testing errors should be prevented by quality assurance efforts.

8.6.3 Hazards Specific to FPGA-based systems

The hazards mentioned in the former section can be classified into two groups: hazards specific to FPGA-based systems and general hazards (i.e., not specific to FPGA-based systems). In this section, the hazards specific to FPGA-based systems and countermeasures to be taken against them are described.

Figure 8.18 summarizes possible errors that are considered as hazards of FPGA-based systems in the development process.

8.6.3.1 Small Logic Errors

Small logic errors are specific to the use of FEs proposed by Toshiba [6]. Although the use of FEs is one of the important points to minimize the possibility of logic errors in the FPGAs, and this use provides adequate assurance to the quality of FPGA logic, if there are errors in FEs, the approach will be unsuccessful. To eliminate small logic errors, full pattern testing for FEs is performed. For an FE, the engineers prepare a test procedure including test cases that cover all possible

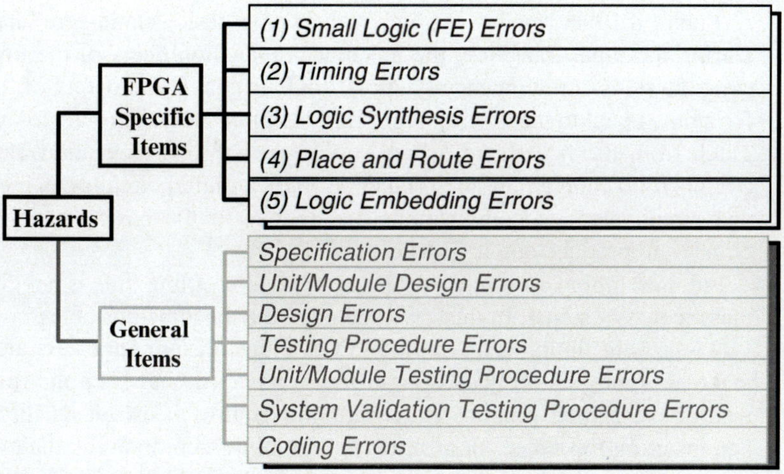

Fig. 8.18 Hazards to FPGA-based systems

combinations of input signals to the FE. The FE testing includes both simulations and actual FPGA testing.

Once the test procedure is prepared, simulation is performed to check whether the FE operates correctly. This check compares the output signals with the predetermined expected values. If the output signals are not equal to the expected values, then the FE designed will be corrected. This design correction will continue until the FE operates correctly in the simulation. After the FE is verified in the simulation, a test FPGA is programmed as a QIC, and the full pattern testing is performed for the actual FPGA chip for the same test cases used in the simulation to verify that the FPGA operates correctly.

8.6.3.2 Timing Errors

Timing errors are another hazard specific to FPGA-based systems. They are derived from the electrical signal propagation delay in an FPGA chip. As mentioned above, there are thousands of logic cells in an array in the FPGA chip. Each cell consists of logic gates. In addition to the logic cells, the FPGA chip has many signal lines running vertically and horizontally. In the FPGA programming, sets of signal lines are selected and connected by anti-fusing according to the fuse map. Therefore, the length that a signal propagates from one cell to another cell depends on the selection of signal lines in the programming.

Glitches are one type of timing errors. They are an unwanted fast "spike" in an electronic signal that is produced by timing hazards inherent in a poorly designed circuit. As such, they are undesirable switching activities that occur before a signal settles to its intended value. Glitches can cause incorrect values to be latched by asynchronous circuits within the electronic device. In particular, glitches can cause improper registering of memory values. Therefore, flip-flops are inserted in the logic to implement synchronous logic and avoid timing issues.

Figure 8.19 shows how a glitch occurs on a basic "static-zero" hazard circuit. During the input transition, the inherent propagation delay of the inverter circuit creates a transient, unintended logical "high" signal at the output of the AND gate for a time equal to the inverter signal propagation delay. This creates a short output glitch from the AND gate, which can cause improper operation of downstream circuits if no countermeasure is taken. Complex digital circuits can include embedded circuit element combinations that reduce to the basic "static-zero" hazard circuit and produce output glitches.

Another timing error may occur when enough settling time is not given to a gate that receives a signal. In this case, the gate may become unstable.

To prevent timing errors in the FPGA chips, design engineers are required to make an appropriate FPGA design rule in advance, and keep the rule during the whole development process. The design rule requires the use of synchronous design, i.e., inserting flip-flops, and restricts the use of asynchronous elements that include gates between flip-flops. The use of the synchronous design can prevent the harmful effect of glitches. The flip-flop receiving the output from the AND gate operates,

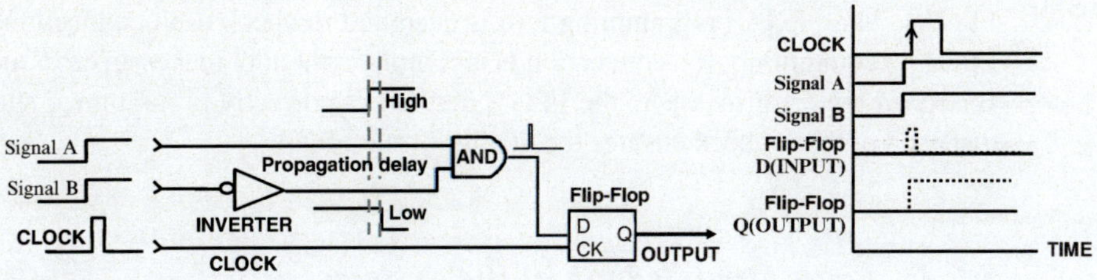

Fig. 8.19 Example of a glitch

synchronizing with the clock signal, and eliminates the spike. The restricted use of asynchronous elements gives an adequate settling time to the gate that receives the signal because a smaller number of gates will not delay the signal.

To confirm that the circuit operates without timing errors, the design engineers perform static analyses and dynamic analyses of the FPGA design, and evaluate the signal propagation time in the FPGAs. Finally, the FPGAs are validated by the unit/module validation testing, where all FPGAs are mounted on the modules and operated in the same conditions as in the actual use.

8.6.3.3 Logic Synthesis Errors and Place and Route Errors

Logic synthesis errors and place and route errors are also FPGA specific, because these processes are not used in the development of computer-based systems. Tool reliability is one of the keys to minimize these errors. However, the tools are provided by the FPGA vendor. By interviews and investigations, the tool reliability should be evaluated, based on information obtained from a vendor. Furthermore, IEEE Standard 7-4.3.2, endorsed by the United States Nuclear Regulatory Commission (USNRC), requires the approach that satisfies one or both of the following methods against software tool issues.

– Method A: A tool validation that guarantees the necessary features of the software tool function as required.
– Method B: Use of software tools in a manner in which V&V activities detect errors that tools cannot detect.

This tools used in the FPGA development are complex software and third party software. Then, method A was thought to be impractical, and method B was chosen in the case of the development in Toshiba [6].

To meet the requirements of method B, an independent check of the netlists is performed. FPGA testing is designed to ensure that every connection among FEs is toggled and is performed. In this way, the results of the testing should be satisfactory.

8.6.3.4 Logic Embedding Errors

The logic is finally embedded into the actual FPGAs. These errors differ from the other errors mentioned above, in that they are not a logic error, but a production error of

FPGAs. The FPGA programming tool is designed to check each connection during FPGA programming. If a connection is incomplete, the tool makes an error message. Based on the error message, the FPGA that is not correctly programmed should be disposed of. This check ensures the quality of the FPGAs.

8.6.4 Logic Qualification of the Systems

For the safety-related systems such as the PRNMs mentioned before, hazard analyses and V&V efforts were performed along with the development process. The hazard analyses were performed to address the potential hazards in the systems. Special attention was paid to the FPGA-specific hazards. The hazard analyses revealed that two concurrent occurrences of a failure in a system were required to get a harmful effect on plant operation, because the safety-related systems have redundant configuration. Then the possibility of a common cause failure occurring due to a logic error was minimal by taking appropriate countermeasures. As a result, the hazard analyses concluded that the risks of the system were within the acceptable level.

The V&V efforts included design review, traceability analyses, and validation testing of the systems. In addition, the V&V efforts included the review of hazard analyses. As a result, the V&V efforts confirmed that all requirements for the systems were fulfilled in the final product, and that the systems were suitable for use in NPPs.

8.7 Next Generation I&C systems; what the Future Holds

8.7.1 Design Scope and Concept

I&C systems and the MCR of ABWR were milestones in technology. As a next step in full digitalization, technology for FPGA-based systems has been developed. The FPGA-based systems are certainly very important and should be one of the core technologies for the next generation I&C systems of NPPs. However, consideration should be given to whether there will be any further innovations. As a possible answer to this question, the authors present a perspective view of the next generation system and they introduce some new developing elemental technologies.

Safety and reliability of instruments and operability were drastically improved in the digitalization of the third generation systems. Sustainability of component supply and design will be certainly guaranteed by FPGA technology, as well. However, the contribution of the I&C systems is still restricted from the viewpoint of plant life cycle (such as construction, test operation, normal operation, periodic inspection, shutdown, and decommissioning).

Figure 8.20 shows design scopes for I&C systems and the MCR. The upper bar chart shows the current design scope. The MCR, even in the third generation, has been designed by mainly focusing on plant operation (test and normal operations). During

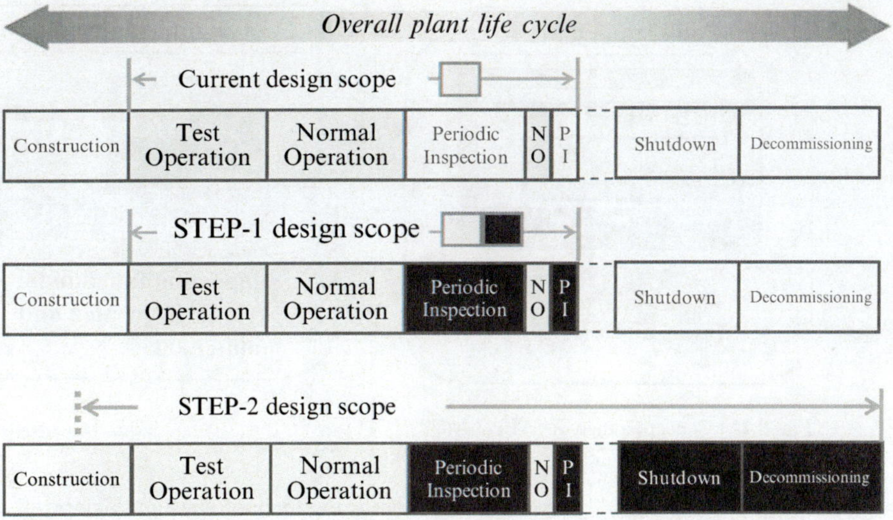

Fig. 8.20 Design scopes for I&C systems and the MCR

an inspection, I&C and MCR are used only for keeping the plant stable and monitoring it. Even in the operation phase, for example, ultimate optimized and flexible automated operation has not been realized yet. The next step of design scope, STEP-1, is shown in the middle chart. In this concept, the design scope of I&C and MCR is extended to periodic inspection phase (periodic inspection). The I&C and MCR are effectively utilized during both operation and inspection, and then the MCR becomes a control station even in inspections. The final design scope is STEP-2, shown in the bottom chart. In the concept of STEP-2, coordination of overall plant life management activities and ultimate safety automated operation are established. In this step, MCR will be an effective management and operating station for plant total life management. At the first phase in a plant construction, various kinds of information, such as specifications, performance data of materials and equipment are systematically stored in a record database system, indexed to the respective installation conditions. After that, each record is utilized in each plant phase by mutual coordination. Records continue to be kept, updated, replaced, and traced during operation and inspection phases, and are inherited until the decommissioning phase.

Figure 8.21 shows a concept image of the STEP-1 MCR configuration. The MCR is accompanied by maintenance functions using a central information board. The central information board has supporting functions for periodic inspections and other maintenance activities. They are based on the condition monitoring tools and monitoring tools of progress and results of the periodic inspection, evaluation and reporting tools coordinating with the above information. By combining a third generation MCR and a central information board, the room is utilized for not only operation, but also for inspection. For maintenance activities, the maintenance plan and construction schedule will be optimized and planned in the office using the data of condition monitoring results, inspection results, prediction tools, and other evaluation results.

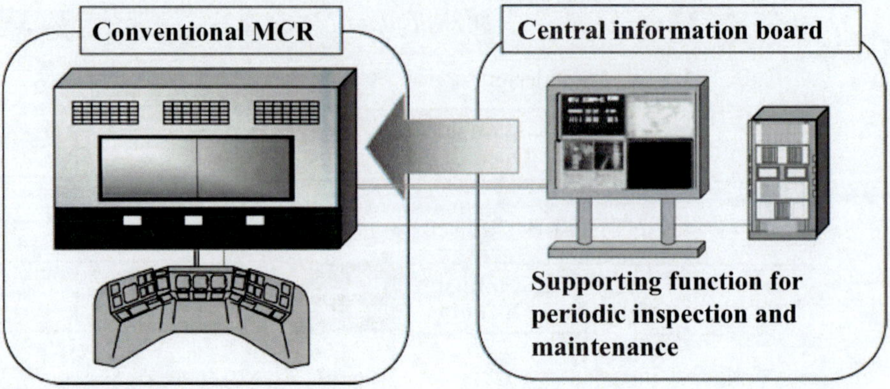

Fig. 8.21 Concept image of STEP-1 MCR

To realize STEP-1, conditions of monitoring technologies and diagnosis technologies should be extended further not only to mechanical equipment, such as pumps and valves, but also to sensors, electronics, water chemistry, etc. Coordinating inspection results and condition monitoring data are important. To maximize the plant availability while keeping plant safety, the plant maintenance decisions (plans and schedules) will be made by coordinating the plant operating records with condition monitoring data and inspection results. Aging trend predictions, optimized overhaul times, degradation predictions for materials and equipment, and other various data related to plant availability will be evaluated and coordinated with each other.

8.7.2 Elemental Technologies

Various measurement and monitoring tools and data mining and diagnosis tools are required to establish the "central information board" described above, and these are now under development. Some examples are described here.

8.7.2.1 Measurement and Monitoring Tools

As an example of a measurement and monitoring tool, a fiber-optic sensing system using fiber Bragg grating (FBG) sensor is shown in Fig. 8.22 [9]. FBG sensors are embedded in optical fiber transmission lines.

Grating pitch defines the characteristic of reflection wavelength in a FBG. Microexpansion and contraction of the FBGs, induced by temperature and mechanical strain, change the grating pitch. As a result, reflection wavelength changes, and physical quantities (temperature and strain) can be obtained by interrogating reflection wavelength of the FBGs. In Fig. 8.22, temperature sensors, static strain sensors, and vibration sensors (dynamic strain) are connected. About 200 FBGs can be connected and 200 points giving physical quantities can be monitored using a light source and a data acquisition system. The sensors and transmission lines of this system can operate without electric power supply. Especially FBGs with single mode fibers using infrared light (1.55 μm) are radiation-tolerant, compared with silicon devices. Currently, this kind of system is still being tested and its functions and performance are being confirmed in laboratories. However, it will be a promising tool for online monitoring of conditions in the primary containment vessel during plant operation in the near future.

The next example is a condition monitoring method of sensors and electronic instrumentations. Conventionally, vibration data of pumps and motors are frequently measured as condition monitoring items. However, from now on, condition monitoring of sensors and electronic instrumentations in the fully digitalized system will become more important. Then, various methods for condition monitoring of sensors and electronic instrumentations are now under development using various algorithms such as statistics methods, neural networks, fuzzy logic, and so on. In the fully digitalized system, almost all sensor readout information is collected and stored as digital data in plant computer systems. Using these data, signal drift and other abnormal status can be flexibly examined and evaluated by computer arithmetic algorithms.

Figure 8.23 shows an example of drift detection results obtained using a statistic method [10]. In this figure, dummy drift data were added to readout the trend data of water level of a tank, and the trend data are evaluated using the sequential probability ratio test (SPRT) method to detect a drift of magnitude less than the required loop accuracy. The early period after reaching full power operation is defined as the base period. In this period, sensors and instrumentations were doubtlessly exposed to a drift phenomenon, because they were accurately calibrated during an inspection. In the base period, a reference distribution of readout values is calculated. On the other hand during a monitoring period, a distribution of that is

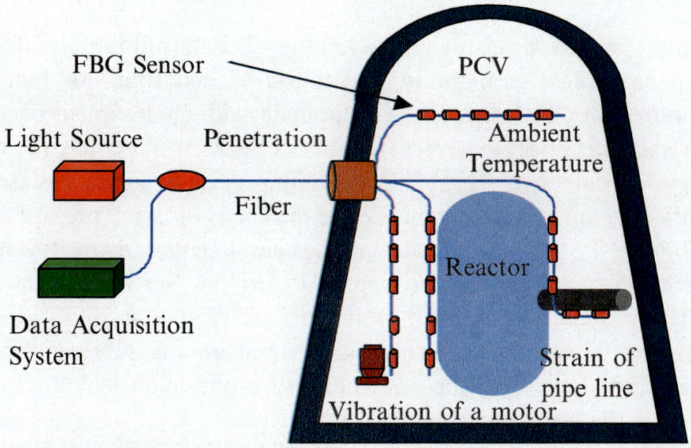

Fig. 8.22 Fiber-optic sensing system as a condition monitoring tool

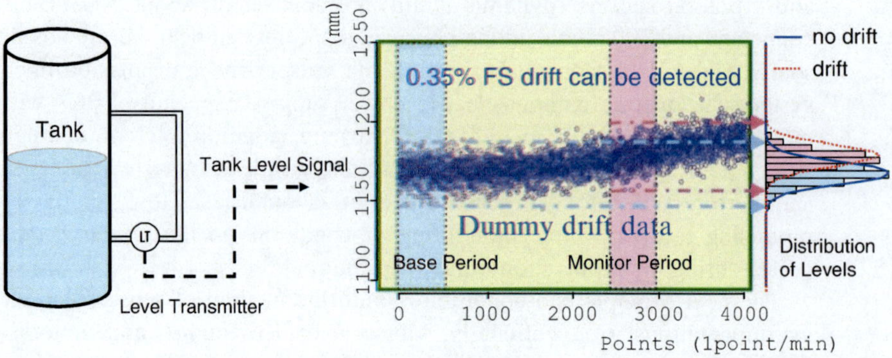

Fig. 8.23 Example of drift detection using SPRT method

newly calculated. A change between these two distributions is examined using SPRT method, based on "null hypothesis: not changed" and "alternative hypothesis: changed." If the alternative hypothesis is accepted, the readout data of the sensor are compared with the sensors of concern to discriminate value change due to a sensor drift from that due to a process state change. Finally, the occurrence of drift phenomenon is recognized.

The detection example of Fig. 8.23 detects a drift of 0.35% (of full-scale). In this case, the sensor loop accuracy is 1%. As shown in this example, this method can detect a drift of magnitude less than the loop accuracy. Using a statistic method such as SPRT, management workload of individual threshold values for the amount of sensors can be minimized, and the reliability and accountability for the determination of drift recognition will be improved. The calibration and maintenance period of sensors can be optimized using these results.

8.7.2.2 Data Mining and Diagnostic Tools

Another important elemental technology is data mining and diagnostic tools. To keep high plant availability, utilization of conditions of monitoring data, and coordination of information of equipment with environment conditions are important and effective. However, amounts of these data are massive and their relationships are complex. Then, a data mining system with databases and evaluation methods relating to maintenance are the focus of much present research.

Figure 8.24 shows an example system, "Device Record Management System (DRMS)" based on this concept [11]. In this software, data are stored in the database. It is like a patient's medical chart in a hospital. Monitoring and data sampling of such items as vibration, temperature, lubricant content, electronics error, and signal drift are equivalent to a physician making an auscultation and entering the information in the patient's chart.

Stored data in the database are handled and data mined by the Device Record Management System. Degradation detection, aging trend prediction, and overhaul

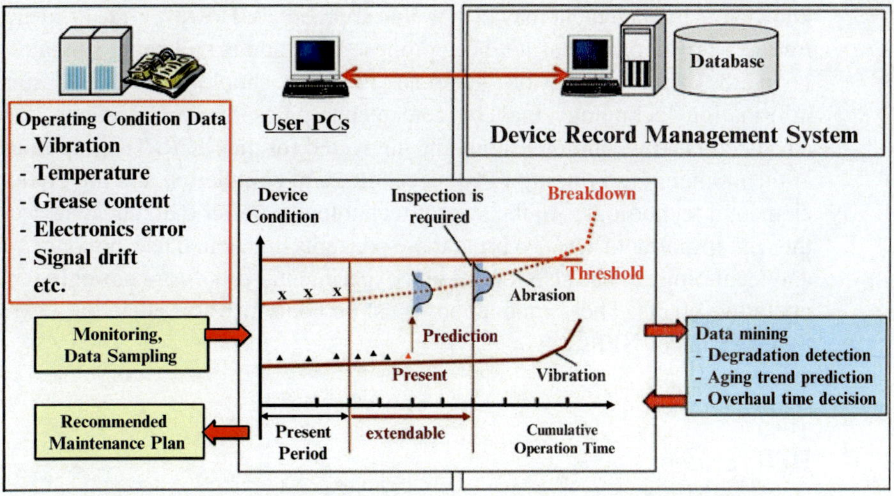

Fig. 8.24 Device Record Management System: statistical analysis tool for maintenance and renovation planning

time decision are reported. The prediction is performed basically by using statistic models. A physical model may be accurate and the best approach in some cases; however, not all equipment and phenomena can be represent by accuracy physical models. In this system, a statistic prediction method using probability calculations was adopted. Maintenance plans as outputs of this system can offer effective prospects for inspections, such as an extendable time without an inspection, a recommended inspection timing, and a breakdown timing. Effective inspection and renovation plans can be elaborated using this information.

8.7.3 Toward the Next Generation

In the ultimate phase, STEP-2, I&C systems, and the MCR will be the core tools and the central station for total plant life management. The concept of total plant life management is that the important values, such as safety, security, economy, and ecology, are optimized and accord with each other throughout the plant life span. Compared with STEP-1, the amount of data is even more massive, and the data coordination is even more complex, because initial information on equipment and materials is inherited throughout the total plant life until its decommissioning. To handle these data, flexible integration and coordination technologies for a vast array of data are required. The conventional video display concept of a HMI and conventional database applications do not seem to be enough. Multidimensional real-time visualization technologies and virtual display technologies may be required. For software technology, agent technology, synthesis of ontology technology, and knowledge management may be possible approaches. However, the critical solution for the STEP-2 phase has not been proposed yet, and is still being sought.

In the next generation systems, FPGA technology and the state-of-art information technology may be core technologies for I&C systems; further an advanced HMI technology should be innovated for the MCR. The important things at the moment are bringing FPGA technology to completion and integration of the elemental technologies to the system technologies. After that, an advanced HMI is the next important issue. To bring these concepts into actual use, not only new ideas and technologies, but also logical and experimental proofs are strongly required at operating plants. These innovations must be continued to realize the next generation systems of NPPs.

References

1. United States Nuclear Regulatory Commission, Fact sheet available via http://www.nrc.gov/reading-rm/doc-collections/fact-sheets/3mile-isle.html
2. Fujii K, Neda T, Takamiya S, Suto O, Ikeda Y, Hayakawa H (1983) BWR plant advanced central control panel – PODIA. IEEE Trans Nucl Sci 30(1):833–837
3. Mori N, Makino M, Naito N (1992) Advanced instrumentation and control technologies for nuclear power plants (in Japanese). Toshiba Rev 11(47):842–844
4. JEAG 4609, Application criteria for programmable digital computer system in safety-related system of nuclear power plants, Japan Electric Association
5. First Workshop on the Applications of Field-Programmable Gate Arrays (FPGA) in Nuclear Power Plants, 8–10 October 2008, EdF R&D, Chatou, France. Presentation files are available via http://entrac.iaea.org/I-and-C/WS_EDF_CHATOU_2008_10/Start.htm
6. Goto Y, Oda N, Igawa S, Odanaka S, Tanaka A, Izumi M (2004) Development of FPGA-based safety-related I&C systems, Proceedings of 14th Annual Joint ISA POWID/EPRI Controls and Instrumentation Conference, No. 015, Colorado Springs, CO, USA, 6–11 June 2004
7. Hayashi T, Oda N, Ito T, Miyazaki T, Haren Y (2009) Logic qualification of FPGA-based safety-related I&C systems. Proceedings of ICAPP '09, No. 9251, Shinjuku Tokyo, Japan, 10–14 May 2009
8. IEEE Standard 1012-1998, IEEE Standard for software verification and validation, IEEE Standard Association
9. Arai R, Sumita A, Makino S, Maekawa T, Morimoto S (2002) Large-scale hybrid monitoring system for temperature, strain, and vibration using fiber Bragg grating sensors. Proceedings of SPIE, vol 4920, 62–72, Shanghai, China, 14 October 2002
10. Hirose Y, Tamaoki T, Hayashi T, Enomoto M, Maekawa T, Masugi T (2008) Online inspection of sensors in nuclear power plants. Trans Am Nucl Soc 99:771–772
11. Sonoda Y, Hirose Y (2003) Inspection and condition monitoring service on the web for nuclear power plants. Proceedings of HCI International 2003, 1308-1312, Crete, Greece, 22–27 June 2003

Chapter 9
Advanced 3D-CAD and Its Application to State-of-the-Art Construction Technologies in ABWR Plant Projects

Junichi Kawahata

Since the first nuclear power plant (NPP) was constructed in the 1960s, more than 50 NPPs have been built in Japan. As an active player in the field of NPP construction, Hitachi, Ltd. (now, Hitachi-GE Nuclear Energy, Ltd. (HGNE)) has constructed 22 of these Japanese NPPs till 2009. Through this extensive experience, HGNE has developed and applied its own advanced technologies, including a unique 3D-CAD-based integrated plant engineering environment and streamlined design-to-manufacturing/construction management system. These technologies have been continuously improved with the evolution of HGNE's construction management philosophies, often providing an enabling force to greater achievements in project performance. In addition to the latest ABWR Shika-2 completed in 2006, HGNE is currently leading two more ABWR construction projects, Shimane Unit 3 and Ohma Unit 1, both of which are on target for an "On-Budget and On-Schedule" completion. In this chapter, the state-of-the-art engineering and construction technologies established on the advanced 3D-CAD platform, currently being applied to these projects, are introduced.

9.1 3D Integrated Engineering System

9.1.1 Introduction

The HGNE 3D-CAD system, in which physical plant facilities are visually modeled using 3D computer graphics, is widely applied to the nuclear plant design process. Although there are some general-purpose commercial 3D-CAD systems available, HGNE mainly uses its own in-house system that was developed and customized

J. Kawahata (✉)
Hitachi-GE Nuclear Energy, Ltd, Tokyo, Japan
e-mail: junichi.kawahata.hh@hitachi.com

T. Saito et al. (eds.), *Advances in Light Water Reactor Technologies*,
DOI 10.1007/978-1-4419-7101-2_9, © Springer Science+Business Media, LLC 2011

specifically for HGNE's nuclear plant projects. Described here are its unique features and its application to layout design, manufacturing design, fabrication, and construction planning and management.

HGNE was the primary contractor in the construction of Japan's first domestically produced NPP in 1974, Shimane Unit 1. Since then, HGNE has led the design and construction of over 20 domestic NPPs, which implement its world-leading boiling water reactor (BWR) technologies and nuclear equipment.

In the 1980s, HGNE developed its proprietary computer system, specially adapted to its plant design approach and construction methodologies. However, in the beginning, development of the 3D-CAD system was initiated to simply provide a replacement for plastic models. It has since proven itself a far more valuable tool than originally conceived through the practical application of lessons learned from many actual NPP construction projects. It was through the continuous optimization of this 3D-CAD system over the last 20 years that the engineering database, in combination with the 3D-CAD system, is now regarded as a core engineering tool, integrating the design of both upstream and downstream systems. The entire system, comprising the design systems, the manufacturing support systems, the construction support systems, and the engineering database is called the "Plant Integrated CAE System."

9.1.2 Outline of the Plant Integrated CAE System

The outline of the Plant Integrated CAE System is shown in Fig. 9.1. 3D plant design work, including piping layout, is performed utilizing upstream design information such as the plot plan and system design data. After a careful review

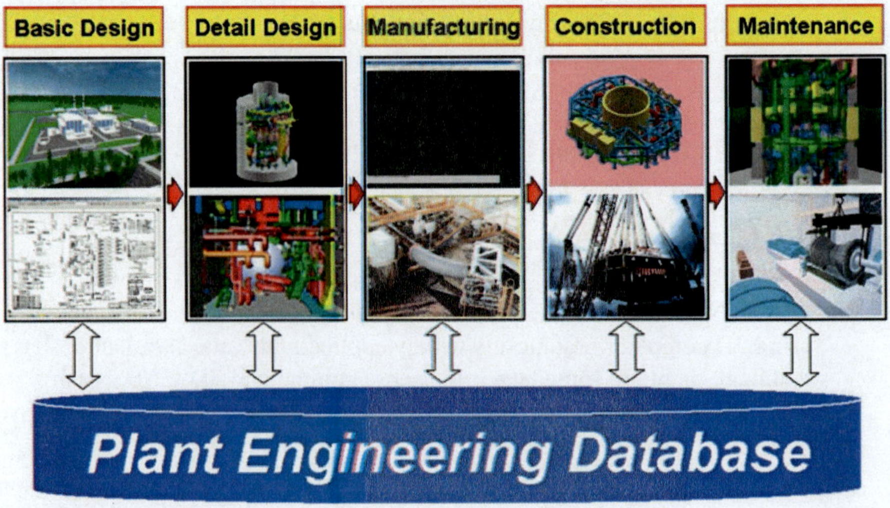

Fig. 9.1 Plant Integrated CAE System

of the 3D plant design, fabrication and installation information, such as shop and field weld data, are added to the 3D data. Fabrication design is performed using the 3D data stored in the database, which is used to produce fabrication drawings, procurement data, and input data for computer-controlled fabrication machines. 3D design data and fabrication data are then available for construction planning, which includes activities like designing temporary facilities, planning a construction schedule, planning a detailed construction sequence, and reviewing the moving range of construction cranes.

To support these activities, the Plant Integrated CAE System is linked with approximately ten computer servers in HGNE's design office and local construction offices. Within the last decade, the rapid improvement in PC performance has enabled the widespread use of graphics workstations. The HGNE system has grown to fully utilize this capability and integrates hundreds of PCs throughout HGNE, connecting the design office, the factory, the construction sites, subvendor offices, and customer offices.

9.1.3 Plant Layout Design Using 3D-CAD

9.1.3.1 3D Design System and Database System

Figure 9.2 shows the plant 3D-CAD system architecture, which includes the 3D-CAD database, the 3D layout CAD program, and the input and animation functions.

The detailed arrangement of equipment and piping is crafted in harmony with the general arrangement of each building. In the process of creating a layout plan with

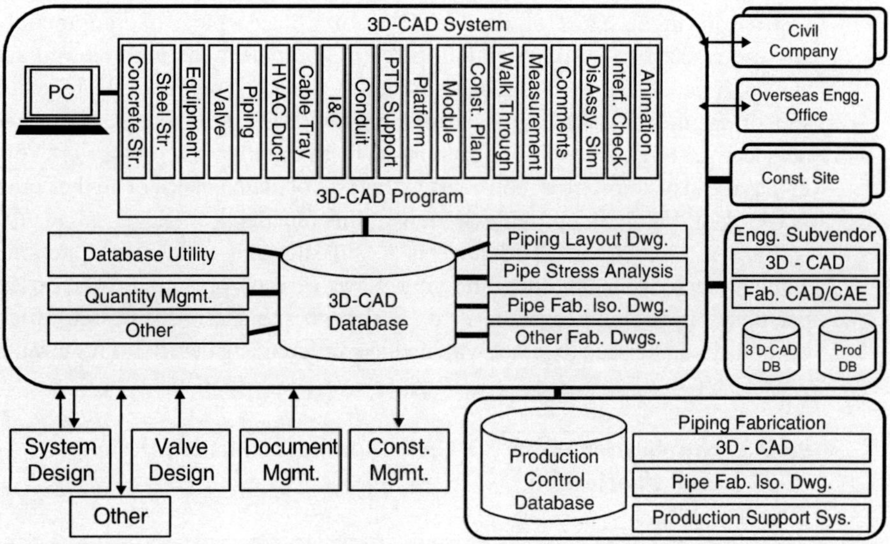

Fig. 9.2 Plant 3D-CAD system architecture

3D-CAD, various checks are made by employing the design support functions included in the software. This ensures an expeditious and high-quality piping layout design. The layout plan is critical to assuring that fundamental characteristics like safety, operability, maintainability, and constructability are fully optimized. Therefore, the plan is reviewed and checked in detail to ensure that it satisfies all requirements of plant equipment and operations. Planning of the building structure and auxiliary facilities is carried out simultaneously with the layout design, and a synthesized arrangement model ("composite coordination") is made to integrate the concrete design requirements and the requirements of all relevant specialist disciplines.

For the manufacturing and fabrication design, detailed information is created using the design system and stored in the 3D database. This information includes items such as material specifications, welding procedures, and inspection requirements. This information is later used for material and parts procurement as well as the production of pipe fabrication drawings in the downstream design process.

Most importantly, the database has improved cooperation between the engineering, manufacturing, and construction specialists by facilitating the sharing of information through the groupware system and improving the organization of data through the document management system.

9.1.3.2 Review and Evaluation System

The "walk-through" function enables an engineer to visually check the plant layout by creating a virtual plant model through which the engineer can navigate as the first person. This graphical representation aids the engineer in reviewing the operability and maintainability of the plant. For example, any interferences or restrictions in the space required to install or replace a piece of equipment or in the space necessary to easily operate a piece of equipment can be reviewed and redesigned, if necessary. Several other simulation functions on 3D-CAD are available to support engineers as well as the customer in reviewing the layout design from various perspectives (e.g., accessibility, equipment disassembly/reassembly, work volume for in-service inspection, etc.), considering the ease of maintenance of all the components in the plant. Figure 9.3 provides a view of this function.

Finally, "remote CAD review" is a walk-through simulation function that is available across the Internet, allowing views of each operation to be observed by personnel at different locations (e.g., at the job site and in the design office). This capability can be used together with a videoconferencing device for review meetings.

9.1.3.3 Application of the 3D-CAD Data to Production, Design, and Fabrication

After reviewing the plant layout and finalizing the information and the specifications that use the information from the 3D-CAD database, pipe fabrication drawings are

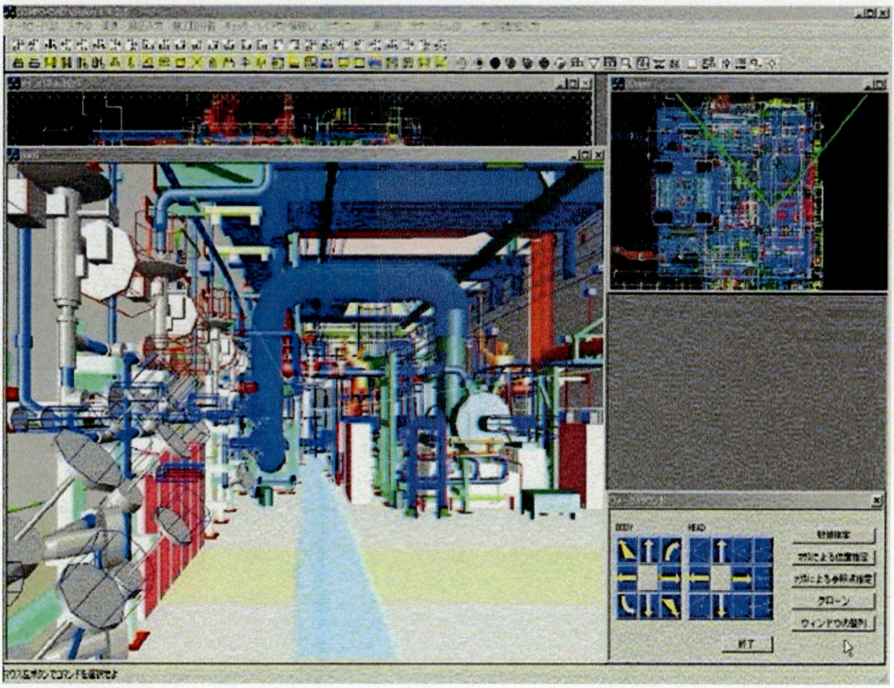

Fig. 9.3 Walk-through simulation

created. Here the automatic design application of the CAD system improves the reliability and quality of the design, while reducing the man-hours to create the drawings. The features of the system are shown below:

1. Automatic specification of pipe parts (fittings, gaskets, bolt-nuts, etc.)
2. Automatic generation of manufacturing information
3. Automatic generation of bill-of-materials information
4. Automatic generation of dimensions and annotations
5. Automatic generation of numerical control (NC) data for pipe processing
6. Automatic generation of shop and field inspection information
7. Cooperation with construction planning and management systems

Production information is added to the piping design data automatically, and is stored in the production control database of the pipe fabrication shop. Based on this information, piping production is optimized in the master schedule and is broken down into weekly and daily schedules. Moreover, a "work instruction" document is sent to a work team and NC data for a production machine are generated and published from the database. Finally, a "work actual result" is fed back into the database for management purposes.

3D-CAD is also an indispensable tool in modularized construction, which is one of HGNE's advanced construction methods. By using 3D-CAD, module design is carried out in parallel with layout design, achieving a more optimal modularization

Fig. 9.4 Construction module

and construction plan. Key information on factors such as module manufacturing and parts delivery schedule is integrated into the database, and can be used at the module factory. An example of the 3D-CAD application to modularization is shown in Figure 9.4.

9.1.3.4 Application to Construction Planning and Management

Construction engineers at the office, or construction supervisors at the construction site, use the 3D-CAD data to review layouts to facilitate construction planning and management. Also, 3D-CAD data are exchanged with civil companies, enabling them to create optimal and comprehensive construction plans using the visual engineering environment.

The construction planning of a nuclear plant can be categorized into schedule plan, shipping plan, installation plan, and temporary facility plan. The construction planning begins at the conceptual planning phase of the plant, and proceeds from a general master plan to a detailed plan, taking into account customer requirements, engineering requirements, civil company variables, and the capabilities of subcontractors.

Construction Planning System

The construction planning system helps construction engineers in the office and supervisors at the site create the submaster schedule, detailed schedule, carry-in plan, installation sequence, temporary facility plan, yard plan, crane and lifting device plan, and so on. Figure 9.5 shows an example of a typical construction animation. With respect to the design and manufacturing processes, site schedules are intended to be highly accurate to allow for Just-in-Time (JIT) delivery of drawings and products, while maintaining the flexibility required to deal with the

Fig. 9.5 Construction animation example

unexpected. This system, together with the interconnected database system, helps us create the JIT plan.

Construction Management System

The construction management system helps people at a construction site to control schedules and resources, together with documents and other necessary information. The following are the major purposes of the construction management system:

1. Effective use of the product data created with 3D-CAD in installation and inspection work
2. Efficient control of products and parts delivery
3. Rapid response to any sudden changes
4. Real-time support for construction management
5. Quick feedback to the design and the manufacturing specialists of construction status and changes

The construction management system uses product data, such as equipment lists, valve lists, pipe spool lists, welding numbers in each pipe fabrication drawing, pipe support numbers, inspection requirements and schedules, etc., generated by 3D-CAD and other systems. A "work instruction sheet" is created by the system for each job, such as welding or inspection. When a work team finishes the job, results are collected and stored in the database. This is done on a daily basis and managers can check the updated construction status in real time. At our construction sites, cooperation between HGNE and all construction subcontractors is achieved using this system with a PC network.

9.2 Advanced Construction Technologies

9.2.1 Introduction

Over the last few decades, Japan's nuclear plant construction environment has changed dramatically. For example, the number of construction workers has decreased,

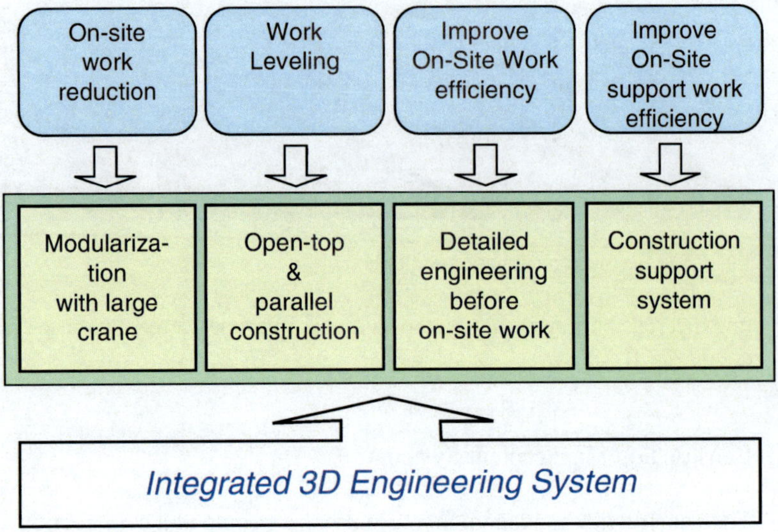

Fig. 9.6 Concepts of construction strategies

while the average age of workers has increased. However, customer demand for cost reduction and shorter construction duration has continued to grow stronger. Therefore, the optimization of construction is a key issue in the power plant business. To answer these demands for improved performance, HGNE has focused on construction strategies that are based on feedback gained from many years of NPP construction experience in Japan. HGNE has developed technology-based solutions to each strategy that have provided significant gains in overall project performance. Our key strategies are shown below and summarized in Figure 9.6:

1. Reduce on-site work volume
2. Level on-site manpower
3. Improve on-site work efficiency
4. Improve on-site support work efficiency

These concepts are quite simple in principle; however, their effectiveness has been proven through the successes of past projects.

One of the world's latest new-build projects, the 1,358 MW ABWR Shika Unit 2 (Shika-2) of Hokuriku Electric Power Company, was constructed "On-Budget and On-Schedule" applying these technologies. Shika-2 was the first ABWR plant in which all the major pieces of equipment, including the reactor pressure vessel, turbine, and generator, were supplied and installed by one prime contractor, HGNE. HGNE was also responsible for everything from the basic design through commissioning. The construction started with the foundation excavation of the main building in September 1999, and, 58 months after rock inspection, the plant started commercial operation in March 2006.

In Japan, two more ABWR projects, Shimane Unit 3 and Ohma Unit 1 are now under construction. Again, HGNE is taking a primary role and applying more advanced design and construction technologies.

The world energy market currently faces the Nuclear Renaissance, partly fueled by a demand for noncarbon generating energy sources. The result is a strong pull for new NPP construction projects all over the world. How well we can manage the construction is not only crucial for any one project's success, but also the extent to which nuclear energy will remain the right solution amidst the challenges of the current world economy.

9.2.2 Applied Construction Technologies

Among the advanced construction technologies applied to Shika-2, the following provided the greatest benefits:

1. Broader application of large module/block construction method
2. Open-top and parallel construction method
3. Application of floor packaging construction method
4. Full application of information technology to quality plant engineering and construction

An approximate 25% peak workload reduction at the construction site was achieved through the implementation of these technologies. These technologies are briefly introduced in the following sections.

9.2.2.1 Broader Application of Large Module/Block Construction Method

The large module/block construction method is one of HGNE's hallmark construction strategies. This method utilizes a heavy-lift crane for lifting and installing large-scale modules/blocks.

HGNE has employed this method on the construction of NPPs since the early 1980s. More than 1,000 modules have been manufactured and installed so far. During the design stage, a 3D computer-aided engineering (CAE) system that has special features for module engineering is fully utilized. In 2000, HGNE established a dedicated module factory and has since assembled and shipped all shop-manufactured modules from this factory. The factory is fully enhanced and integrated with the CAE and project control systems.

For the Shika-2 construction effort, about 200 mechanical and electrical modules were applied. Among them, the RCCV upper drywell module was one of the greatest achievements. The RCCV upper drywell module consisted of a large steel structure, radiation shields, pipes, valves, and other components contained in the BWR drywell. The total weight of the module was approximately 650 metric tons (see Figure 9.4.)

9.2.2.2 Open-Top and Parallel Construction Method

The open-top and parallel construction method is currently applied to NPP construction in Japan. When applying this method, construction activities from both the civil and mechanical engineering disciplines are conducted in parallel with detailed coordination. Major components must be placed before the ceiling concrete work starts. After the concrete cures in the walls and ceiling, mechanical installation work within the room or area starts. At the same time, wall concrete work and then major component carry-in is performed on the level above. This process continues until the top floor is reached. The method demonstrates how building construction and mechanical/electrical installation work proceed on simultaneous paths, while offering the additional benefit of leveling the peak labor requirements at the construction site. Since various activities are going in parallel,

Fig. 9.7 Open-top and parallel construction method

this method requires very detailed coordination between civil engineering and mechanical/electrical contractors through installation procedures, schedules, tight delivery controls, etc. Figure 9.7 offers a view of an actual application of the open top and parallel construction method.

9.2.2.3 Application of Floor Packaging Construction Method

Traditionally, hydrostatic pressure testing is conducted after the system is completely assembled, including components. The purpose is to demonstrate the integrity of the entire system; however, this approach also results in a large concentration of labor hours for testing toward the end of construction. HGNE has developed a new concept called the floor packaging method which resolves this issue. In this method, partial hydrostatic pressure testing (floor by floor) of a system is performed prior to completion of the whole system. This method enables us to close-out all the work in each construction area from the bottom floor up and offers the benefit of leveling the maximum workload typically experienced toward the end of construction. Figure 9.8 depicts this benefit.

9.2.2.4 Full Application of Information Technology to Quality Plant Engineering and Construction

The application of the advanced 3D-CAD system, which was introduced in the prior section, has made the plant layout design more efficient and accurate. In addition,

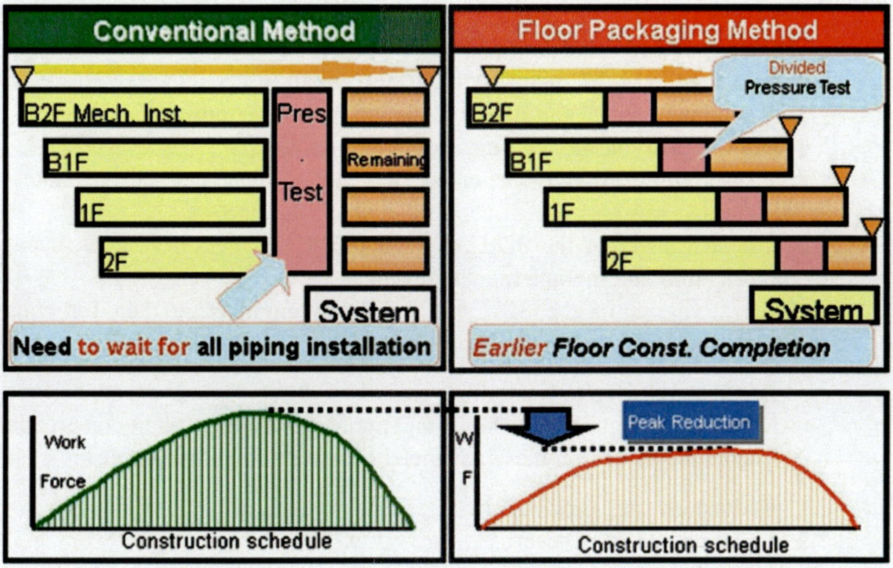

Fig. 9.8 Floor packaging method

its simulation function has helped to improve the engineering approach for plant operability, such as accessibility, maintainability, and constructability. The simulation functions have also made it easier to confirm the transport paths for component installation and to optimize crane usage.

9.2.3 *Development of Advanced Technologies*

In the current on-going project, more advanced technologies are being developed and applied. Application of radio frequency identification (RFID) can be noted as one such advanced technology. More than 100,000 RFIDs are utilized in the areas of manufacturing, transportation, and installation of components in HGNE's latest ABWR construction project, Shimane-3.

9.3 Conclusion

In this chapter, an advanced 3D-CAD system and its application to the latest ABWR were introduced.

The Plant Integrated CAE System (3D-CAD) facilitates efficient construction from the earliest design phase through the completion of the construction effort. Specifically, the following are the advantages and benefits of the system that significantly strengthen HGNE's design and construction capability:

1. Comprehensive software architecture to support planning, engineering, manufacturing, procurement, and construction
2. Extended 3D-CAD plant layout design capability
3. Visual review and simulation of design including remote CAD review
4. Information sharing with interconnected databases
5. Proper engineering management with target date and design status control systems
6. Direct transfer of the 3D-CAD data to the production systems, such as pipe fabrication and module manufacturing
7. Effective use of the 3D-CAD and other engineering data for construction planning and construction management

Based on the system capabilities described above, various advanced construction technologies have been developed and applied to ABWR plant construction.

The concepts behind these advanced construction technologies are as follows:

1. Reduce on-site work volume
2. Level on-site manpower
3. Improve on-site work efficiency
4. Improve on-site support work efficiency

HGNE has demonstrated and proven these practical engineering concepts and construction technologies through a long history of NPP construction projects. HGNE is confident that these concepts and technologies can be effectively applied to any NPP construction project in the world with expectations of similar results.

References

1. Yamanari S et al (2006) The development of a comprehensive integrated nuclear power plant construction management system. Hitachi Rev 88(2):173–178
2. Morita K, Akagi K et al (2006) Advanced construction technology for Shika Nuclear Power Station Unit No. 2 of the Hokuriku Electric Power Company. Proceedings of the 15th Pacific Basing Nuclear Conference (PBNC), Sydney, Australia, 2006
3. Akagi K, Akahori S et al (2008) The latest application of Hitachi's state-of-the-art construction technology and further evolution towards new build NPP projects. Proceedings of the 29th annual conference of the Canadian Nuclear Society (CNS), 2008

Chapter 10
Progress in Seismic Design and Evaluation of Nuclear Power Plants

Shohei Motohashi

In Japan, seismic design of nuclear power reactor facilities is examined according to the "Regulatory Guide for Examining Seismic Design of Nuclear Power Reactor Facilities" [1]. Therefore, the seismic design of a nuclear power plant (NPP) is conducted based on this guide.

The latest guide was revised in September 2006 by the Nuclear Safety Commission. The previous guide was established in 1978 and was partially revised in 1981. The major reason for revising the guide has been to take into account new knowledge about seismology and earthquake engineering, which were accumulated in recent several years through the experience of several big earthquakes such as the Hyogo-ken Nambu Earthquake in 1995.

The main revised items are as follows:

(a) Taking into account new information from topographical and physical surveys regarding investigation of active faults.
(b) Taking into account uncertainties of parameters for determination of the design basis earthquake ground motions (DBEGMs).
(c) Using the strong motion simulation method using a fault model in addition to the conventional empirical method to estimate the earthquake ground motion.
(d) Improving the assessment of a DBEGM whose source cannot be identified.
(e) Improving the classification of importance in seismic design by using three classes instead of four.
(f) Taking into account accompanying events of an earthquake such as the slope stability and the tsunami safety.
(g) Taking into account the concept of residual risk.

S. Motohashi (✉)
Japan Nuclear Energy Safety Organization, Tokyo, Japan
e-mail: motohashi-shohei@ines.go.jp

T. Saito et al. (eds.), *Advances in Light Water Reactor Technologies*,
DOI 10.1007/978-1-4419-7101-2_10,

In particular, the concept of the residual risk is newly added to the revised guide. It is based on the premise that the possible occurrence of stronger earthquake ground motions exceeding the determined DBEGMs cannot be denied. One of the valid methods to evaluate the residual risk is a seismic probabilistic safety assessment (SPSA).

In this chapter, outlines are presented for the seismic design procedures of nuclear reactor facilities based on the revised guide [1] and for the SPSA to evaluate residual risk. Recent seismic topics from the big earthquake which occurred in July 2007 near the Kashiwazaki–Kariwa NPP are also briefly described.

10.1 Outline of Seismic Design of NPPs

An outline of the seismic design flow for NPPs is shown in Fig. 10.1. First, an investigation of previous earthquakes around a site is conducted. Then DBEGM is determined based on the investigated earthquakes. Next, seismic design of buildings and equipment is conducted against the DBEGM considering seismic classification of the facilities. On the other hand, accompanying events of earthquake, such as land slope stability and tsunami safety, are also confirmed for the DBEGM. At the last stage of the seismic design, residual risk (seismic risk for over the DBEGM) of the NPP to earthquake hazards is assessed.

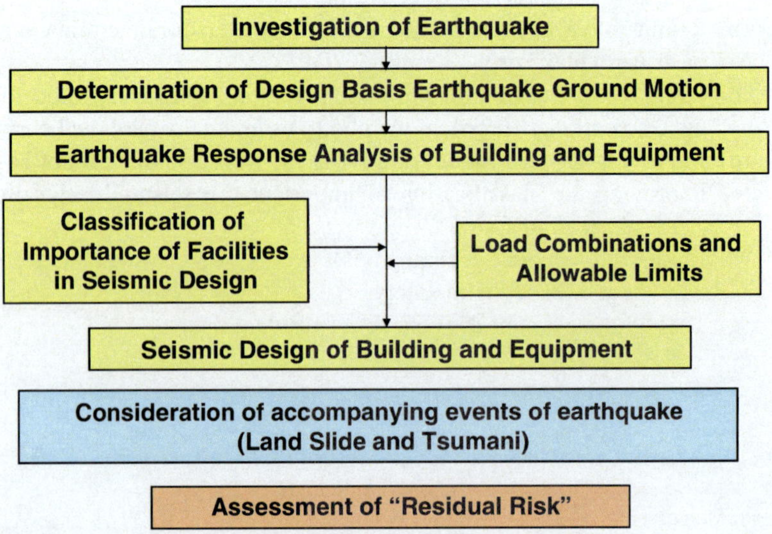

Fig. 10.1 Outline of seismic design of NPPs

10.1.1 Investigation of Earthquakes

10.1.1.1 Plate Tectonics Around Japan

Plate tectonics around the Japanese archipelago is shown in Fig. 10.2. The islands are on the Eurasian Plate and the North American Plate; and these plates are a continuous plate with East Asia and North America. At the same time, the Philippine Sea Plate and the Pacific Ocean Plate are moving and they push and sink under the Eurasian Plate and the North American Plate at the southern and eastern sides of the Japanese archipelago. These plate movements cause earthquakes in and around Japan.

10.1.1.2 Earthquake Source Patterns

In Japan, three types of earthquake source patterns are considered as shown in Fig. 10.3. The first is an inland (plate) earthquake which is caused by fault slips in the inland plate under Japanese island where there is a strong compressive stress from the movement of the Philippine Sea Plate and the Pacific Ocean Plate. The second is an interplate earthquake which is caused by fault slips at the boundary of the Eurasian Plate and Philippine Sea Plate or North American Plate and Pacific Ocean Plate, and this type ordinarily results in earthquakes of large magnitude. The last one is an intraplate earthquake which is caused by a fault in the sinking Philippine Sea Plate or the Pacific Ocean Plate.

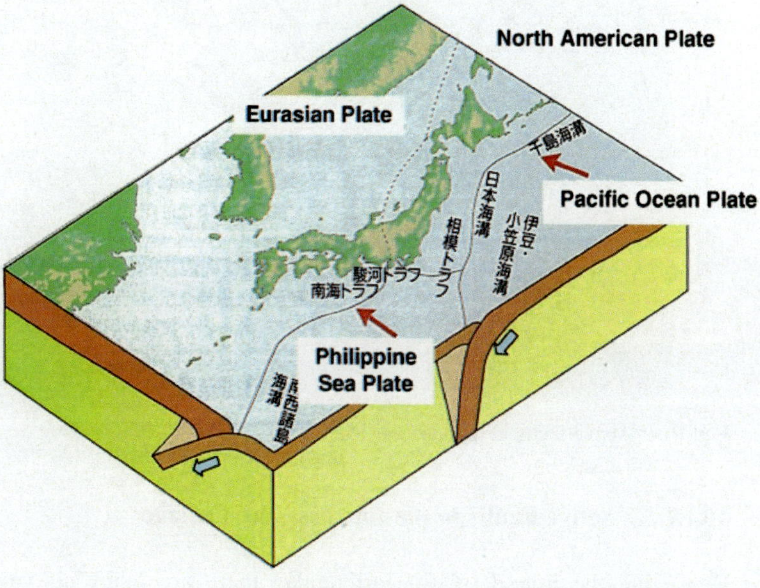

Fig. 10.2 Plate tectonics around the Japanese archipelago

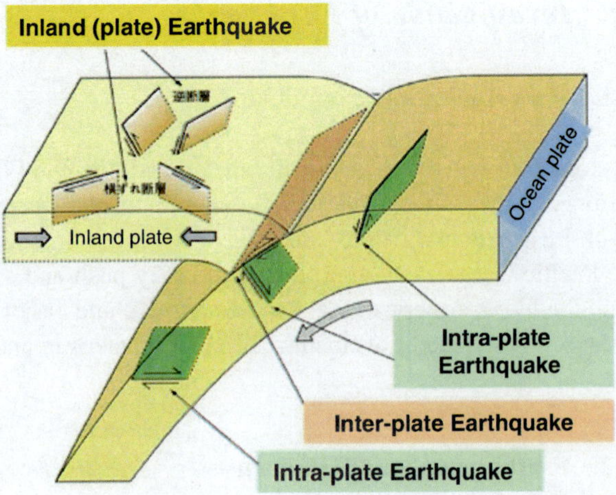

Fig. 10.3 Types of earthquake source patterns

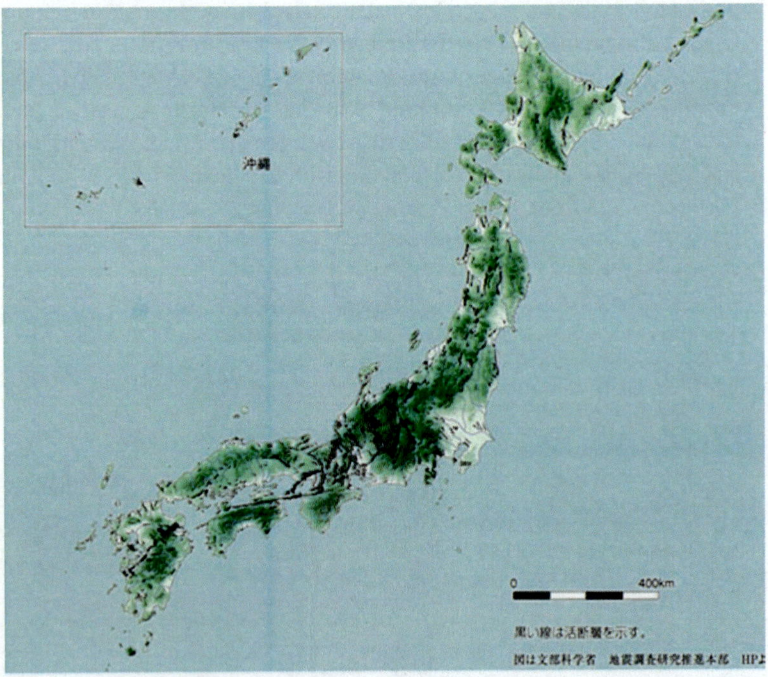

Fig. 10.4 Active faults in the Japanese archipelago

10.1.1.3 Active Faults in the Japanese Archipelago

Regarding the inland (plate) earthquake, there are many active faults in and around the Japanese archipelago which are shown in Fig. 10.4. The black lines show active faults. Generally, the magnitude of an inland earthquake is rather small, but earthquake motion at the ground surface is sometimes big at a local site because the fault is at a shallow depth. The return period (occurrence frequency) of an inland earthquake is generally very long with the interval being about 1,000–10,000 years.

10.1.1.4 Surveys of Active Faults

Earthquake sources (faults) around a proposed NPP site are thoroughly investigated by various surveys as described in the following.

1. Literature and document surveys. Literature and document surveys are conducted to investigate past and historical earthquakes around a proposed NPP site. These surveys may include books and maps (Fig. 10.5) and old documents (Fig. 10.6). An example representation of a geographical distribution of past earthquakes investigated around a site is shown in Fig. 10.7.
2. Topography and lineament surveys. Near the site, a detailed active fault survey is conducted by topography and/or lineament surveys as shown in Figs. 10.8 and 10.9. The lineament is the gap line appearing on the ground surface. An active fault is usually slipping repeatedly over a long period and some of the slips will become visible as the gap on the ground surface. Topographical surveys investigate undulation and/or winding of the ground surface and estimate the location

Fig. 10.5 Books and maps describing active faults in Japan

Fig. 10.6 Old documents about earthquakes

Fig. 10.7 Example of past earthquakes around a site

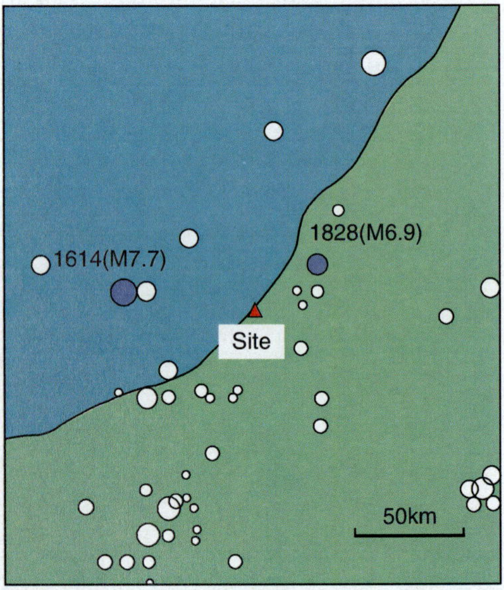

of the active fault under the ground. These surveys are done by using geological maps and/or aerial photographs.

3. Trench surveys. Trench surveys are conducted to confirm a fault that is currently active. This survey is done by digging up the ground near a proposed NPP site

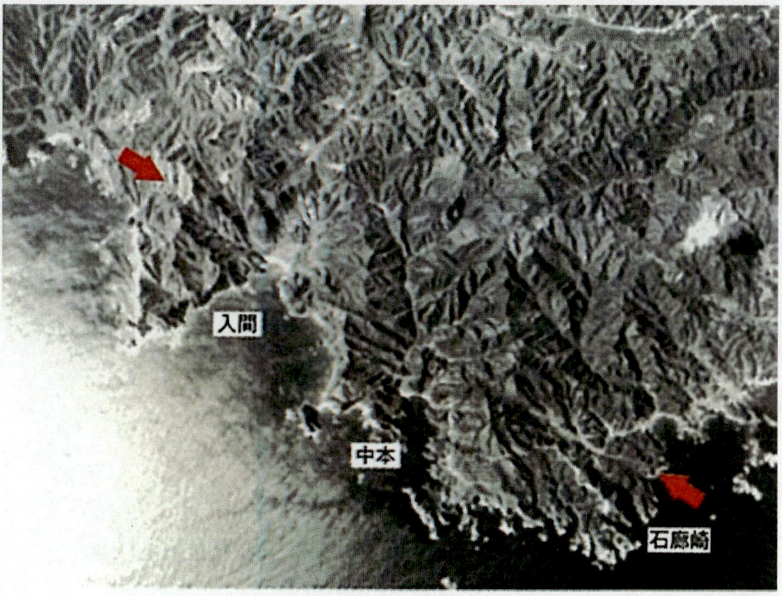

Fig. 10.8 Fault shown in an aerial photograph

Fig. 10.9 Image of a gap line on the ground surface

and investigating any slip traces of the fault directly as shown in Figs. 10.10 and 10.11. Trench surveys are also done to identify the latest activity of the fault as estimated by its geological age.

4. Physical surveys. Physical surveys such as sound wave surveys are also conducted to investigate an active fault as shown in Fig. 10.12. A soil stratum under

Fig. 10.10 Trench survey in the field

Fig. 10.11 Trench survey for soil strata

the ground is obtained by the sound wave survey and a fault is estimated by a gap or deformation of the soil stratum. Sound wave surveys are also applied to investigate the underwater grounds as shown in Fig. 10.13.

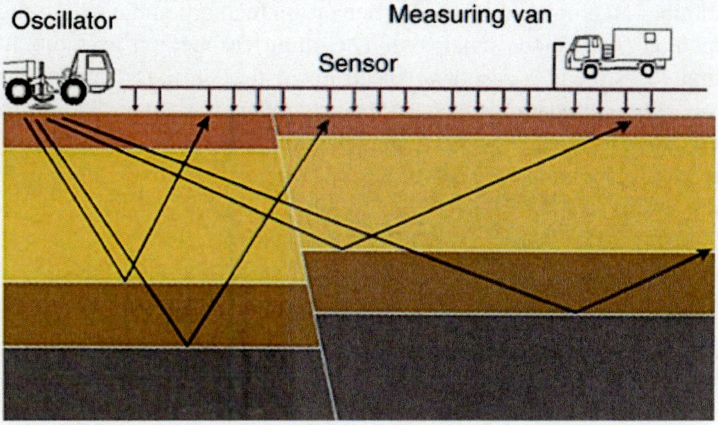

Fig. 10.12 Sound wave survey in the ground

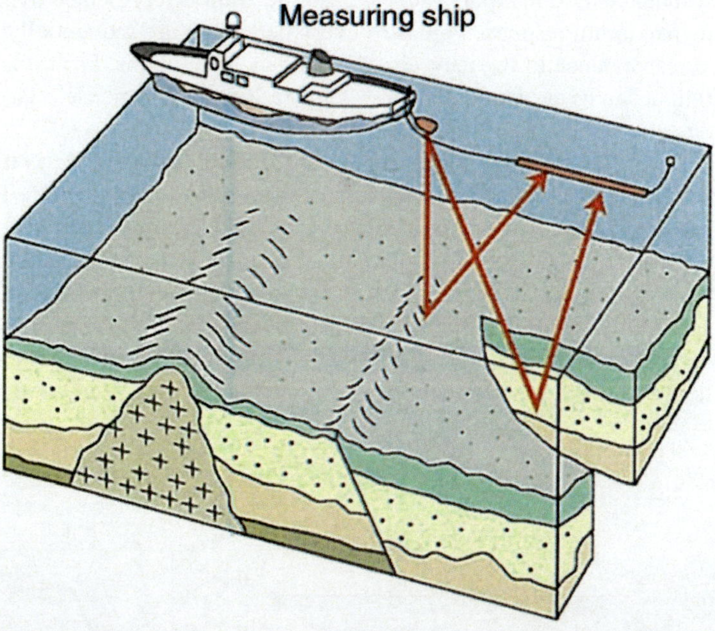

Fig. 10.13 Sound wave survey of underwater ground

10.1.2 Determination of Design Basis Earthquake Ground Motions

10.1.2.1 Estimation of Earthquake Ground Motion

DBEGMs at the site are estimated from the active faults around the site determined by the above surveys. There are two methods to estimate the earthquake ground motion: an empirical method (attenuation method) and a strong motion simulation method (fault model analysis). The empirical method uses the attenuation equation to estimate a response spectrum at the site based on parameters of earthquake magnitude and hypocenter distance which is the distance between the site and the earthquake source treated as a point source. The fault model analysis is a more precise method to calculate an earthquake motion at the site based on treating the earthquake source as the fault plane.

An image of the empirical method is shown in Fig. 10.14. Response spectra at the site are calculated for all earthquake sources around the site.

The response spectrum is a useful graph to represent the frequency characteristic of an earthquake motion which is a random wave, and it is calculated from an earthquake time history as shown in Fig. 10.15. On the other hand, all structures and components have natural periods (natural frequencies) like a pendulum. Many pendulums with various natural periods are arranged on the same baseplate and earthquake responses of every pendulum on the baseplate are analyzed against the earthquake motion input. The response spectrum is represented by a graph in which the maximum response values of every pendulum are sequentially plotted by their correspondence to the natural periods of the pendulums. From the response spectrum, it can be easily recognized whether a structural response is large or small from a natural period of the structure.

Regarding the fault model analysis, the earthquake source is treated as a fault plane which is divided into small elements as shown in Fig. 10.16. The fault rupture is assumed to start at a certain point in the fault plane and it is transferred in the fault plane which is modeled to rupture the fault elements one after another according to fault rupture speed. Earthquake motions at the site from each fault element are calculated considering the wave propagation in the ground and the time lag of each element rupture, and total earthquake motions at the site are obtained by composing the waves from each element. It is known that there exist a few regions in the fault

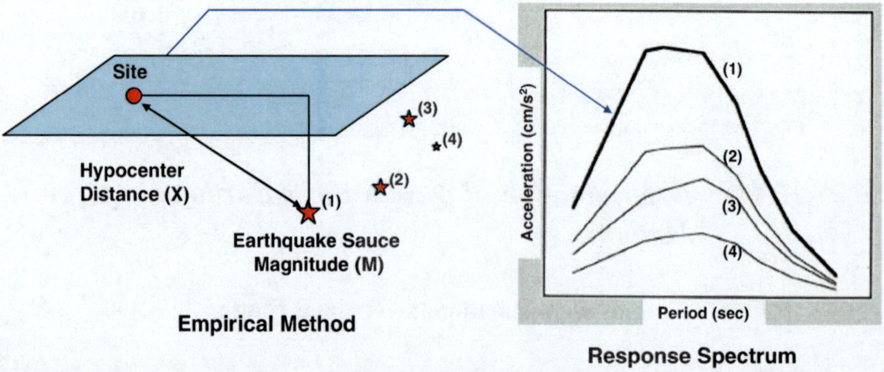

Fig. 10.14 Image of an empirical method for estimating response spectrum of an earthquake

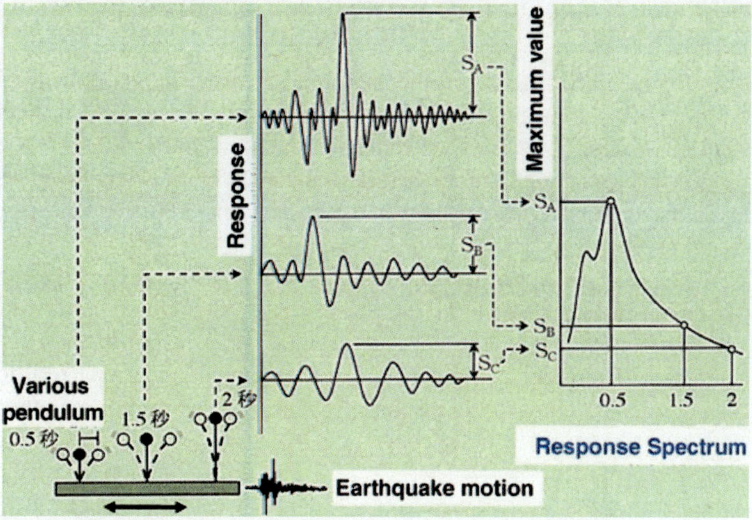

Fig. 10.15 Calculation flow of response spectrum from earthquake motions

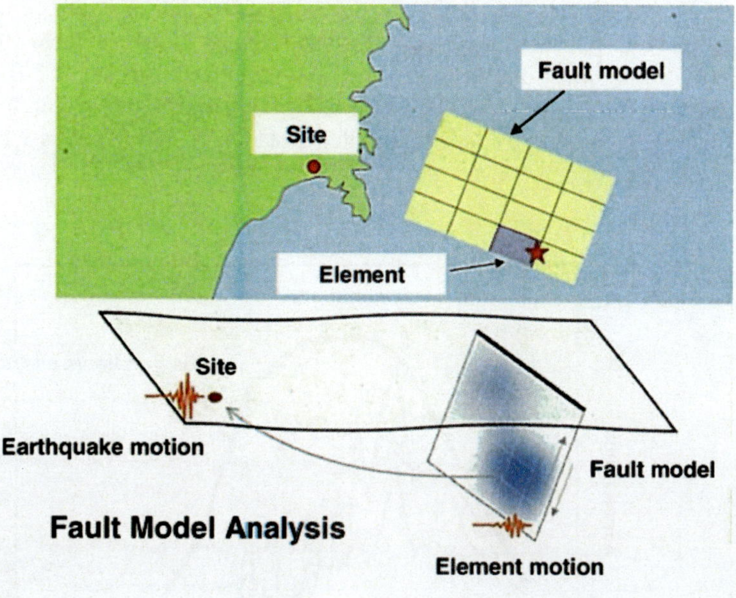

Fig. 10.16 Image of fault model analysis

plane which generate strong waves (energy); such a region is known as an asperity and it is usually considered in the fault plane model.

A time history of the earthquake motions at the site is directly obtained by the fault model analysis. The response spectrum is calculated from the time history using the above method.

10.1.2.2 Estimation of Design Basis Earthquake Ground Motions

Many response spectra of the earthquake motions are obtained from various earthquake sources around the site, calculated by the empirical method and/or the fault model analysis. They are compared as shown in Fig. 10.17. The most effective (largest) response spectrum is selected as the DBEGM. Sometimes several response spectra are selected for the DBEGM in case one spectrum does not cover other spectra.

The DBEGM is determined for the horizontal direction and the vertical direction as shown in Fig. 10.18.

When applying the empirical method, the response spectrum is directly obtained. Therefore, a time history of earthquake motions (artificial waves) is usually generated to fit the response spectrum by an iteration method.

10.1.3 Classification of Importance in Seismic Design

Classification of importance of the facilities in the seismic design uses three classes, S, B, and C classes, which are determined by the safety importance regarding radioactive discharge for an accident condition as shown in Table 10.1. The S class includes the most important facilities and they have to bear the DBEGMs (Ss) and the static seismic force of 3.0 Ci. Here 1.0 Ci is the static seismic force for general buildings provided by Japanese building standards. The B class includes facilities second in importance and they have to bear the static seismic force of 1.5 Ci. The

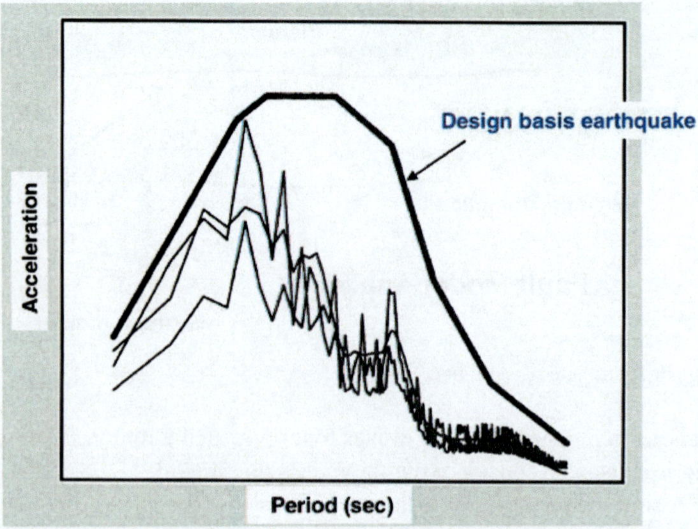

Fig. 10.17 Response spectra compared for various sources

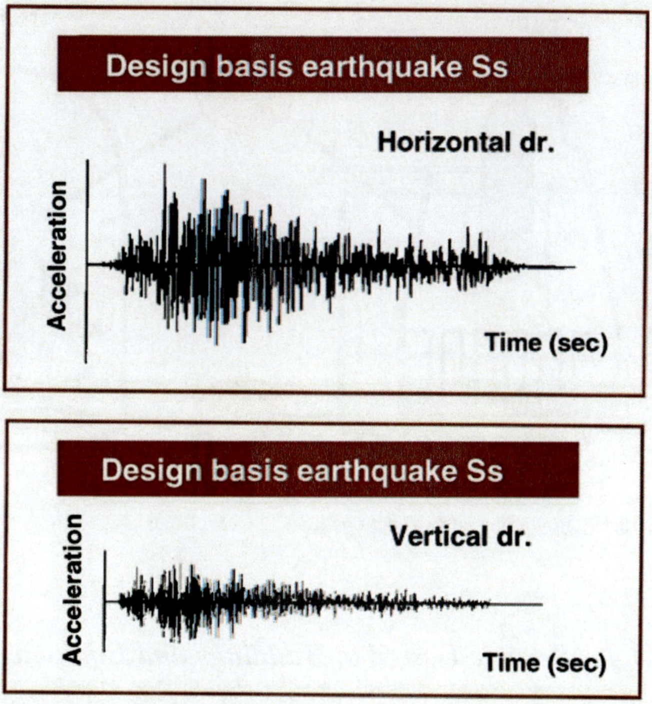

Fig. 10.18 Horizontal and vertical DBEGMs

Table 10.1 Classification of seismic importance

Class	Definition	Seismic force
S	Facilities containing radioactive materials by themselves or related directly to facilities containing radioactive materials, and whose influences are very significant	Ss (performance design) Sd*, 3.0 Ci** (elastic design)
B	Facilities of the same functional categories as above S Class, however whose influences are relatively small	1.5 Ci** (elastic design)
C	Facilities except for S or B Class, and ones required to ensure equal safety as general industrial facilities	1.0 Ci** (elastic design)

Sd*: Design earthquake motion for elastic design which is greater than 1/2 Ss
Ci**: Static seismic force for general building provided by Japanese building standard

C class facilities are not related to radioactive material safety and they have to bear the static seismic force of 1.0 Ci which is the same seismic force as general buildings.

The most important functions for safety of NPPs are shutdown, cooling, and confinement of the reactor core and the facilities related to these functions are put in S class. Image of the structures and the components related S class is shown in Fig. 10.19.

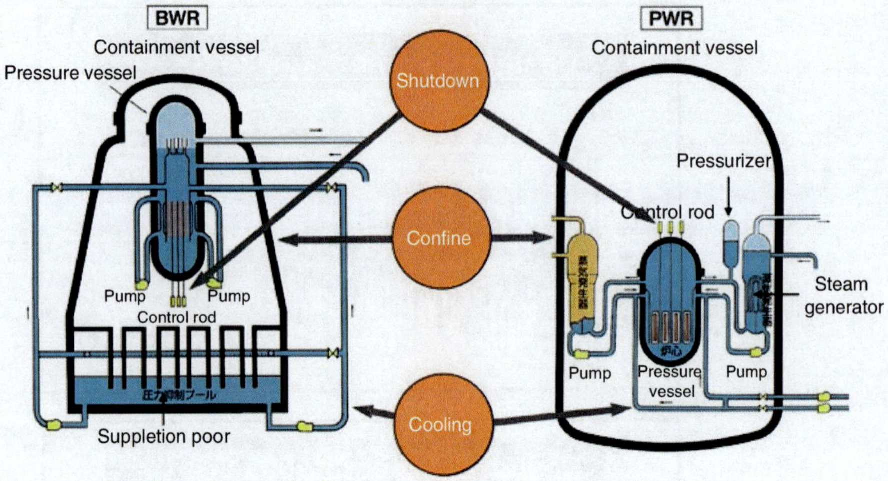

Fig. 10.19 Important safety functions for S class facilities

10.1.4 Seismic Design of Buildings and Equipment

The reactor building for an advanced boiling water reactor (ABWR) is shown in Fig. 10.20. The building is composed of rigid seismic walls made of reinforced concrete. Therefore, an earthquake response analysis model is usually constructed by modeling the seismic walls as a stick lumped-mass and beam model as shown in Fig. 10.21.

An earthquake response analysis is conducted for the DBEGMs using the building analysis model considering the soil–structure interactions as shown in Fig. 10.22. A model for large and heavy equipment is sometimes connected to the building model and is analyzed together with the building model considering the equipment and building interactions.

Maximum response values such as accelerations, displacements, shear forces, and bending moments of the building and the components are obtained by the earthquake response analysis. Using these response values, the structures and components are designed to bear the seismic loads and other loads such as dead load, thermal load, pressure load, and so on. Figure 10.23 shows an example of a stress analysis model of a component for the seismic design.

For the piping system, an analytical model is constructed for a beam model or a finite element model, and the earthquake response analysis of the piping system is conducted by input using building response values as shown in Fig. 10.24.

The earthquake response analysis of the building also confirms the ground stability under the base mat of the building. The ground has to bear a sliding force and a pressure force which are applied by the building during an earthquake as shown in Fig. 10.25.

Fig. 10.20 Overview of ABWR NPP

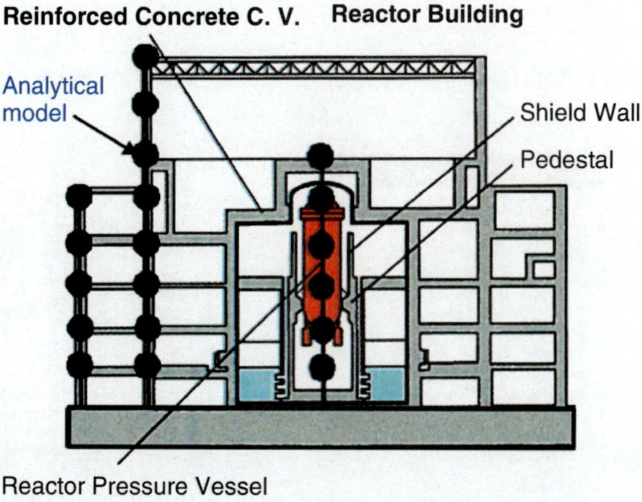

Fig. 10.21 Analytical model of reactor building

10.1.5 Events Accompanying an Earthquake

Attention must also be paid to events accompanying an earthquake. If there is a land slope close to the reactor building or other facilities, the ground stability regarding sliding and rupturing of the slope must be confirmed as shown in Fig. 10.26.

Safety against a tsunami must also be confirmed. A tsunami is generated by a fault slip or land slide at the sea bottom as shown in Fig. 10.27. This tsunami is sometimes amplified by the shapes of the coast line and of the sea bottom. So a

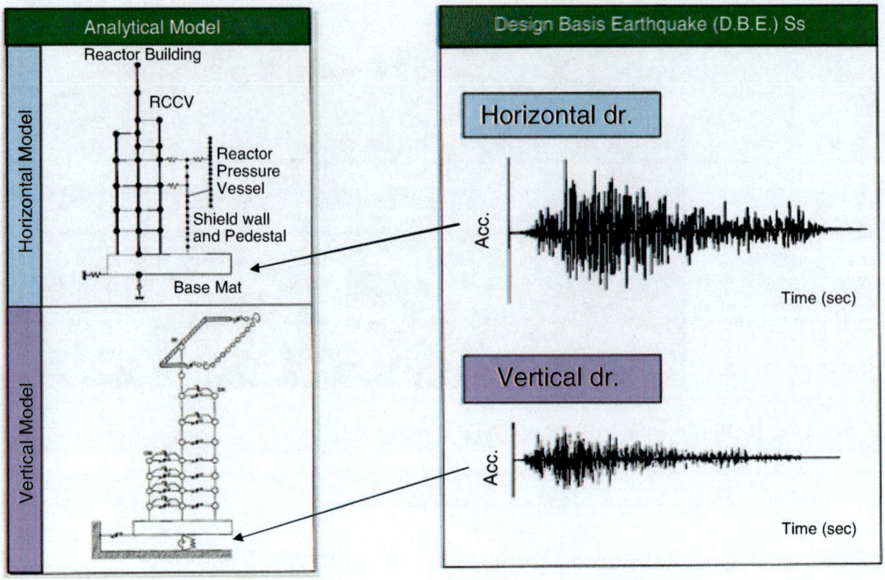

Fig. 10.22 Earthquake response analysis for horizontal and vertical motions

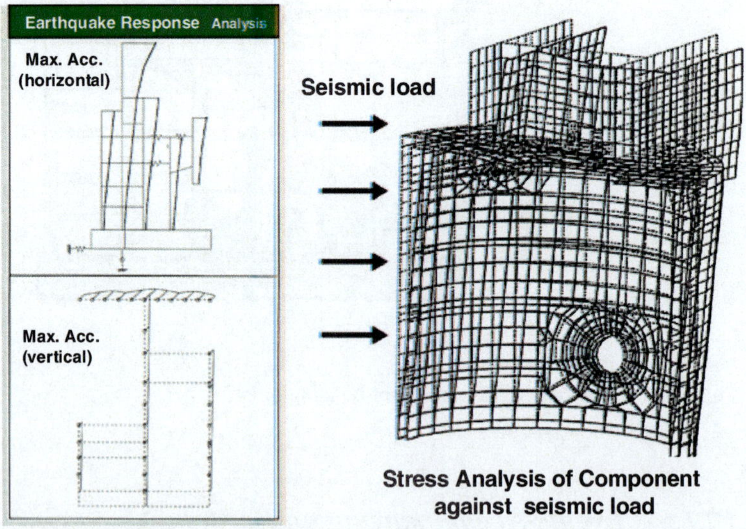

Fig. 10.23 Earthquake response and seismic design of component

tsunami propagation analysis using topographical data around the site is conducted. It should be confirmed that the highest level of the tsunami does not exceed the ground level of the site and that the lowest level is not below the intake of cooling sea water as shown in Fig. 10.28.

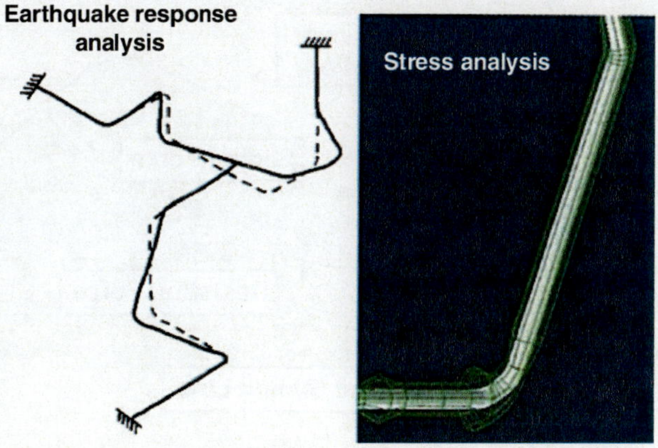

Fig. 10.24 Analysis model for piping system

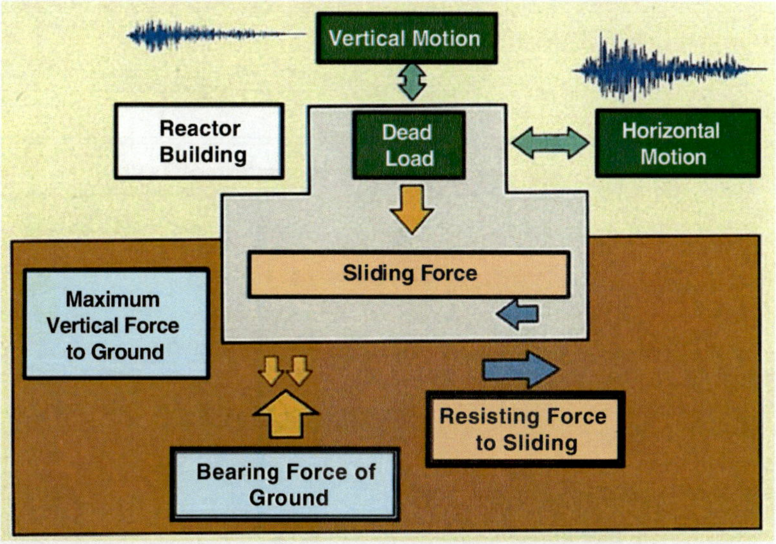

Fig. 10.25 Ground stability to sliding and pressure forces from the building

10.1.6 Ensuring Safety in the Seismic Design

Next, attention is directed to the seismic safety in construction and operational stages.

Important structures and components have to be supported on sufficiently stiff soil and a firmly reinforced concrete base mat. Figure 10.29 shows a scene during an inspection of the base foundation before the construction of a reactor building and Fig. 10.30 shows the carefully arranged reinforcing bars of the base mat of the reactor building.

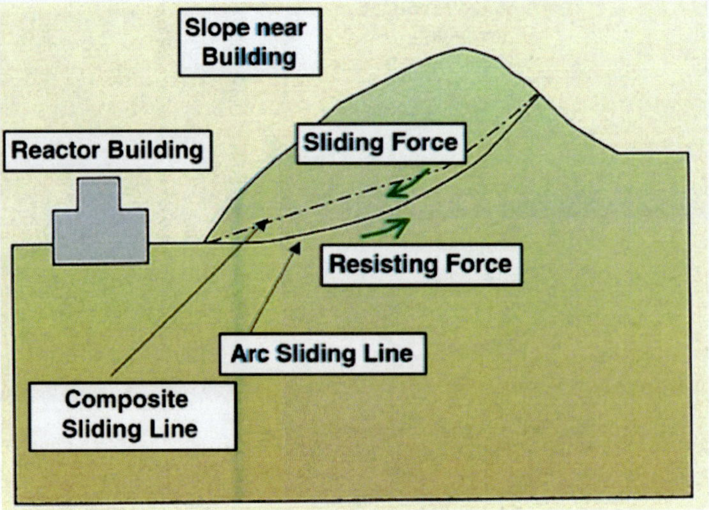

Fig. 10.26 Slope stability near reactor building

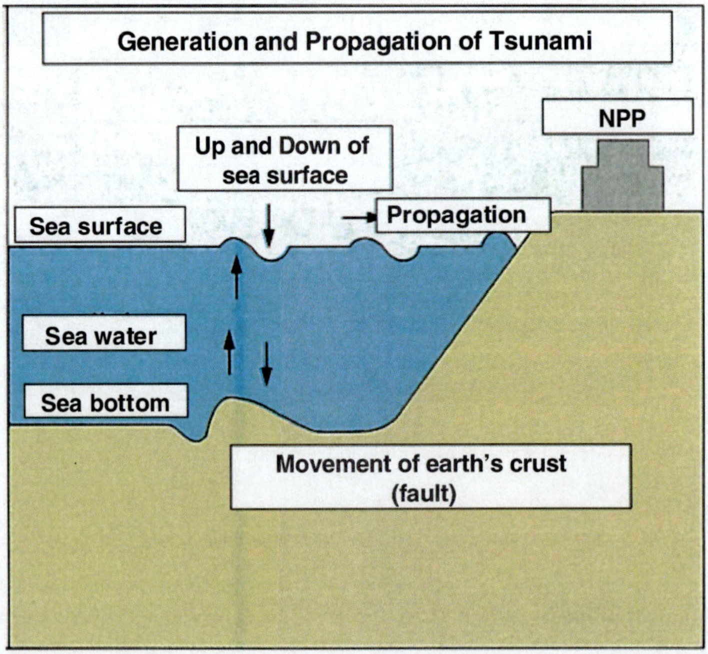

Fig. 10.27 Tsunami generation and propagation

Figure 10.31 shows a schematic drawing of an automatic shutdown system. Such an automatic shutdown system against earthquakes is installed in all Japanese NPPs. When strong earthquake motions occur at an NPP, sensors installed in the

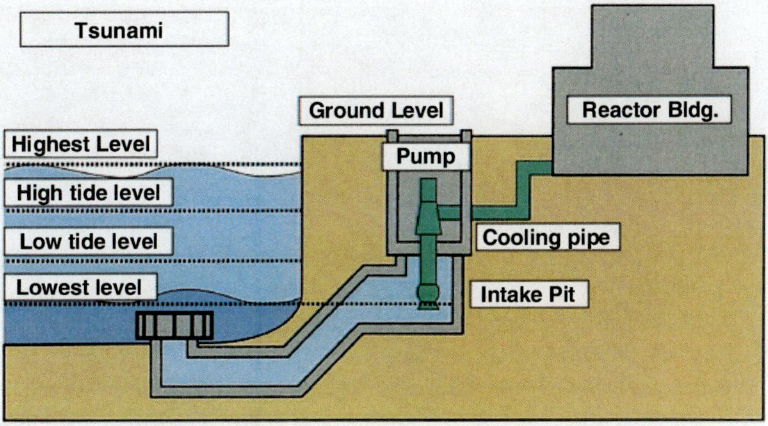

Fig. 10.28 Safety of NPP against tsunami

Fig. 10.29 Inspection of base foundation

reactor building detect them and the reactor is automatically shut down by a signal from the sensors.

Seismic performance and capacities of safety-related structures and components are confirmed by shaking table tests. Figure 10.32 shows the shaking table test of a reinforced concrete containment vessel where a scale model is used. Figure 10.33 shows the shaking table test of actual electric panels.

Fig. 10.30 Reinforcing bars arranged in the base mat

10.2 Assessment of Residual Risk

In the revised guide [1], the residual risk is mentioned as follows:

- From a seismological standpoint, the possible occurrence of stronger earthquake motions which exceed the determined DBEGMs cannot be denied.
- At the design stage of facilities, appropriate attention should be paid to this possible occurrence of exceeding the determined DBEGMs.
- Recognizing the existence of residual risk, every effort should be made to minimize it to the as low as practically possible level, not only in the design stage but also in the following stages.

SPSA is one of the most prominent methods to evaluate the residual risk. Figure 10.34 outlines the SPSA, which is composed of a seismic hazard analysis, a fragility analysis of facilities, and an accident sequence analysis including an event tree and a fault tree analysis. Core damage frequency against an earthquake considering all of the systems in the NPP is estimated by combining these analyses.

The basic concept of the SPSA is that damage frequency of the facilities is represented by the probabilistic seismic response of the facility exceeding the probabilistic seismic capacity of the facility, which is represented as the gray area in Fig. 10.35.

Figure 10.36 shows the flow of the SPSA. The seismic hazard is represented as a hazard curve derived from the magnitudes, hypocenter distance, and occurrence

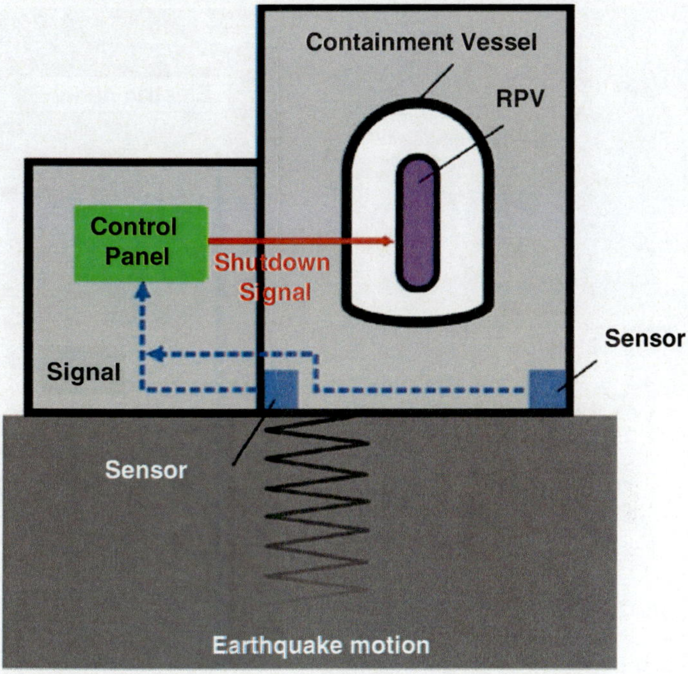

Fig. 10.31 Automatic shutdown system installed in a reactor building

Fig. 10.32 Shaking table test of RCCV

frequencies of all the earthquakes around the site. The fragility curves are derived from seismic capacities of facilities considering their means and deviations. The accident sequence evaluation is conducted through scenario analysis and system

Fig. 10.33 Shaking table test of electric panels

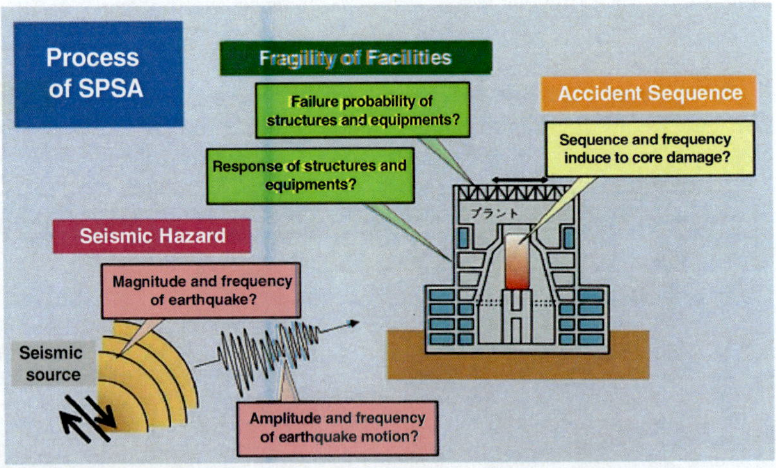

Fig. 10.34 Outline of the seismic probabilistic safety assessment (SPSA)

reliability evaluation. The core damage frequency is estimated by combining these analyses.

The goal of the core damage frequency is specified as 10^{-4}/year/unit by the Nuclear Safety Committee in Japan. The goal of the containment failure frequency is specified as 10^{-5}/year/reactor.

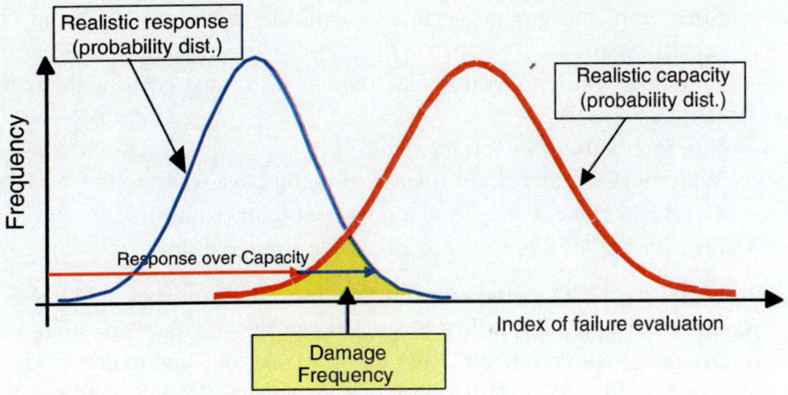

Fig. 10.35 Basic concept of SPSA

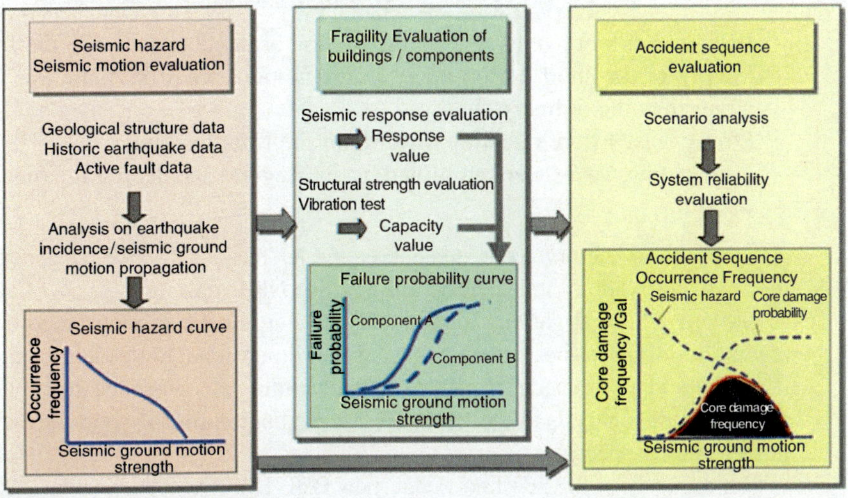

Fig. 10.36 Flow chart of SPSA

10.3 Recent Seismic Topics at Kashiwazaki–Kariwa NPP

On July 16, 2007, the Niigataken Chuetsu-Oki (NCO) Earthquake occurred in the vicinity of the Kashiwazaki–Kariwa NPP. Very large earthquake motions which were two to three times over the DBEGMs were observed at the site. This was the first time in the world that an NPP experienced such a large earthquake. There are seven units at the Kashiwazaki–Kariwa NPP site, and all of the units operating at the time were shut down safely by the automatic shutdown system against earthquakes. But some things occurred and facilities not related to nuclear safety were damaged as follows:

- Sinkage of the ground surface around the reactor building and cracking of a service road
- Flooding by fire protection pipe rupture with water entering the reactor building (Unit 1)
- Fire in a house transformer (Unit 3)
- Water leakage and slight release of radioactive water to the sea (Unit 6)
- Overhead crane damaged at a universal joint (Unit 6)
- Overturning of a low level radioactive storage drums

Just after the NCO earthquake, the Nuclear and Industrial Safety Agency (NISA) required the operating utility to investigate the reasons why such a large unpredicted earthquake occurred in the Kashiwazaki area and to check for the safety of all the facilities. The NISA also requested cross-check analyses by the Japan Nuclear Safety Organization (JNES) to represent the government side.

Not only the utility and JNES but also researchers at many universities and institutes started studies about the earthquake source and motions just after the earthquake. The findings of their investigations are summarized below:

- Pulse waves were generated by the rupture of three asperities in the fault.
- Energy of the short period range in the motions from the fault was 1.5 times larger than the ordinary energy.
- Strong waves were radiating toward the site from the fault.
- Propagating waves were amplified by an irregular and deep soil structure under the site.

All the above factors were considered to be reasons for the large earthquake motions observed at the Kashiwazaki–Kariwa NPP site.

Moreover, investigations about observed earthquake motions recorded in the reactor building was performed using a three-dimensional finite element model, and it was elucidated that deformation of the building floor needed to be taken into account in the analysis model for a precise evaluation for the short period range of floor response spectra.

Reflecting the above new knowledge, new DBEGMs (Ss) at the Kashiwazaki–Kariwa site which are three to five times larger than the old values were determined by a committee of the NISA. Unit 7 and Unit 6 were checked and the safety of their facilities was reviewed against the NCO earthquake and the DBEGMs at first. Check and review work by the utility of the other units is being prepared to begin in September 2009.

References

1. Nuclear Safety Committee, Regulatory Guide for Examining Seismic Design of Nuclear Power Reactor Facilities, Sept 2006 (in Japanese)
2. Nuclear and Industrial Safety Agency and Japan Nuclear Safety Organization, Pamphlet of Seismic Safety of Nuclear Power Plants, 2nd edn. Sept 2007 (in Japanese)

Index